T0299083

Structure and Randomness in
Computability and Set Theory

Structure and Randomness in Computability and Set Theory

Edited by

Douglas Cenzer
University of Florida, USA

Christopher Porter
Drake University, USA

Jindrich Zapletal
University of Florida, USA

World Scientific

EW JERSEY · LONDON · SINGAPORE · BEIJING · SHANGHAI · HONG KONG · TAIPEI · CHENNAI · TOKYO

Published by

World Scientific Publishing Co. Pte. Ltd.

5 Toh Tuck Link, Singapore 596224

USA office: 27 Warren Street, Suite 401-402, Hackensack, NJ 07601

UK office: 57 Shelton Street, Covent Garden, London WC2H 9HE

Library of Congress Cataloging-in-Publication Data

Names: Cenzer, Douglas, editor. | Porter, Christopher (Christopher P.), editor. |
    Zapletal, Jindrich, editor.
Title: Structure and randomness in computability and set theory / edited by
    Douglas Cenzer, University of Florida, Christopher Porter, Drake University,
    Jindrich Zapletal, University of Florida.
Description: New Jersey : World Scientific, [2021] | Includes
    bibliographical references and index.
Identifiers: LCCN 2020034089 (print) | LCCN 2020034090 (ebook) |
    ISBN 9789813228221 (hardcover) | ISBN 9789813228238 (ebook for institutions) |
    ISBN 9789813228245 (ebook for individuals)
Subjects: LCSH: Set theory. | Computable functions.
Classification: LCC QA248 .S892 2021 (print) | LCC QA248 (ebook) |
    DDC 511.3/22--dc23
LC record available at https://lccn.loc.gov/2020034089
LC ebook record available at https://lccn.loc.gov/2020034090

British Library Cataloguing-in-Publication Data
A catalogue record for this book is available from the British Library.

For any available supplementary material, please visit
https://www.worldscientific.com/worldscibooks/10.1142/10661#t=suppl

Desk Editors: Kwong Lai Fun/Vishnu Mohan

Typeset by Stallion Press
Email: enquiries@stallionpress.com

Printed in Singapore

Preface

The goal of this collection of review chapters is to provide an in-depth overview of work in three areas that have emerged on the frontier of research in set theory and computability in recent years: (1) infinitary combinatorics and ultrafilters; (2) algorithmic randomness and algorithmic information; and (3) computable structure theory.

The unifying themes of these areas are an emphasis on structure, randomness, and the interplay between them. Taking the above-listed three research areas in reverse order, these themes appear as follows: First, we have an emphasis on the use of the tools of computability theory to analyze the complexity of various mathematical structures in computable structure theory. Second, we can study the extent to which effective methods allow us to classify objects as random or nonrandom, a central theme in algorithmic randomness, or, in the case of algorithmic information theory, measuring the amount of algorithmic information in an object, placing it on a scale that at the lower levels corresponds to more structure, while at the higher levels corresponds to more randomness. Third, in infinitary combinatorics, one can study the extent to which order emerges from disorder, particularly in the case of Ramsey theory for infinite sets, thereby yielding an interesting balance between structure and a certain kind of randomness.

We now layout the specifics of the book. Part I "Infinitary Combinatorics and Ultrafilters" deals with a remarkably persistent theme in set theory. Nonprincipal ultrafilters on natural numbers are well-known to be difficult to analyze and treat in detail. At the same time, they are exceptionally useful in many directions, as one can see from the proof of van der Waerden theorem via the compactification of the semigroup of natural numbers or (more recently) Malliaris's and Shelah's proof of $\mathfrak{p} = \mathfrak{t}$. Over the years, set theorists isolated certain critical properties of ultrafilters which completely determine their combinatorial features.

As the oldest example of such a critical property, consider selective ultrafilters: the ultrafilters U which, for every coloring $c : [\omega]^2 \to 2$, contain a set homogeneous for such a coloring. It turns out that it is impossible to find further distinctions between selective ultrafilters using formulas from a certain broad syntactically identified class. In other words, if we know that a certain ultrafilter is selective, we probably know most of its other combinatorial properties as well. One way to precisely formulate this intuition is found in a result of Todorcevic: granted sufficiently large cardinals, every selective ultrafilter U is generic over the canonical model $L(\mathbb{R})$ (the smallest model of ZF set theory containing all reals and all ordinals) for the partially ordered set P of all infinite subsets of ω ordered by inclusion. Thus, the study of combinatorial properties of U expressible in the generic extension $L(\mathbb{R})[U]$ reduces to the study of the partial ordering P and is completely independent of U. The model $L(\mathbb{R})[U]$ attracted plenty of attention over the last 30 years from authors such as Todorcevic, Di Prisco, Dobrinen, Paul Larson and Zapletal; it is one of the canonical and best understood objects in transfinite set theory.

Given the success story of Ramsey ultrafilters, one can ask whether it is possible to find other properties playing similar critical role. The answer is affirmative and the list of such critical properties is continually growing. Most of the examples found so far can be restated in terms of partition calculus. For each such critical property ϕ, there is a definition of a partial ordering P_ϕ such that every ultrafilter satisfying ϕ is in fact generic over the model $L(\mathbb{R})$ for the poset P_ϕ. This converts the study of an ultrafilter with the property ϕ to the study of the ordering P_ϕ. The comparison of the models $L(\mathbb{R})[U]$ for ultrafilters U of various critical types then slowly dissects the unwieldy set of all ultrafilters into smaller, manageable units. This open-ended, bold program has seen continual progress over the years, and it is intimately connected with the tools of abstract partition calculus.

Part I consists of two chapters. Chapter 1 "Topological Ramsey Spaces Dense in Forcings", by Natasha Dobrinen, is an extensive survey of this area. Dobrinen frames this traditional field of inquiry using the theory of topological Ramsey spaces of Stevo Todorcevic, which support infinite-dimensional Ramsey theory similarly to the Ellentuck space. Each topological Ramsey space is endowed with a partial ordering which can be modified to a σ-closed "almost reduction" relation analogously to the partial ordering of "mod finite" on $[\omega]^\omega$. Such forcings add new ultrafilters satisfying weak partition relations and have complete combinatorics. In cases where a forcing turned out to be equivalent to a topological Ramsey space, the strong Ramsey-theoretic techniques have aided in a fine-tuned analysis of the Rudin–Keisler and Tukey structures associated with the forced ultrafilter and in discovering new ultrafilters with complete combinatorics. This original perspective allows her to organize the wealth of existing research in a particularly incisive way and to prove a number of new results as well. Dobrinen's exposition provides a long dictionary of critical combinatorial properties of ultrafilters and pointers for extending this dictionary further. Readers interested in using techniques using topological Ramsey spaces to study ultrafilters with various partition relations should find Dobrinen's survey to be instructive.

Chapter 2 "Infinitary Partition Properties of Sums of Selective Ultrafilters", by Andreas Blass, is more specific and deals with a particular pair of critical combinatorial ultrafilter properties. It concerns two kinds of ultrafilters on ω^2, the first kind given by the sums of nonisomorphic selective ultrafilters that are indexed by another selective ultrafilter, and the second kind given by ultrafilters that are generic with respect to the forcing the conditions of which are subsets of ω^2 that have an infinite intersection with $\{n\} \times \omega$ for infinitely many $n \in \omega$. Although these two kinds of ultrafilters share a number of properties, such as being Q-points but not P-points and not being at the top of the Tukey ordering, they also differ in several respects, as only ultrafilters of the first kind are basically generated while only ultrafilters of the second kind are weak P-points. Blass first shows that the infinitary partition property of ultrafilters of the first kind is of the same strength with what has been previously shown about the infinite partition property of ultrafilters of the second kind. This, in turn, leads to Blass's second main result, obtained via an application of complete combinatorics, that the two kinds of ultrafilters are the same when viewed in different models of set theory. Lastly, Blass uses this result to account for how both the similarities and differences between the two kinds of ultrafilter arise.

Part II "Algorithmic Randomness and Information" concerns two different research strands, namely, the study of algorithmically random sequences under Turing reductions, and the study of effective notions of Hausdorff dimension defined in terms of Kolmogorov complexity, the central concept in algorithmic information theory.

One of the primary aims in the study of algorithmic randomness is to study various definitions of algorithmically random sequences and the properties of such sequences. The most well-studied definition of algorithmic randomness is Martin-Löf randomness. Martin-Löf's original idea behind his definition was to formalize the notion of an effective statistical test, given by a sequence of effectively generated open sets the measures of which are effectively converging to zero. A sequence that is not in the intersection of any such test is a Martin-Löf random sequence. Alternative definitions of algorithmic randomness can be obtained, for instance, by modifying the underlying notion of an effective statistical test, although Martin-Löf has proven to be, in certain respects, more well-suited to being studied from a computability-theoretic point of view compared to alternative definitions of randomness.

One respect in which Martin-Löf randomness is amenable to study using tools from computability theory, namely the behavior of random sequences under Turing reductions, is the subject of the first randomness-theoretic chapter in this book, "Limits of the Kučera–Gács Coding Method", by George Barmpalias and Andrew Lewis-Pye. This chapter discusses an improvement of methods used independently by Kučera and Gács to prove what is now considered to be a classical result in the field concerning the computational power of Martin-Löf random sequences.

In 1985, Kučera proved that for every sequence $A \in 2^\omega$, there is some Martin-Löf random sequence B such that $A \leq_T B$. Kučera's proof involved a method of coding according to which one encodes an arbitrary sequence $A \in 2^\omega$ by a member of a fixed Π_1^0 class of positive Lebesgue measure. If we use this method of coding on a Π_1^0 class consisting of Martin-Löf random sequences (such as the complement of some level of the universal Martin-Löf test), we obtain the desired reduction to a random sequence.

Independently, in 1986 Gács proved the same result using an alternative method of coding, which allowed him to provide a more fine-grained analysis of the reduction in question, which he shown can be given by a wtt-functional. That is, Gács proved that there is a computable function $f : \omega \to \omega$ such that every sequence A is computable from a Martin-Löf random sequence B with the function f bounding the use of this reduction. In fact, Gács proved that we can take this function to be given by $f(n) = n + \sqrt{n} \log(n)$.

In their chapter, Barmpalias and Lewis-Pye first carefully layout a modular argument for the theorem that encompasses the coding methods employed by both Kučera and Gács. Next, they identify a key limitation of this approach which involves the redundancy of the reduction, i.e., which, for each n, is the number of additional input bits beyond n that are needed to yield n output bits (for instance, the proof due to Gács establishes the result with redundancy $g(n) = \sqrt{n}\log(n)$). The authors then provide an alternative coding method, which allows them to prove that every sequence can be computed from a Martin-Löf random sequence with an optimal logarithmic redundancy, thereby significantly improving the original Kučera–Gács result.

One of the most useful tools in the study of algorithmic randomness, namely Kolmogorov complexity, can be used to measure the amount of algorithmic information in a given sequence (for this reason, this area of research is sometimes referred to as algorithmic information theory). Informally, the Kolmogorov complexity of a finite string is the number of bits required of a universal machine to output that string; in the case that the Kolmogorov complexity of a string exceeds the length of a string, such a string is called *incompressible* (it is standard to consider incompressibility up to a fixed additive constant, but we will suppress that detail here). One can then consider the Kolmogorov complexity of the initial segments of an infinite sequence. It was recognized fairly early in the development of the theory of algorithmic randomness that Martin-Löf random sequences are precisely those sequences with sufficiently incompressible initial segments (at least with respect to certain modifications of the originally formulated notion of Kolmogorov complexity, such as prefix-free complexity and monotone complexity).

It was later discovered that studying the initial segment complexity of initial segments of nonrandom sequences can still be quite fruitful. This leads to the following chapter, "Information vs. Dimension: An Algorithmic Perspective", by Jan Reimann, which is a self-contained survey that covers the work on the interactions between algorithmic information theory and various notions of fractal dimension dating back over 30 years. Informally, the fractal dimension of a set, for instance a subset of $\mathbb{R}^2$, measures the complexity of the set in a way that corresponds to the extent that it fills spaces. Moreover, fractal dimension is not integer-valued, as it can provide intermediate values between the standard spatial dimensions.

Beginning in the 1980s, results due to Ryabko and Staiger indicated a strong connection between the classical notion of the Hausdorff dimension of a subset of a sequential space and the Kolmogorov complexity of initial segments of its members. Lutz later extended this work by defining effective notions of Hausdorff dimension and packing dimension that are defined for individual binary sequences, a development that marked a departure from the classical versions of these notions, which always assign dimension zero to an individual object. These definitions of effective dimension proved to be robust: Lutz's definitions of these two notions of dimension were initially given in terms of certain betting strategies known as gales, but it was later shown by various authors that these two notions can be given formulated in terms of Kolmogorov complexity: the effective Hausdorff dimension of an infinite sequence X is the limit infimum of the ratio $K(X \restriction n)/n$ (where $K(X \restriction n)$ is the prefix-free Kolmogorov complexity of the first n bits of the sequence X), while the effective packing dimension of X is the limit supremum of this same ratio.

These developments, as well as more recent developments, are covered in Reimann's survey. After reviewing the basics of classical information theory, Kolmogorov complexity, and Hausdorff dimension, Reimann details various aspects of effective Hausdorff dimension, including its formulation in terms of effective null sets and its relationship to Kolmogorov complexity. He then highlights the role effective Hausdorff plays in the effective version of the Shannon–McMillan–Breiman theorem from classical information theory, a result that underwrites the informal identity "entropy = complexity = dimension". Finally, Reimann discusses some of his own results on multifractal measures that draw upon effective Hausdorff dimension.

Part III, "Computable Structure Theory", contains three chapters. There are some unifying themes. One is definability in the arithmetic and hyperarithmetic hierarchy, that is, the effective Borel hierarchy. A second theme is computable reducibility. This leads to notions of completeness of sets and relations at various levels of the hierarchy. Index sets are used to characterize properties of structures frequently by showing that the index sets are complete at a certain level.

Chapter 5 by Russell Miller "Computable Reducibility for Cantor Space" is a contribution to the quickly growing field of computable descriptive set theory, evaluating the computational content of various objects and relations appearing in classical descriptive set theory. The field identifies distinctions which are too fine for descriptive set theory to detect, and opens a new perspective in each major theme of descriptive set theory.

Descriptive set theorists often build hierarchies of objects found in mathematical analysis. There is the Wadge hierarchy of definable subsets of the Baire space, which is in a certain direction the ultimate complexity measurement. However, one can find other similar tools with different aims. Large parts of several decades of research in the theory of cardinal invariants of the continuum can be viewed as the study of σ-ideals on Polish spaces ordered by Borel–Katětov reductions. A somewhat younger research direction in similar vein is the comparison of Borel equivalence relations via Borel reducibility. In this case, a detailed map of Borel equivalence relations has been drawn, encompassing and rating many known equivalence problems. This has been a very successful enterprise; several long-standing classification research programs in mathematical analysis went down in flames after it was shown that they can never capture the inherent complexity of the underlying objects.

As soon as workers in computability found a satisfactory notion of a computable function between Polish spaces, they started wondering how complex the reduction functions in these descriptive set theoretic hierarchies must be from computability standpoint. A parallel to the theory of cardinal invariants was discovered in the study of Turing degrees, and interesting finesses appeared in the computational complexity of Katětov reductions between various σ-ideals. The theory of computable reductions between Borel equivalence relations has not received as much attention, and Miller's paper aims to rectify this situation.

Miller surveys a number of possible ways of restricting the comparison to computable reducibility of equivalence relations on the Cantor space. Different versions are obtained by modifying some aspect of the reduction. For instance, one can vary the number of jumps of the input sequence that the functional is allowed to use, one need not require that the reduction succeed for all members of Cantor space but only for arbitrary finite or countable subsets of Cantor space, or one can add an additional oracle to be used in the reduction. As Miller shows, these refinements of Borel reducibility allow one to classify Borel reductions in terms of a level of difficulty or to account for why a Borel reduction fails to exist.

Chapter 6 "Logic Programming and Effectively Closed Sets", by Douglas Cenzer, Victor W. Marek and Jeffrey B. Remmel, surveys results on index problems for effectively closed sets and their applications to models of logic programs. Effectively closed sets are the foundational level of the effective Borel hierarchy. They play a key role in the area of algorithmic randomness. Effectively closed sets, or Π^0_1 classes, arise naturally in the study of computable structures. For many mathematical problems, such as finding a prime ideal of a Boolean algebra or finding a zero of a continuous function, the set of solutions to a given problem may be viewed as a closed set under some natural topology. For a computable problem, the set of solutions may be viewed as a Π_1^0 class. Thus, completeness results for properties of Π_1^0 classes may be used to obtain completeness results for many types of computable problems. For example, the set of complete consistent extensions of an axiomatizable theory may be represented as a Π_1^0 class, and the property of having a computable complete consistent extension can be shown to be Σ_3^0 complete.

The authors introduce a new notion of boundedness for trees and examine the complexity of index sets for the corresponding closed sets. This classification is in turn used to study the so-called recognition problem in the metaprogramming of finite normal predicate logic programs. In particular, for any property $\mathcal{P}$ of finite normal predicate logic programs over a fixed computable first-order language, the associated index set $I_{\mathcal{P}}$ is classified in the arithmetical hierarchy. Here, the authors' classification of the index sets of closed sets associated with the above-described bounded trees serves as the primary tool. For example, the authors determine the complexity of the index sets relative to all finite predicate logic programs and relative to certain special classes of finite predicate logic programs of properties such as (i) having no stable models, (ii) having no recursive stable models, (iii) having exactly c stable models for any given positive integer c, (iv) having exactly c recursive stable models for any given positive integer c, (v) having only finitely many stable models, and (vii) having only finitely many recursive stable models.

Chapter 7 "Computability and Definability", by Trang Ha, Valentina Harizanov, Leah Marshall and Hakim Walker provides a comprehensive survey of recent results in computable structure theory, including much recent work that has not been covered in previous surveys on the field. One important theme is degree spectrum of a structure and of a theory. Another key topic is the notion of a Scott sentence of a structure, in particular for finitely generated structures. Other topics covered include

effective aspects of theories, diagrams, and models, computable
infinitary formulas and Scott rank, the index sets of structures
and classes of structures, strongly minimal structures, relatively
Δ_α^0-categorical structures, and aspects of definability and complexity of relations on structures. Particular structures studied
include graphs, trees, linear orders, groups and fields.

Recent results of Knight, Harrison-Trainor and others characterize those structures which have a d-Σ_2^0 Scott sentence and
establish that any finitely generated field has such a sentence.
Computable structures of Scott rank ω_1^{CK} lead up to the solution
by Harrison-Trainor, Igusa and Knight, of a long-standing question of Sacks by showing that there such a computable structure
such that the computable infinitary theory is not $\aleph_0$-categorical
extensive treatment of computable and Δ_α^0 categoricity includes
characterization of such categoricity in terms of Scott families;
the distinction between Δ_α^0 categoricity and relative categoricity. Results of Downey, Lempp, Montalban, Turetsky and others
show that the index set for computably categorical structures
is shown to be Π_1^1 complete whereas the index set for relatively computable structures is only Σ_3^0 complete. There are
new results of Adams and Cenzer on weakly ultrahomogeneous
structures. There is a detailed presentation of the work of Hari-
izanov, Knight and others on intrinsically Σ_α^0 relations. This
includes very interesting new results about the degree spectrum
of a relation. Index sets for classes of structures are examined in
detail. For example, the index set of computable prime models
is $\Pi_{\omega+2}^0$-complete, the index set of computable structures with
noncomputable Scott rank is Σ_1^1-complete, and the set of finitely
generated free groups is Σ_3^0 complete within the class of all free
groups.

The topics covered in this book were inspired by a sequence
of meetings in Gainesville, FL, known as the Southeastern Logic
Symposium (SEALS). Experts from computability and set theory have been meeting since 1985 at a number of institutions

in the Southeast US, with support from the National Science Foundation.

The editors would like to thank the authors and the referees for their hard work and patience.

Douglas Cenzer, Chris Porter and Jindra Zapletal
Gainesville, FL

© 2021 World Scientific Publishing Company
https://doi.org/10.1142/9789813228238_fmatter

About the Editors

Douglas Cenzer is Professor of Mathematics at the University of Florida, where he was Department Chair from 2013 to 2018. He has more than 100 research publications, specializing in computability, complexity, and randomness. He joined the University of Florida in 1972 after receiving his Ph.D. in mathematics from the University of Michigan.

Christopher Porter is Assistant Professor of Mathematics at Drake University, specializing in computability theory, algorithmic randomness, and the philosophy of mathematics. He received his Ph.D. in mathematics and philosophy from the University of Notre Dame in 2012, was an NSF international postdoctoral fellow at Université Paris 7 from 2012 to 2014, and a postdoctoral associate at the University of Florida from 2014 to 2016, before joining Drake University in 2016.

Jindrich Zapletal is Professor of Mathematics at University of Florida, specializing in mathematical logic and set theory. He received his Ph.D. in 1995 from the Pennsylvania State University, and held postdoctoral positions at MSRI Berkeley, Cal Tech and Dartmouth College, before joining the University of Florida in 2000.

Contents

Part I
Infinitary Combinatorics and Ultrafilters

Chapter 1

Topological Ramsey Spaces Dense in Forcings

Natasha Dobrinen

*Department of Mathematics, University of Denver,
C.M. Knudson Hall 302, 2290 S. York St., Denver,
CO 80208, USA*
natasha.dobrinen@du.edu

http://cs.du.edu/~ndobrine

Topological Ramsey spaces are spaces which support infinite dimensional Ramsey theory similarly to the Ellentuck space. Each topological Ramsey space is endowed with a partial ordering which can be modified to a σ-closed "almost reduction" relation analogously to the partial ordering of "mod finite" on $[\omega]^\omega$. Such forcings add new ultrafilters satisfying weak partition relations and have complete combinatorics. In cases where a forcing turned out to be equivalent to a topological Ramsey space, the strong Ramsey-theoretic techniques have aided in a fine-tuned analysis of the Rudin–Keisler and Tukey structures associated with the forced ultrafilter and in discovering new ultrafilters with complete combinatorics. This expository paper provides an overview of this collection of results and an entry point for those interested in using topological Ramsey space techniques to gain finer insight into ultrafilters satisfying weak partition relations.

1. Overview

Topological Ramsey spaces are essentially topological spaces which support infinite-dimensional Ramsey theory. The prototype of all topological Ramsey spaces is the Ellentuck space. This is the space of all infinite subsets of the natural numbers equipped with the Ellentuck topology, a refinement of the usual metric, or equivalently, product topology. In this refined topology, every subset of the Ellentuck space which has the property of Baire is Ramsey. This extends the usual Ramsey Theorem for pairs or triples, etc., of natural numbers to infinite dimensions, meaning sets of infinite subsets of the natural numbers, with the additional requirement that the sets be definable in some sense.

Partially ordering the members of the Ellentuck space by almost inclusion yields a forcing which is equivalent to forcing with the Boolean algebra $\mathcal{P}(\omega)/\text{fin}$. This forcing adds a Ramsey ultrafilter. Ramsey ultrafilters have strong properties: They are Rudin–Keisler minimal, Tukey minimal, and have complete combinatorics over $L(\mathcal{R})$, in the presence of large cardinals.

These important features are not unique to the Ellentuck space. Rather, the same or analogous properties hold for a general class of spaces called topological Ramsey spaces. The class of such spaces were defined by abstracting the key properties from seminal spaces of Ellentuck, Carlson–Simpson, and Milliken's space of block sequences, and others. Building on the work of Carlson and Simpson [8], the first to form an abstract approach to such spaces, Todorcevic presented a more streamlined set of axioms guaranteeing a space is a topological Ramsey space in [35]. This is the setting that we work in.

Topological Ramsey spaces come equipped with a partial ordering. This partial ordering can be modified to a naturally defined σ-closed partial ordering of *almost reduction*, similarly to how the partial ordering of inclusion modulo finite is defined from the partial ordering of inclusion. This almost reduction ordering was defined for abstract topological Ramsey spaces by Mijares in [27]. He showed that forcing with a topological Ramsey space partially ordered by almost reduction adds a new ultrafilter on the countable base set of first approximations. Such ultrafilters inherit some weak partition relations from the fact that they were forced by a topological Ramsey space; they behave like weak versions of a Ramsey ultrafilter.

When an ultrafilter is forced by a topological Ramsey space, one immediately has strong techniques at one's disposal. The Abstract Ellentuck Theorem serves to both streamline proofs and helps to clarify what exactly is causing the particular properties of the forced ultrafilter. The structure of the topological Ramsey space aids in several factors of the analysis of the behavior of the ultrafilter. The following are made possible by knowing that a given forcing is equivalent to forcing with some topological Ramsey space.

(1) A simpler reading of the Ramsey degrees of the forced ultrafilter.
(2) Complete Combinatorics.
(3) Exact Tukey and Rudin–Keisler structures, as well as the structure of the Rudin–Keisler classes inside the Tukey classes.
(4) New canonical equivalence relations on fronts — extensions of the Erdős–Rado and Pudlák–Rödl Theorems.
(5) Streamlines and simplifies proofs, and reveals the underlying structure responsible for the properties of the ultrafilters.

This chapter focuses on studies of ultrafilters satisfying weak partition relations and which can be forced by some σ-closed partial orderings in [10, 11, 13, 15, 16], and other work. These works concentrate on weakly Ramsey ultrafilters and a family of ultrafilters with increasingly weak partition properties due to Laflamme in [23]; P-points forced *n-square forcing* by Blass in [3] which have Rudin–Keisler structure below them a diamond shape; the k-arrow, not $k + 1$-arrow ultrafilters of Baumgartner and Taylor in [1], as well as the arrow ultrafilters; new classes of P-points with weak partition relations; non-P-points forced by $\mathcal{P}(\omega \times \omega)/(\text{Fin} \otimes \text{Fin})$ and the natural hierarchy of forcings of increasing complexity, $\mathcal{P}(\omega^\alpha)/\text{Fin}^{\otimes \alpha}$.

It turned out that the original forcings adding these ultrafilters actually contain dense subsets which form topological Ramsey spaces. The Ramsey structure of these spaces aided greatly in the analysis of the properties of the forced ultrafilters. In the process some new classes of ultrafilters with weak partition properties were also produced. Though there are many other classes of ultrafilters not yet studied in this context, the fact that in all these cases dense subsets of the forcings forming topological Ramsey spaces were found signifies a strong connection between ultrafilters satisfying some partition relations and topological Ramsey spaces. Thus, we make the following conjecture.

Conjecture 1. *Every ultrafilter which satisfies some partition relation and is forced by some σ-closed forcing is actually forced by some topological Ramsey space.*

While this is a strong conjecture, so far there is no evidence to the contrary, and it is a motivating thesis for using topological Ramsey spaces to find a unifying framework for ultrafilters satisfying some weak partition properties.

Finally, a note about attributions: We attribute work as stated in the papers quoted.

2.   A Few Basic Definitions

Most definitions used will appear as needed throughout this chapter. In this section, we define a few notions needed throughout.

Definition 2. A *filter* $\mathcal{F}$ on a countable base set B is a collection of subsets of B which is closed under finite intersection and closed under superset. An *ultrafilter* $\mathcal{U}$ on a countable base set

B is a filter such that each subset of B or its complement is in $\mathcal{U}$.

We hold to the convention that all ultrafilters are proper ultrafilters; thus $\emptyset$ is not a member of any ultrafilter.

Definition 3. An ultrafilter $\mathcal{U}$ on a countable base set B is

(1) *Ramsey* if for each $k, l \geq 1$ and each coloring $c : [B]^k \to l$, there is a member $U \in \mathcal{U}$ such that $c \upharpoonright [U]^k$ is constant.
(2) *selective* if for each function $f : \omega \to \omega$, there is a member $X \in \mathcal{U}$ such that f is either constant or one-to-one on U.
(3) *Mathias-selective* if for each collection $\{U_s : s \in [\omega]^{<\omega}\}$ of members in $\mathcal{U}$, there is an $X \in \mathcal{U}$ such that for each $s \in [\omega]^{<\omega}$ for which $\max(s) \in X$, $X \setminus (\max(s) + 1) \subseteq U_s$.

The three definitions above are equivalent. Booth proved in [7] that (1) and (2) are equivalent, and Mathias proved in [26] that (1) and (3) are equivalent.

Definition 4. An ultrafilter $\mathcal{U}$ on a countable base set B is a *P-point* if for each sequence U_n, $n < \omega$, of members of $\mathcal{U}$, there is an X in $\mathcal{U}$ such that for each $n < \omega$, $X \subseteq^* U_n$. Equivalently, $\mathcal{U}$ is a *P-point* if for each function $f : \omega \to \omega$ there is a member X in $\mathcal{U}$ such that f is either constant or finite-to-one on X.

$\mathcal{U}$ is *rapid* if for each strictly increasing function $f : \omega \to \omega$, there is a member $X \in \mathcal{U}$ such that for each $n < \omega$, $|X \cap f(n)| < n$.

Using the function definition of P-point, it is clear that a selective implies P-point, which in turn implies rapid.

A different hierarchy of ultrafilters may be formed by weakening the Ramsey requirement to only require some bound on the number of colors appearing, rather than requiring homogeneity on a member in the ultrafilter. The first of this type of weakening is a *weakly Ramsey* ultrafilter, which is an ultrafilter $\mathcal{U}$ such that for each $l \geq 3$ and coloring $f : [\omega]^2 \to l$, there is a member

$X \in \mathcal{U}$ such that the restriction of f to $[X]^2$ takes no more than two colors. The usual notation to denote this statement is

$$\mathcal{U} \to (\mathcal{U})^2_{l,2}. \tag{1}$$

This idea can be extended to any $k \geq 2$, defining a *k-Ramsey* ultrafilter to be one such that for each $l > k$ and $f : [\omega]^2 \to l$, there is a member $X \in \mathcal{U}$ such that the restriction of f to $[X]^2$ takes no more than k colors. This is denoted

$$\mathcal{U} \to (\mathcal{U})^2_{l,k}. \tag{2}$$

As we shall review later, Laflamme forced a hierarchy of ultrafilters $\mathcal{U}_k$, $k \geq 1$, such that $\mathcal{U}_k$ is $k + 1$-Ramsey but not k-Ramsey [23].

Next, we present one of the most useful ways of constructing new ultrafilters from old ones.

Definition 5 (Fubini Product). Let $\mathcal{U}$ and $\mathcal{V}_n$, $n < \omega$, be ultrafilters on ω. The *Fubini product* of $\mathcal{U}$ and $(\mathcal{V}_n)_{n<\omega}$ is the ultrafilter $\lim_{n\to\mathcal{U}} \mathcal{V}_n$ on base set $\omega \times \omega$ such that a set $A \subseteq \omega \times \omega$ is in $\lim_{n\to\mathcal{U}} \mathcal{V}_n$ if and only if

$$\{n < \omega : \{j < \omega : (n,j) \in A\} \in \mathcal{V}_n\} \in \mathcal{U}. \tag{3}$$

In other words, a subset $A \subseteq \omega \times \omega$ is in $\lim_{n\to\mathcal{U}} \mathcal{V}_n$ if and only if for $\mathcal{U}$ many n, the nth fiber of A is a member of $\mathcal{V}_n$. If all $\mathcal{V}_n$ are equal to the same ultrafilter $\mathcal{V}$, then the Fubini product is written as $\mathcal{U} \cdot \mathcal{V}$.

The Fubini product construction can be continued recursively any countable ordinal many times. In particular, for any countable ordinal α, the αth Fubini iterate of $\mathcal{U}$ is denoted by $\mathcal{U}^\alpha$. The importance of this fact will be seen in later sections.

3. The Prototype Example: Ramsey Ultrafilters and the Ellentuck Space

The connections and interactions between Ramsey ultrafilters and the Ellentuck space provide the fundamental example of the phenomena we are illustrating in this chapter. Recall that a *Ramsey ultrafilter* is an ultrafilter $\mathcal{U}$ on a countable base set, usually taken to be ω, which contains witnesses of Ramsey's theorem: For each $k, l \geq 1$ and each coloring $c : [\omega]^k \to l$, there is an $X \in \mathcal{U}$ such that $c \restriction [X]^k$ is constant. Ramsey ultrafilters are forced by $\mathcal{P}(\omega)/\text{Fin}$, which is a σ-closed forcing. This is forcing equivalent to the partial ordering $([\omega]^\omega, \subseteq^*)$, where for $X, Y \in [\omega]^\omega$, $Y \subseteq^* X$ if and only if $Y \setminus X$ is finite. This section reviews the complete combinatorics of Ramsey ultrafilters, the exact Rudin–Keisler and Tukey structures connected with Ramsey ultrafilters, and the roles played by the Ellentuck space, either implicitly or explicitly in these results, and the crucial theorems of Nash–Williams, Ellentuck, and Pudlák–Rödl. This provides the groundwork from which to understand the more general results.

3.1. Complete combinatorics of Ramsey ultrafilters

Saying that an ultrafilter has *complete combinatorics* means that there is some forcing and some well-defined combinatorial property such that any ultrafilter satisfying that property is generic for the forcing over some well-defined inner model. There are two main formulations of complete combinatorics for Ramsey ultrafilters. The first has its inception in work of Mathias in [26] and was formulated by Blass in [5]: Any Ramsey ultrafilter in the model $V[G]$ obtained by Lévy collapsing a Mahlo cardinal to $\aleph_1$ is $\mathcal{P}(\omega)/\text{Fin}$-generic over $\text{HOD}(\mathbb{R})^{V[G]}$. Thus, we say that Ramsey ultrafilters have *complete combinatorics* over $\text{HOD}(\mathbb{R})^{V[G]}$, where $V[G]$ is obtained by Lévy collapsing a Mahlo cardinal to $\aleph_1$. This form of complete combinatorics does not take place in the original model V, but only presupposes the existence of a Mahlo cardinal.

The second formulation of complete combinatorics is due to Todorcevic (see [18, Theorem 4.4]) building on work of Shelah and Woodin [32]. It presupposes the existence of large cardinals stronger than a Mahlo but has the advantage that the statement is with respect to the canonical inner model $L(\mathbb{R})$ inside V rather than in a forcing extension of V collapsing a Mahlo cardinal. If V has a supercompact cardinal (or somewhat less), then any Ramsey ultrafilter in V is generic for the forcing $\mathcal{P}(\omega)/\mathrm{Fin}$ over the Solovay model $L(\mathbb{R})$ inside V. Thus, we say that in the presence of certain large cardinals, each Ramsey ultrafilter in V has *complete combinatorics* over $L(\mathbb{R})$. This second formulation lends itself to natural generalizations to forcing with abstract topological Ramsey spaces, as we shall review later.

3.2.   *Rudin–Keisler order*

The well-studied Rudin–Keisler order on ultrafilters is a quasi-ordering in which "stronger" ultrafilters are smaller. Given ultra-filters $\mathcal{U}$ and $\mathcal{V}$, we say that $\mathcal{V}$ is *Rudin–Keisler reducible to* $\mathcal{U}$ if there is a function $f : \omega \to \omega$ such that

$$\mathcal{V} = f(\mathcal{U}) := \{X \subseteq \omega : f^{-1}(X) \in \mathcal{U}\}. \qquad (4)$$

Two ultrafilters are Rudin–Keisler (RK) equivalent if and only if each is RK-reducible to the other. In this case, we write $\mathcal{U} \equiv_{\mathrm{RK}} \mathcal{V}$. It turns out that two ultrafilters are RK equivalent if and only if there is a bijection between their bases taking one ultrafilter to the other (see [2] or [7]). Thus, we shall use the terminology *RK equivalent* and *isomorphic* interchangeably. The collection of all ultrafilters RK equivalent to a given ultrafilter $\mathcal{U}$ is called the *RK class* or *isomorphism class* of $\mathcal{U}$.

Recall that an ultrafilter $\mathcal{U}$ is *selective* if for each function $f : \omega \to \omega$, there is a member $X \in \mathcal{U}$ such that f is either one-to-one or constant on X. If f is constant on some member of $\mathcal{U}$, then the ultrafilter $f(\mathcal{U})$ is principal. If f is one-to-one on some member of $\mathcal{U}$, then $f(\mathcal{U})$ is isomorphic to $\mathcal{U}$. Hence, selective ultrafilters are Rudin-Keisler minimal among nonprincipal ultrafilters.

3.3. *Tukey order on ultrafilters*

The Tukey order between partial orderings was defined by Tukey in order to study convergence in Moore–Smith topology. In recent decades it has found deep applications in areas where isomorphism is too fine a notion to reveal useful information. In the setting of ultrafilters partially ordered by reverse inclusion, the Tukey order is a coarsening of the Rudin–Keisler order and provides information about the cofinal types of ultrafilters.

Given ultrafilters $\mathcal{U}$ and $\mathcal{V}$, we say that $\mathcal{V}$ is *Tukey reducible to* $\mathcal{U}$ if there is a function $f : \mathcal{U} \to \mathcal{V}$ such that for each filter base $\mathcal{B}$ for $\mathcal{U}$, $f''\mathcal{B}$ is a filter base for $\mathcal{V}$. Such a map is called a *cofinal* map or a *convergent* map. Equivalently, $\mathcal{V}$ is Tukey reducible to $\mathcal{U}$ if there is a function $g : \mathcal{V} \to \mathcal{U}$ such that for each unbounded subset $\mathcal{X} \subseteq \mathcal{V}$, the image $g''\mathcal{X}$ is unbounded in $\mathcal{U}$. It is worth noting that whenever $\mathcal{V} \leq_T \mathcal{U}$, then there is a *monotone* cofinal map witnessing this; that is, a map $f : \mathcal{U} \to \mathcal{V}$ such that whenever $X \supseteq Y$ are members of $\mathcal{U}$, then $f(X) \supseteq f(Y)$.

If both $\mathcal{U}$ and $\mathcal{V}$ are Tukey reducible to each other, then we say that they are *Tukey equivalent*. For directed partial orderings, Tukey equivalence is the same as cofinal equivalence: There is some other directed partial ordering into which they both embed as cofinal subsets. Since, for any ultrafilter $\mathcal{U}$, the partial ordering $(\mathcal{U}, \supseteq)$ is directed, Tukey equivalence between ultrafilters is the same as cofinal equivalence. The collection of all ultrafilters Tukey equivalent to $\mathcal{U}$ is called the *Tukey type* or *cofinal type* of $\mathcal{U}$.

Each Rudin–Keisler map induces a monotone cofinal map. If $h : \omega \to \omega$ and $\mathcal{V} = h(\mathcal{U})$, then the map $f : \mathcal{U} \to \mathcal{V}$ given by $f(X) = \{h(n) : n \in X\}$, for $X \in \mathcal{U}$, is a cofinal map witnessing that $\mathcal{V} \leq_T \mathcal{U}$. Thus, Tukey types form a coarsening of the RK classes of ultrafilters.

Todorcevic proved in [31] that, analogously to the Rudin–Keisler order, Ramsey ultrafilters are minimal among nonprincipal ultrafilters in the Tukey ordering. His proof uses a theorem that P-points carry continuous cofinal maps and a theorem of Pudlák and Rödl regarding canonical equivalence relations on barriers of the Ellentuck space. These will be discussed below, after which we will return to an outline of the proof of this theorem.

3.4. *Continuous cofinal maps from P-points*

It was proved in [14] that every P-point carries continuous cofinal maps. The members of a given ultrafilter $\mathcal{U}$ are subsets of ω. Using the natural correspondence between a subset of ω and its characteristic function as an infinite sequence of 0's and 1's, each ultrafilter on ω may be seen as a subspace of the Cantor space 2^ω, endowed with the product topology. A continuous map from 2^ω into itself is a function such that the preimage of any open set is open. This amounts to continuous functions having initial segments of their images being decided by finite amounts of information. Thus, $f : 2^\omega \to 2^\omega$ is continuous if and only if there is a finitary function $\hat{f} : 2^{<\omega} \to 2^{<\omega}$ such that $\hat{f}$ preserves end-extensions and reproduces f. Precisely, for $s \sqsubseteq t$, $\hat{f}(s) \sqsubseteq \hat{f}(t)$, and for each $X \in 2^\omega$, $f(X) = \bigcup_{n<\omega} \hat{f}(X \upharpoonright n)$. (For finite sets of natural numbers, the notation $s \sqsubseteq t$ is used to denote that s is an initial segment of t, meaning that $s = \{n \in t : n \le \max(s)\}$. $s \sqsubset t$ denotes that s is a proper initial segment of t: s is an initial segment of t and s is not equal to t.)

Recall that an ultrafilter $\mathcal{U}$ is a *P-point* if whenever X_n, $n < \omega$, are members of $\mathcal{U}$ such that each $X_{n+1} \subseteq^* X_n$, then there is a member $U \in \mathcal{U}$ such that for each $n < \omega$, $U \subseteq^* X_n$. Such a set U is called a *pseudointersection* of the sequence of $\{U_n : n < \omega\}$.

Theorem 6 (Dobrinen and Todorcevic [14]). *For each P-point $\mathcal{U}$, if $f : \mathcal{U} \to \mathcal{V}$ is a monotone cofinal map, then there is a member $X \in \mathcal{U}$ such that f is continuous when restricted to the set $\mathcal{U} \upharpoonright X := \{U \in \mathcal{U} : U \subseteq X\}$. Moreover, there is a monotone continuous function f^* from $\mathcal{P}(\omega)$ into $\mathcal{P}(\omega)$ such that $f^* \upharpoonright (\mathcal{U} \upharpoonright \tilde{X}) = f \upharpoonright (\mathcal{U} \upharpoonright \tilde{X})$.*

3.5.   *The Ellentuck space*

The Ellentuck space has as its points the infinite subsets of the natural numbers, $[\omega]^{\omega}$. For $a \in [\omega]^{<\omega}$ and $X \in [\omega]^{\omega}$, $a \sqsubset X$ denotes that $a = \{n \in X : n \leq \max(a)\}$. The basic open sets inducing the Ellentuck topology are of the following form: Given a finite set $a \in [\omega]^{<\omega}$ and an infinite set $X \in [\omega]^{\omega}$, define

$$[a, X] = \{Y \in [\omega]^{\omega} : a \sqsubset Y \subseteq X\}. \tag{5}$$

The *Ellentuck topology* is the topology on the space $[\omega]^{\omega}$ induced by all basic open sets of the form $[a, X]$, for $a \in [\omega]^{<\omega}$ and $X \in [\omega]^{\omega}$. Notice that this topology refines the usual metric or equivalently product topology on $[\omega]^{\omega}$.

It is this topology which is the correct one in which to understand infinite-dimensional Ramsey theory. Infinite-dimensional Ramsey theory is the extension of Ramsey theory from finite dimensions, that is, colorings of $[\omega]^{k}$ where k is some positive integer, to colorings of $[\omega]^{\omega}$. Assuming the Axiom of Choice, the following statement is false: "Given a function $f : [\omega]^{\omega} \to 2$, there is an $M \in [\omega]^{\omega}$ such that f is constant on $[M]^{\omega}$". However, if the coloring is sufficiently definable, then Ramsey theorems hold. This is the content of the progression from the Nash–Williams Theorem [29] through the work of Galvin–Prikry [19], Mathias [26], Silver [33] and Louveau [25] up to the theorem of Ellentuck showing that the Ellentuck topology is the correct topology in which to obtain optimal infinite-dimensional Ramsey theory.

Theorem 7 (Ellentuck [17]). *If $\mathcal{X} \subseteq [\omega]^{\omega}$ has the property of Baire in the Ellentuck topology, then for any basic open set $[a, X]$, there is a member $Y \in [a, X]$ such that either $[a, Y] \subseteq \mathcal{X}$ or else $[a, Y] \cap \mathcal{X} = \emptyset$.*

Those familiar with Mathias forcing will notice the strong correlation between Ellentuck's basic open sets and conditions in Mathias forcing. Forcing with the collection of basic open sets in the Ellentuck topology, partially ordered by inclusion, is in fact equivalent to Mathias forcing.

3.6. *Fronts, barriers, and the Nash-Williams theorem*

Fronts and barriers are collections of finite sets which approximate all infinite sets and are minimal in some sense.

Definition 8. A set $\mathcal{F} \subseteq [\omega]^{<\omega}$ is a *front* if

(1) For each $X \in [\omega]^{\omega}$, there is an $a \in \mathcal{F}$ such that $a \sqsubset X$.
(2) Whenever $a, b \in \mathcal{F}$ and $a \neq b$, then $a \not\sqsubset b$.

$\mathcal{F}$ is a *barrier* if it satisfies (1) and also (2′) holds:

(2′) Whenever $a, b \in \mathcal{F}$ and $a \neq b$, then $a \not\subset b$.

A family satisfying (2) is called *Nash–Williams* and a family satisfying (2′) is called *Sperner*. The notions of front and barrier may be relativized to any infinite subset of ω. By a theorem of Galvin, for each front $\mathcal{F}$ on some infinite $M \subseteq \omega$, there is an infinite subset $N \subseteq M$ such that $\mathcal{F}|N := \{a \in \mathcal{F} : a \subseteq N\}$ is a barrier on N.

Notice that for each $k < \omega$, the set $[\omega]^k$ is both a front and a barrier on ω. The set $[\omega]^k$ is in fact the uniform barrier of rank k. Uniform barriers, and fronts, are defined by recursion on the rank. Given uniform barriers $\mathcal{B}_n$ on $\omega \setminus (n+1)$, with the rank of $\mathcal{B}_n$ being α_n, where either all α_n are the same or else they are strictly increasing, the barrier

$$\mathcal{B} = \{\{n\} \cup a : a \in \mathcal{B}_n,\ n < \omega\} \tag{6}$$

is a uniform barrier of rank $\sup\{\alpha_n + 1 : n < \omega\}$. The *Shreier barrier* $\mathcal{S}$ is the fundamental example of a uniform barrier of rank ω.

$$\mathcal{S} = \{a \in [\omega]^{<\omega} : |a| = \min(a) + 1\}. \tag{7}$$

This is the same as letting $\mathcal{B}_n = [\omega \setminus (n+1)]^n$ and defining $\mathcal{S}$ as in Eq. (6).

The Nash–Williams theorem shows that every clopen subset of the Baire space with the metric topology has the Ramsey property. Though this follows from the Ellentuck theorem, we state it here since it will be useful in several proofs which do not require the full strength of the Ellentuck theorem.

Theorem 9 (Nash–Williams). *Given any front $\mathcal{F}$ on an infinite set $M \subseteq \omega$ and any partition of $\mathcal{F}$ into finitely many pieces, $\mathcal{F}_i$, $i < l$ for some $l \geq 1$, there is an infinite $N \subseteq M$ such that $\mathcal{F}|N \subseteq \mathcal{F}_i$ for one $i < l$.*

Remark. Any ultrafilter $\mathcal{U}$ generic for $([\omega]^\omega, \subseteq^*)$ satisfies a generic version of the Nash–Williams theorem. By a density argument, for each $k \geq 1$, given any partition of $[\omega]^k$ into finitely many sets, there is a member $X \in \mathcal{U}$ such that $[X]^k$ is contained in one piece of the partition. Thus, we may see the Ramsey property for the generic ultrafilter as instances of the Nash–Williams theorem for uniform barriers of finite rank.

3.7. *Canonical equivalence relations on barriers*

Letting the number of colors increase, we see that coloring with infinitely many colors corresponds to forming an equivalence relation on a collection of finite sets, where two finite sets are equivalent if and only if they have the same color. The first Ramsey-like theorem for infinite colorings of finite sets of natural numbers is due to Erdős and Rado. Though it is not always possible to find a large homogeneous set in one color, it is possible to find a large set on which the equivalence relation is in some sense *canonical*. This term is used to refer to some simple equivalence relation which once achieved, is inherited by all further infinite subsets.

Given $k \geq 1$, the canonical equivalence relations on $[\omega]^k$ are of the form E_I, where $I \subseteq k$, defined by

$$a \, E_I \, b \longleftrightarrow \{a_i : i \in I\} = \{b_i : i \in I\}, \tag{8}$$

where $a, b \in [\omega]^k$ and $\{a_0, \ldots, a_{k-1}\}$ and $\{b_0, \ldots, b_{k-1}\}$ are the strictly increasing enumerations of a and b.

Theorem 10 (Erdős–Rado). *Given $k \geq 1$ and an equivalence relation E on $[\omega]^k$, there is some infinite $M \subseteq \omega$ and some $I \subseteq k$ such that the restriction of E to $[M]^k$ is exactly $E_I \restriction [M]^k$.*

It is often quite useful to think of canonical equivalence relations in terms of projection maps. For $I \subseteq k$ and $a \in [\omega]^k$, let $\pi_I(a) = \{a_i : i \in I\}$. Then $a \, E_I \, b$ if and only if $\pi_I(a) = \pi_I(b)$.

The notion of canonical equivalence relations may be extended to all uniform barriers of any countable ordinal rank. This is the content of the next theorem of Pudlák and Rödl. Since the lengths of members of a uniform barrier of infinite rank are not bounded, the notion of canonical becomes a bit less obvious at the start. However, just as the canonical equivalence relations on $[\omega]^k$ can be thought of as projections to the ith members of a, for i in a given index set, so too it can be instructive to think of irreducible functions as projections to certain indexed members of a given finite set.

Definition 11 (irreducible function). Let $\mathcal{B}$ be a barrier on ω. A function $\varphi : \mathcal{B} \to [\omega]^{<\omega}$ is *irreducible* if

(1) For each $a \in \mathcal{B}$, $\varphi(a) \subseteq a$.
(2) For all $a, b \in \mathcal{B}$, if $\varphi(a) \neq \varphi(b)$, then $\varphi(a) \not\subseteq \varphi(b)$.

Property (2) implies that the image set $\{\varphi(a) : a \in \mathcal{B}\}$ is a Sperner set, and thus forms a barrier on some infinite set.

Theorem 12 (Pudlák–Rödl). *Given an equivalence relation E on a barrier $\mathcal{B}$ on ω, there is an infinite subset $M \subseteq \omega$ and an irreducible function φ which canonize E on $\mathcal{B} \restriction M$: For all $a, b \in \mathcal{B}|M$, $a \, E \, b$ if and only if $\varphi(a) = \varphi(b)$.*

The Erdős–Rado theorem follows from the Pudlák–Rödl theorem, and it may be instructive for the reader to prove this.

Remark. We stated that it can be instructive to think of irreducible functions as projection maps. This view provides intuition for understanding irreducible maps canonizing equivalence relations on barriers for topological Ramsey spaces which are more complex than the Ellentuck space.

3.8. *Fubini iterates of ultrafilters and the correspondence with uniform fronts*

New ultrafilters may be constructed from given ultrafilters using the process of Fubini product and iterating it countably many times. Recall that the *Fubini product* $\lim_{n \to \mathcal{U}} \mathcal{V}_n$ of $\mathcal{U}$ with $\mathcal{V}_n$, $n < \omega$, is

$$\{A \subseteq \omega \times \omega : \{n < \omega : \{j < \omega : (n,j) \in A\} \in \mathcal{V}_n\} \in \mathcal{U}\}. \quad (9)$$

Note that $[\omega]^2$ is in one-to-one correspondence with the upper triangle of $\omega \times \omega$. As long as $\mathcal{U}$-many of the ultrafilters $\mathcal{V}_n$ are nonprincipal, the upper triangle $\{(n,j) : n < j < \omega\}$ is a member of $\lim_{n \to \mathcal{U}} \mathcal{V}_n$. Thus, $\lim_{n \to \mathcal{U}} \mathcal{V}_n$ is isomorphic to the ultrafilter $\mathcal{W}$ defined on base set $[\omega]^2$ by $B \subseteq [\omega]^2 \in \mathcal{W}$ if and only if for $\mathcal{U}$ many $n < \omega$, the set $\{j > n : \{n,j\} \in B\} \in \mathcal{V}_n$. Hence we may assume that the base set for the ultrafilter is $[\omega]^2$ rather than $\omega \times \omega$. It thus is natural to let $\lim_{n \to \mathcal{U}} \mathcal{V}_n$ denote this ultrafilter $\mathcal{W}$ on base set $[\omega]^2$.

This connection continues when we iterate the Fubini product construction. If we have ultrafilters $\mathcal{W}_n$ each of which is a Fubini product of some ultrafilters and $\mathcal{X}$ is another ultrafilter, then $\lim_{n \to \mathcal{X}} \mathcal{W}_n$ is an ultrafilter on base set $\omega \times \omega \times \omega$. As long as the ultrafilters are nonprincipal, this is isomorphic to an ultrafilter on base set $[\omega]^3$, since each ultrafilter $\mathcal{W}_n$ is (modulo some set in $\mathcal{W}_n$) an ultrafilter on base set $[\omega]^2$.

This recursive construction continues onward so that to each ultrafilter $\mathcal{W}$ which is obtained via a countable iteration of Fubini products of ultrafilters, there corresponds a barrier $\mathcal{B}$ of the same rank as the recursive rank of the construction of $\mathcal{W}$

so that the base set of $\mathcal{W}$ may without any loss of information be assumed to be $\mathcal{B}$. This idea of using uniform barriers as the base sets for iterated Fubini products is due to Todorcevic. It is delineated in more detail in [12].

3.9. *Ramsey ultrafilters are Tukey minimal, and the RK structure inside its Tukey type is exactly the Fubini powers of the Ramsey ultrafilter*

Given an ultrafilter $\mathcal{U}$ and $1 \leq \alpha < \omega_1$, $\mathcal{U}^\alpha$ denotes the αth Fubini power of $\mathcal{U}$. This is formed by the Fubini product where all the $\mathcal{V}_n = \mathcal{U}$ at each iteration of the Fubini product. For limit α, given $\mathcal{U}^\beta$ for all $\beta < \alpha$, $\mathcal{U}^\alpha$ denotes $\lim_{n \to \mathcal{U}} \mathcal{U}^{\beta_n}$, where $(\beta_n)_{n<\omega}$ is any increasing sequence cofinal in α. The following theorem is attributed to Todorcevic in [31].

Theorem 13 (Todorcevic [31]). *Each Ramsey ultrafilter is Tukey minimal. Moreover, if $\mathcal{U}$ is Ramsey and $\mathcal{V} \leq_T \mathcal{U}$, then $\mathcal{V} \equiv_{\mathrm{RK}} \mathcal{U}^\alpha$ for some $\alpha < \omega_1$.*

The structure of his proof is as follows: Suppose that $\mathcal{U}$ is a Ramsey ultrafilter, $\mathcal{V} \leq_T \mathcal{U}$, and $f : \mathcal{U} \to \mathcal{V}$ is a monotone cofinal map.

(1) By Theorem 6, there is a member $X \in \mathcal{U}$ such that f is continuous when restricted to $\mathcal{U} \upharpoonright X := \{U \in \mathcal{U} : U \subseteq X\}$. Without loss of generality, f may be assumed to be continuous on all of $\mathcal{U}$.

(2) The finitary map $\hat{f} : [\omega]^{<\omega} \to [\omega]^{<\omega}$ approximating f on $\mathcal{U}$ is used to define the set $\mathcal{F}$ of all $U \cap n$ where $U \in \mathcal{U}$ and n is minimal such that $\hat{f}(U \cap n) \neq \emptyset$. The set $\mathcal{F}$ forms a *front on* $\mathcal{U}$ since each member of $\mathcal{U}$ end-extends some member of $\mathcal{F}$, and the minimality of the members of $\mathcal{F}$ imply that this set is Nash-Williams.

(3) The set $\mathcal{F}$ is the countable base set for a new ultrafilter. For each set $U \in \mathcal{U}$,

$$\mathcal{F}|U = \{a \in \mathcal{F} : a \subseteq U\}. \tag{10}$$

$\mathcal{U} \upharpoonright \mathcal{F}$ denotes the filter on base set $\mathcal{F}$ generated by the sets $\mathcal{F}|U$, for $U \in \mathcal{U}$. This filter turns out to be an ultrafilter on base set $\mathcal{F}$. This follows since $\mathcal{U}$ being Ramsey implies, with a bit of work, that for any partition of $\mathcal{F}$ into two pieces $\mathcal{F}_0$, $\mathcal{F}_1$, there is an $X \in \mathcal{U}$ such that $\mathcal{F}|X \subseteq \mathcal{F}_i$ for one $i < 2$.

(4) Define a function $g : \mathcal{F} \to \omega$ by $g(a) = \min(\hat{f}(a))$. Then g is a Rudin–Keisler map from the countable base set $\mathcal{F}$ to ω. It turns out that the ultrafilter $\mathcal{V}$ is equal to the Rudin–Keisler image $g(\mathcal{U} \upharpoonright \mathcal{F})$. Thus, to compare $\mathcal{V}$ with $\mathcal{U}$, it is sufficient to compare $g(\mathcal{U} \upharpoonright \mathcal{F})$ with $\mathcal{U}$.

(5) g colors $\mathcal{F}$ into infinitely many colors, thus inducing an equivalence relation on $\mathcal{F}$: For $a, b \in \mathcal{F}$, a and b are equivalent if and only if $g(a) = g(b)$. By the Pudlák–Rödl theorem applied to this equivalence relation on $\mathcal{F}$, there is a member $U \in \mathcal{U}$ such that below U, the equivalence relation is canonical witnessed by some irreducible map φ. The image of $\mathcal{F}$ under φ also is a barrier (possibly restricting below some smaller set in $\mathcal{U}$); let $\mathcal{B} = \{\varphi(a) : a \in \mathcal{F}\}$. Thus,

$$\mathcal{V} = g(\mathcal{U} \upharpoonright \mathcal{F}) \cong \varphi(\mathcal{U} \upharpoonright \mathcal{F}) = \mathcal{U} \upharpoonright \mathcal{B}, \tag{11}$$

where $\varphi(\mathcal{U} \upharpoonright \mathcal{F})$ is the ultrafilter generated by the sets $\{\varphi(a) : a \in \mathcal{F}|Y\}$, $Y \in \mathcal{U}|U$, and $\mathcal{U} \upharpoonright \mathcal{B}$ is the ultrafilter on base set $\mathcal{B}$ generated by the sets $\mathcal{B}|Y := \{b \in \mathcal{B} : b \subseteq Y\}$, where $Y \in \mathcal{U}$. This is a form of being a Fubini power of $\mathcal{U}$, with rank equal to the rank of the barrier $\mathcal{B}$.

This proof outline turns out to work for many other cases of ultrafilters associated to topological Ramsey spaces, as was first discovered in [15]. The structures become more complex as we move towards ultrafilters with weaker partition relations, but for the cases of P-points investigated so far, we can find the exact structure of the RK classes inside the Tukey types.

4.   Forcing with Topological Ramsey Spaces

In the previous section, we outlined some key properties of forcing with the Ellentuck space partially ordered by almost inclusion, which is forcing equivalent to the partial ordering $\mathcal{P}(\omega)/\text{Fin}$. We saw that the Ramsey property is a sufficient combinatorial property to completely characterize when an ultrafilter is forced by $\mathcal{P}(\omega)/\text{Fin}$ over certain special inner models of the forms $\text{HOD}(\mathbb{R})^{V[G]}$ or $L(\mathbb{R})$. We outlined how theorems regarding canonical equivalence relations on barriers can be applied using continuous cofinal maps to classify the ultrafilters which are Tukey reducible to a Ramsey ultrafilter. Moreover, the methods employed made clear the exact structure of the Rudin–Keisler classes inside the Tukey type of a Ramsey ultrafilter: These are the isomorphism classes of the countable Fubini powers of the Ramsey ultrafilter. These results turn out to be special cases of more general phenomena arising when one forces ultrafilters using topological Ramsey spaces partially ordered by the σ-closed almost reduction. In this section, we provide an overview of these phenomena for abstract topological Ramsey spaces.

We first begin with the abstract definition of a topological Ramsey space from Todorcevic's book [35].

4.1.   *Basics of general topological Ramsey spaces*

Building on earlier work of Carlson and Simpson in [8], Todorcevic distilled key properties of the Ellentuck space into four axioms, **A.1**–**A.4**, which guarantee that a space is a topological Ramsey space. The axioms are defined for triples $(\mathcal{R}, \leq, r)$ of objects with the following properties. $\mathcal{R}$ is a nonempty set, $\leq$ is a quasi-ordering on $\mathcal{R}$, and $r : \mathcal{R} \times \omega \to \mathcal{AR}$

is a mapping giving us the sequence $(r_n(\cdot) = r(\cdot, n))$ of approximation mappings, where $\mathcal{AR}$ is the collection of all finite approximations to members of $\mathcal{R}$. For $a \in \mathcal{AR}$ and $A, B \in \mathcal{R}$,

$$[a, B] = \{A \in \mathcal{R} : A \le B \text{ and } (\exists n) \, r_n(A) = a\}. \quad (12)$$

For $a \in \mathcal{AR}$, let $|a|$ denote the length of the sequence a. Thus, $|a|$ equals the integer k for which $a = r_k(a)$. For $a, b \in \mathcal{AR}$, $a \sqsubseteq b$ if and only if $a = r_m(b)$ for some $m \le |b|$. $a \sqsubset b$ if and only if $a = r_m(b)$ for some $m < |b|$. For each $n < \omega$, $\mathcal{AR}_n = \{r_n(A) : A \in \mathcal{R}\}$.

A.1 (a) $r_0(A) = \emptyset$ for all $A \in \mathcal{R}$.
   (b) $A \neq B$ implies $r_n(A) \neq r_n(B)$ for some n.
   (c) $r_n(A) = r_m(B)$ implies $n = m$ and $r_k(A) = r_k(B)$ for all $k < n$.

A.2 There is a quasi-ordering $\le_{\text{fin}}$ on $\mathcal{AR}$ such that

   (a) $\{a \in \mathcal{AR} : a \le_{\text{fin}} b\}$ is finite for all $b \in \mathcal{AR}$.
   (b) $A \le B$ if and only if $(\forall n)(\exists m) \, r_n(A) \le_{\text{fin}} r_m(B)$.
   (c) $\forall a, b, c \in \mathcal{AR}[a \sqsubset b \wedge b \le_{\text{fin}} c \to \exists d \sqsubset c \, a \le_{\text{fin}} d]$.

The number $\text{depth}_B(a)$ is the least n, if it exists, such that $a \le_{\text{fin}} r_n(B)$. If such an n does not exist, then we write $\text{depth}_B(a) = \infty$. If $\text{depth}_B(a) = n < \infty$, then $[\text{depth}_B(a), B]$ denotes $[r_n(B), B]$.

A.3 (a) If $\text{depth}_B(a) < \infty$, then $[a, A] \neq \emptyset$ for all $A \in [\text{depth}_B(a), B]$.
   (b) $A \le B$ and $[a, A] \neq \emptyset$ imply that there is $A' \in [\text{depth}_B(a), B]$ such that $\emptyset \neq [a, A'] \subseteq [a, A]$.

If $n > |a|$, then $r_n[a, A]$ denotes the collection of all $b \in \mathcal{AR}_n$ such that $a \sqsubset b$ and $b \le_{\text{fin}} A$.

A.4 If $\text{depth}_B(a) < \infty$ and if $\mathcal{O} \subseteq \mathcal{AR}_{|a|+1}$, then there is $A \in [\text{depth}_B(a), B]$ such that $r_{|a|+1}[a, A] \subseteq \mathcal{O}$ or $r_{|a|+1}[a, A] \subseteq \mathcal{O}^c$.

The *Ellentuck topology* on $\mathcal{R}$ is the topology generated by the basic open sets $[a, B]$; it extends the usual metrizable topology on $\mathcal{R}$ when we consider $\mathcal{R}$ as a subspace of the Tychonoff cube $\mathcal{AR}^{\mathbb{N}}$. Given the Ellentuck topology on $\mathcal{R}$, the notions of nowhere dense, and hence of meager are defined in the natural way. We say that a subset $\mathcal{X}$ of $\mathcal{R}$ has the *property of Baire* if and only if $\mathcal{X} = \mathcal{O} \cap \mathcal{M}$ for some Ellentuck open set $\mathcal{O} \subseteq \mathcal{R}$ and Ellentuck meager set $\mathcal{M} \subseteq \mathcal{R}$. A subset $\mathcal{X}$ of $\mathcal{R}$ is *Ramsey* if for every $\emptyset \neq [a, A]$, there is a $B \in [a, A]$ such that $[a, B] \subseteq \mathcal{X}$ or $[a, B] \cap \mathcal{X} = \emptyset$. $\mathcal{X} \subseteq \mathcal{R}$ is *Ramsey null* if for every $\emptyset \neq [a, A]$, there is a $B \in [a, A]$ such that $[a, B] \cap \mathcal{X} = \emptyset$.

Definition 14 ([35]). A triple $(\mathcal{R}, \leq, r)$ is a *topological Ramsey space* if every subset of $\mathcal{R}$ with the property of Baire is Ramsey and if every meager subset of $\mathcal{R}$ is Ramsey null.

The following result can be found as Theorem 5.4 in [35].

Theorem 15 (Abstract Ellentuck Theorem). *If $(\mathcal{R}, \leq, r)$ is closed (as a subspace of $\mathcal{AR}^{\mathbb{N}}$) and satisfies axioms **A.1**, **A.2**, **A.3**, and **A.4**, then every subset of $\mathcal{R}$ with the property of Baire is Ramsey, and every meager subset is Ramsey null; in other words, the triple $(\mathcal{R}, \leq, r)$ forms a topological Ramsey space.*

Example 16 (The Ellentuck Space). Putting the Ellentuck space into the notation just defined, $\mathcal{R} = [\omega]^{\omega}$ and the partial ordering $\leq$ is simply $\subseteq$. Given $X \in [\omega]^{\omega}$, listing the members of X in increasing order as $\{x_i : i < \omega\}$, the nth approximation to X is $r_n(X) = \{x_i : i < n\}$. In particular, $r_0(X) = \emptyset$. $\mathcal{AR}_n = [\omega]^n$ and $\mathcal{AR} = [\omega]^{<\omega}$. For $a, b \in [\omega]^{<\omega}$, one may define $a \leq_{\text{fin}} b$ if and only if $a \subseteq b$. With these definitions, $([\omega]^{<\omega}, \subseteq, r)$ can be seen to satisfy **Axioms A.1–A.4**, and it is recommended that the reader check this to build a basis for understanding of how the axioms work.

For the Ellentuck space, the definition of $\leq_{\text{fin}}$ is a bit flexible: An alternate definition, $a \leq_{\text{fin}} b$ if and only if $a \subseteq b$ and

$\max(a) = \max(b)$, also satisfies the **Axioms A.1–A.4**. In some spaces, the definition of $\leq_{\text{fin}}$ can be more particular, for instance Milliken's space of strong trees. However, for all the spaces in this chapter, except for Sec. 8.3, the definition of $\leq_{\text{fin}}$ is also flexible in a similar manner as for the Ellentuck space.

4.2. Almost reduction, forced ultrafilters, and complete combinatorics

Let $(\mathcal{R}, \leq, r)$ be any topological Ramsey space. The ordering $\leq$ can be weakened to form a σ-closed order as follows.

Definition 17 (Mijares [27]). For $X, Y \in \mathcal{R}$, write $Y \leq^* X$ if there is an $a \in \mathcal{AR}|Y$ such that $[a, Y] \subseteq [a, X]$. In this case, we say that Y is an *almost reduction* of X.

In the Ellentuck space, $\leq^*$ is simply $\subseteq^*$ where $A \subseteq^* B$ if and only if $A \setminus B$ is finite. Mijares proved in [27] that $(\mathcal{R}, \leq^*)$ is a σ-closed partial ordering. Forcing with $(\mathcal{R}, \leq^*)$ yields a generic filter $\mathcal{U}$ on $\mathcal{R}$. This filter induces an ultrafilter on the countable base set of first-approximations, $\mathcal{AR}_1 := \{r_1(X) : X \in \mathcal{R}\}$, as follows.

Definition 18 (The ultrafilter on first approximations). Let $\mathcal{U}$ be a generic filter for the forcing $(\mathcal{R}, \leq^*)$, where $(\mathcal{R}, \leq, r)$ is some topological Ramsey space. $\mathcal{U} \upharpoonright \mathcal{AR}_1$ denotes the filter on base set $\mathcal{AR}_1$ generated by the sets

$$\mathcal{AR}_1|X := \{r_1(Y) : Y \leq X\}, \quad X \in \mathcal{U}. \tag{13}$$

Fact 19 (Mijares [27]). *For $\mathcal{U}$ generic for $(\mathcal{R}, \leq^*)$, $\mathcal{U} \upharpoonright \mathcal{AR}_1$ is an ultrafilter.*

Since $\mathcal{U}$ is generic and since any subset of $\mathcal{F} \subseteq \mathcal{AR}_1$ induces the clopen (and hence property of Baire) set $\{Y \in \mathcal{R} : \exists a \in \mathcal{F} (a \sqsubset Y)\}$, the Abstract Ellentuck Theorem along with a density argument shows that there is a member $X \in \mathcal{U}$ such that either $\mathcal{AR}_1|X \subseteq \mathcal{F}$ or else $\mathcal{AR}_1|X \cap \mathcal{F} = \emptyset$. This shows that $\mathcal{U} \upharpoonright \mathcal{AR}_1$ is an ultrafilter.

Such generic $\mathcal{U}$ and their induced ultrafilters $\mathcal{U} \upharpoonright \mathcal{AR}_1$ satisfy strong properties. First, they are essentially P-points, but with respect to the space with its finite approximation structure from which they arise rather than with respect to ω. By this, we mean that for any sequence $\langle X_i : i < \omega \rangle$ in $\mathcal{U}$ such that $X_i \geq^* X_{i+1}$ for all i, there is an $X \in \mathcal{U}$ such that $X \leq^* X_i$ for each $i < \omega$. Such an X is pseudo-intersection of the sequence. This often, though not always, will imply that $\mathcal{U} \upharpoonright \mathcal{AR}_1$ is a P-point in the traditional sense, meaning that for each countable set of members $U_i \in \mathcal{U} \upharpoonright \mathcal{AR}_1$, there is a member $X \in \mathcal{U} \upharpoonright \mathcal{AR}_1$ such that for each $i < \omega$, all but finitely many members of X are contained in U_i.

The second and stronger property satisfied by ultrafilters forced by $(\mathcal{R}, \leq^*)$ is the analogue of selectivity, with respect to the topological Ramsey space. In [9], the general definition of semiselective coideal is presented. Here we shall only concentrate on when the generic filter $\mathcal{U}$ is selective. The following definition is the natural abstraction of the formulation of selectivity due to Mathias in [26].

Definition 20 ([9]). Let $\mathcal{U} \subseteq \mathcal{R}$ be a generic filter forced by $(\mathcal{R}, \leq^*)$. Given a family $\mathcal{A} = \{A_a\}_{a \in \mathcal{AR}|A} \subseteq \mathcal{R}$, we say that $Y \in \mathcal{R}$ is a *diagonalization* of $\mathcal{A}$ if for each $a \in \mathcal{AR}|Y$ we have $[a, Y] \subseteq [a, A_a]$.

A set $\mathcal{D} \subseteq$ is said to be *dense open* in $\mathcal{U} \cap [\text{depth}_X(a), X]$ if

(1) (Density) for all $A \in \mathcal{U} \cap [\text{depth}_X(a), X]$, there is a $B \in \mathcal{D}$ such that $B \leq A$.
(2) (Open) for all $A \in \mathcal{U} \cap [\text{depth}_X(a), X]$, for all $B \in \mathcal{D}$, $A \leq B$ implies $A \in \mathcal{D}$.

Given $X \in \mathcal{U}$ and a collection $\vec{\mathcal{D}} = \{\mathcal{D}_a\}_{a \in \mathcal{AR}|X}$ such that each $\mathcal{D}_a$ is open dense in $\mathcal{U} \cap [\text{depth}_X(a), X]$, we say that $Y \in \mathcal{R}$

is a *diagonalization* of $\vec{\mathcal{D}}$ if there is a family $\mathcal{A} = \{A_a\}_{a \in \mathcal{AR}|X}$, with $A_a \in \mathcal{D}_a$, such that Y is a diagonalization of $\mathcal{A}$.

Definition 21 (Abstract Selectivity [9]). A maximal filter $\mathcal{G} \subseteq \mathcal{R}$ is *selective* if for each $A \in \mathcal{G}$ and each collection $\vec{\mathcal{D}} = \{\mathcal{D}_a\}_{a \in \mathcal{AR}|A}$ such that each $\mathcal{D}_a$ is dense open in $\mathcal{G} \cap [\mathrm{depth}_A(a), A]$ and each $B \in \mathcal{G}|A$, there is a $C \in \mathcal{G}$ such that C is a diagonalization of $\mathcal{D}$ and $C \leq B$.

The following may be shown by standard arguments, using genericity.

Fact 22. *Each filter $\mathcal{U} \subseteq \mathcal{R}$ generic for $(\mathcal{R}, \leq^*)$ is selective.*

We now present the version of complete combinatorics for topological Ramsey spaces. This result of Di Prisco, Mijares and Nieto is in fact more general, pertaining to all semiselective coideals. Here we only mention the case for generic filters.

Theorem 23 (Di Prisco, Mijares and Nieto [9]). *If there exists a supercompact cardinal and $\mathcal{G} \subseteq \mathcal{R}$ is generic for $(\mathcal{R}, \leq^*)$, then all definable subsets of $\mathcal{R}$ are $\mathcal{G}$-Ramsey. Hence, each filter in V on base set $\mathcal{R}$ which is selective is generic for the forcing $(\mathcal{R}, \leq^*)$ over the Solovay model $L(\mathbb{R})$.*

For certain topological Ramsey spaces, the Ellentuck space in particular, it is known that for a maximal filter $\mathcal{G}$ on the space, the strong form of selectivity above is equivalent to being Ramsey, meaning that for each $n < \omega$, for any partition of $\mathcal{AR}_n$ into two sets there is a member in $\mathcal{G}$ homogenizing the partition. However, it is still open whether these two notions are the same for any topological Ramsey space. Some equivalents of Ramsey are proved in [36, Chapter 2] and in [13, Sec. 7] for certain collections of topological Ramsey spaces.

We close this subsection by pointing out one more important similarity of general topological Ramsey spaces with the Ellentuck space.

Theorem 24 (Di Prisco, Mijares and Nieto [9]). *Let* $(\mathcal{R}, \leq, r)$ *be a topological Ramsey space and* $\mathcal{U} \subseteq \mathcal{R}$ *be a selective ultrafilter in a transitive model* M *of* $ZF + DCR$. *Let* $\mathbb{M}_{\mathcal{U}}$ *be the Mathias-like forcing consisting of basic open sets* $[a, X]$, *where* $X \in \mathcal{U}$. *Then forcing over* M *with* $\mathbb{M}_{\mathcal{U}}$ *adds a generic* $g \in \mathcal{R}$ *with the property that* $g \leq^* A$ *for each* $A \in \mathcal{U}$. *Moreover,* $B \in \mathcal{R}$ *is* $\mathbb{M}_{\mathcal{U}}$-*generic over* M *if and only if* $B \leq^* A$ *for all* $A \in \mathcal{U}$. *Furthermore,* $M[\mathcal{U}][g] = M[g]$.

For many further forcing properties similar Mathias forcing, see [9].

4.3.   *Continuous cofinal maps*

We saw in Theorem 6 that P-points have continuous Tukey reductions, by which we mean that any monotone cofinal map from a P-point to another ultrafilter is continuous when restricted below some member of the P-point. This is actually a special case of a more general phenomenon, which is seen in the topological Ramsey spaces in [10, 13, 15, 16, 36]. Recall that the metric topology on $\mathcal{R}$ is the one induced by the basic open cones $\{X \in \mathcal{R} : a \sqsubset X\}$, for $a \in \mathcal{AR}$. This is the same as topology inherited by $\mathcal{R}$ viewed as a subspace of $\mathcal{AR}^{\mathbb{N}}$ with the product topology.

The terminology *basic* was introduced by Solecki and Todorcevic in [34] in a study of Tukey structures of analytic posets. When treating an ultrafilter as poset partially ordered by reverse inclusion, it turns out that P-points are exactly ultra-filters which are basic, as was proved in [14]. Since *P-point* has a very well-established meaning, in order to avoid any confusion, we will use the terminology *closed under almost reduction* to refer to a filter $\mathcal{C}$ on $\mathcal{R}$ such that for each sequence $\langle X_i : i < \omega \rangle$ of members of $\mathcal{C}$ such that $X_i \geq^* X_{i+1}$ for each $i < \omega$, there is a diagonalization $Y \in \mathcal{C}$ such that for each $i < \omega$, $Y \leq^* X_i$. This property was called "selective" in [27] and for the next five years following that paper. Work of Trujillo in [37] served to distinguish the two notions for a large class of topological Ramsey spaces, and now when we use the word *selective* we are always referring to Definition 21.

Definition 25 ([13]). Assume that $C \subseteq \mathcal{R}$ is a filter on $(\mathcal{R}, \leq)$. C *has basic Tukey reductions* if whenever $\mathcal{V}$ is a nonprincipal ultrafilter on ω and $f : C \to \mathcal{V}$ exist a monotone cofinal map, there is an $X \in C$, a monotone map $f^* : C \to \mathcal{V}$ which is continuous with respect to the metric topology on $\mathcal{R}$, and a function $\check{f} : \mathcal{AR} \to [\omega]^{<\omega}$ such that

(1) $f \restriction (C \restriction X)$ is continuous with respect to the metric topology on $\mathcal{R}$;
(2) f^* extends $f \restriction (C \restriction X)$ to C;
(3) (a) $s \sqsubseteq t \in \mathcal{AR}$ implies that $\check{f}(s) \sqsubseteq \check{f}(t)$;
    (b) for each $Y \in C$, $f^*(Y) = \bigcup_{k<\omega} \check{f}(r_k(Y))$; and
    (c) $\check{f}$ is monotone: If $s, t \in \mathcal{AR}$ with $s \leq_{\text{fin}} t$, then $\check{f}(s) \subseteq \check{f}(t)$.

The following general theorem encompasses all known examples and provides a weak condition under which a maximal filter $\mathcal{U}$ on $\mathcal{R}$ has basic Tukey reductions. In the following, a member A of $\mathcal{U}$ is fixed, and for each $X \in \mathcal{U}$ such that $X \leq A$, $d(X)$ denotes the set $\{\text{depth}_A(r_n(X)) : n < \omega\}$, the collection of the depths of the finite approximations to X with respect to the fixed A. In the case of the Ellentuck space, each $d(X)$ is simply X. It very well may turn out to be the case that the requirement that $\{d(X) : X \in \mathcal{G}\}$ generates a nonprincipal ultrafilter on ω is simply true for all topological Ramsey spaces. The following theorem extends analogous theorems in [15, 16], and is attributed to Trujillo in [13], where a more general statement of the following can be found.

Theorem 26 (Trujillo [13]). *Let $(\mathcal{R}, \leq, r)$ be any topological Ramsey space. If $\mathcal{U}$ is generic for the forcing $(\mathcal{R}, \leq^*)$ and $\{d(X) : X \in \mathcal{U}\}$ generates a nonprincipal ultrafilter on ω, then $\mathcal{U}$ has basic Tukey reductions.*

4.4. *Barriers and canonical equivalence relations*

The notions of fronts and barriers on the Ellentuck space can be extended to topological Ramsey spaces. The Abstract Nash–Williams Theorem (Theorem 5.17 in [35]) follows from the Abstract Ellentuck Theorem, just as the Nash–Williams theorem follows from Ellentuck's theorem. For many arguments involving the Tukey and Rudin–Keisler structures of forced ultrafilters, the full strength of the Abstract Ellentuck Theorem is not used, but rather Abstract Nash–Williams Theorem suffices.

Definition 27 ([35]). A family $\mathcal{F} \subseteq \mathcal{AR}$ of finite approximations is

(1) *Nash–Williams* if $a \not\sqsubseteq b$ for all $a \neq b$ in $\mathcal{F}$.
(2) *Sperner* if $a \not\leq_{\text{fin}} b$ for all $a \neq b$ in $\mathcal{F}$.
(3) *Ramsey* if for every partition $\mathcal{F} = \mathcal{F}_0 \cup \mathcal{F}_1$ and every $X \in \mathcal{R}$, there are $Y \leq X$ and $i \in \{0, 1\}$ such that $\mathcal{F}_i | Y = \emptyset$.

Definition 28. Suppose $(\mathcal{R}, \leq, r)$ is a closed triple that satisfies **A.1**–**A.4**. Let $X \in \mathcal{R}$. A family $\mathcal{F} \subseteq \mathcal{AR}$ is a *front* on $[0, X]$ if

(1) for each $Y \in [0, X]$, there is an $a \in \mathcal{F}$ such that $a \sqsubset Y$; and
(2) $\mathcal{F}$ is Nash–Williams.

A family $\mathcal{B} \subseteq \mathcal{AR}$ is a *barrier* on $[0, X]$ if it satisfies (1) and also satisfies

(2′) $\mathcal{B}$ is *Sperner*.

Theorem 29 (Abstract Nash–Williams Theorem). *Suppose $(\mathcal{R}, \leq, r)$ is a closed triple that satisfies **A.1**–**A.4**. Then every Nash–Williams family of finite approximations is Ramsey.*

It is also proved in [35] that whenever the quasi-ordering $\leq_{\text{fin}}$ is a partial ordering, then an abstract version of Galvin's lemma holds: For each front $\mathcal{F}$ on $\mathcal{R}$, there is a member $Y \in \mathcal{R}$ such that $\mathcal{F}|Y := \{a \in \mathcal{F} : a \in \mathcal{AR}|Y\}$ is a barrier on $[0, Y]$. In all spaces that we have worked on so far, the quasi-order $\leq_{\mathrm{fin}}$ actually is a partial ordering. Thus, we are free to interchange usages of fronts with barriers as they do not affect any results. Analogously to how $[\omega]^k$ is the uniform barrier of rank k on the Ellentuck space, for any topological Ramsey space, $\mathcal{AR}_k$ is the uniform front of rank k. One can define by recursion fronts of all countable ranks on abstract topological Ramsey spaces, though the definition requires more care in the abstract setting.

The theorem of Pudlák and Rödl in Sec. 3.7 canonizing equivalence relations on fronts on $[\omega]^\omega$ generalizes to a large class of topological Ramsey spaces. This has been seen in [24] and more generally in [22] for the Milliken space of infinite block sequences and more recently in [10, 11, 13, 15, 16], where new topological Ramsey spaces were constructed which are dense inside certain σ-closed forcings producing ultrafilters satisfying weak partition relations. A sampling of the exact formulations of these canonical equivalence relations will be explicated in the following sections. The rest are left for the reader to find in the original sources, as reproducing them here would require too much space. The point we want to make is that such theorems for certain spaces and, in all these cases, have the general form that the canonical equivalence relations are essentially given by projections to substructures. If the Ramsey space has members which have a tree-like structure, then the canonical equivalence relations are defined by projections to subtrees. If the members of the Ramsey space are sequences of ordered structures, then the canonical equivalence relations are defined by projections to substructures. This behavior is what allows for the Rudin–Keisler structure inside the Tukey types to be deduced.

4.5.   Initial Tukey and Rudin–Keisler structures, and Rudin–Keisler structures inside Tukey types

Recall from Sec. 3.9 that if $\mathcal{U}$ is a Ramsey ultrafilter and $\mathcal{V} \leq_T \mathcal{U}$, then there is a front $\mathcal{F}$ and a function $g : \mathcal{F} \to \omega$ such that $\mathcal{V}$ is actually equal to the ultrafilter $g(\langle \mathcal{U} \upharpoonright \mathcal{F} \rangle)$, which is the filter on base set $\{g(a) : a \in \mathcal{F}\}$ generated by the sets $\{g(a) : a \in \mathcal{F}|X\}$, $X \in \mathcal{U}$. From this, one obtains that if $\mathcal{V}$ is nonprincipal, then $\mathcal{V} \equiv_T \mathcal{U}$ and $\mathcal{V} \equiv_{RK} \mathcal{U}^\alpha$ for some $\alpha < \omega_1$. This is a particular instance of a more general phenomenon which has been used successfully in several classes of ultrafilters to classify the structure of those ultrafilters Tukey reducible to an ultrafilter forced by a topological Ramsey space.

The steps (1)–(5) outlined in Sec. 3.9 for Ramsey ultrafilters on ω also provide the outline for obtaining precise results about Tukey and Rudin–Keisler structures below ultrafilters forced by topological Ramsey spaces. A collection of ultrafilters closed under Tukey reduction is called an *initial Tukey structure*. Likewise, a collection of ultrafilters which is closed under Rudin–Keisler reduction is called an *initial RK structure*. Given a topological Ramsey space $\mathcal{R}$ and a filter $\mathcal{U}$ generic for $(\mathcal{R}, \leq^*)$, we are interested in classifying the initial Tukey an initial RK structures below the ultrafilter $\mathcal{U} \upharpoonright \mathcal{AR}_1$. Since $\mathcal{U}$ and $\mathcal{U} \upharpoonright \mathcal{AR}_1$ are Tukey equivalent, we may work with either in the analysis.

The following is a special case of a more general fact shown in [13], stated here only for the case of generic filters. Its proof used Theorem 26 showing that in particular, generic filters have continuous cofinal Tukey reductions. A *front on $\mathcal{U}$* is a family $\mathcal{F} \subseteq \mathcal{AR}$ such that for each $X \in \mathcal{U}$, there is an $a \in \mathcal{F}$ such that $a \sqsubset X$, and no member of $\mathcal{F}$ is a proper initial segment of another member of $\mathcal{F}$. Recall that $g(\mathcal{F}|X) = \{g(a) : a \in \mathcal{F}|X\}$.

Theorem 30 ([13]). *Let $\mathcal{U}$ be generic for $(\mathcal{R}, \leq^*)$, and suppose that $\mathcal{V}$ is a nonprincipal ultrafilter on base set ω such that $\mathcal{V} \leq_T \mathcal{U}$. Then there exist a front $\mathcal{F}$ on $\mathcal{U}$ and a function $g : \mathcal{F} \to \omega$*

such that for each $V \in \mathcal{V}$, there is an $X \in \mathcal{U}$ such that $g(\mathcal{F}|X) \subseteq V$, and moreover each such $g(\mathcal{F}|X)$ is a member of $\mathcal{V}$. Thus, $\mathcal{V}$ equals the ultrafilter on the base set $\{g(a) : a \in \mathcal{F}\}$ generated by the set $\{g(\mathcal{F}|X) : X \in \mathcal{U}\}$.

Since each function $g : \mathcal{F} \to \omega$ induces an equivalence relation on $\mathcal{F}$, once a canonization theorem is proved for equivalence relations on fronts, it will help us understand sets of the form $\{g(a) : a \in \mathcal{F}\}$. Once these are well-understood, the inital Tukey and initial RK structures are well-understood. This will be made clear for a sampling of examples in the following sections.

5. Topological Ramsey Space Theory Applied to Ultrafilters Satisfying Weak Partition Relations: An Overview of the Following Sections

In Sec. 3, we delineated some of the important properties of Ramsey ultrafilters and how these properties are connected with the Ellentuck space. We saw that Ramsey ultrafilters have complete combinatorics, are RK minimal, Tukey minimal, and that the RK classes inside a Ramsey ultrafilter's Tukey type are exactly those of its countable Fubini powers. We also saw how the Nash–Williams theorem provides a quick proof that any ultrafilter forced by $\mathcal{P}(\omega)/\text{Fin}$ is Ramsey. In the following two sections, we will look in-depth at some examples of ultrafilters from the literature which are P-points satisfying weak partition relations. For these examples, we will go through the steps of the results that these ultrafilters are similar to Ramsey ultrafilters, achieving Objectives 1–4 below. In the final section of this expository paper, we will mention similar results for some broad classes of ultrafilters. Some of these ultrafilters are well-known from the literature and some are new, having been produced by the construction of new topological Ramsey spaces. These results were motivated by the following question.

Question 31. What is the structure of the Tukey types of ultrafilters which are close to minimal in the Rudin–Keisler hierarchy?

This is closely related to the question of finding the Tukey structure of ultrafilters satisfying weak partition relations, since in all known examples, partition relations are corollated with increased Rudin–Keisler strength. In investigating these questions, the following objectives were attained for the ultrafilters investigated so far.

Objective 1. Complete combinatorics.

Objective 2. The exact structure of the Tukey types of all ultrafilters Tukey reducible to the forced ultrafilter.

Objective 3. The exact structure of the Rudin–Keisler classes of all ultrafilters Rudin–Keisler reducible to the forced ultrafilter.

Objective 4. The exact structure of the Rudin–Keisler classes inside the Tukey type of each ultrafilter Tukey reducible to the forced ultrafilter.

To attain these goals for ultrafilters which are constructed using some σ-closed partial order by forcing or by using the Continuum Hypothesis, Martin's Axiom, or some weaker cardinal invariant assumption to construct them, it suffices to satisfy the steps (1)–(5) outlined in Sec. 3.9 which were generalized to abstract topological Ramsey spaces in Sec. 4. First, and sometimes challenging, one needs to find a topological Ramsey space which forces the ultrafilter under investigation. It suffices to show that there is a dense set in the partial order which can be structured to form a topological Ramsey space. This attains Objective 1. Second, one needs to prove that cofinal maps from such ultrafilters are continuous, with respect to the correct topology. This has now been done in much generality in Theorem 26. Third, one needs to prove that equivalence relations on fronts for these spaces are canonical when restricted below some member of the space. One needs to understand these canonical equivalence relations very well in order to analyze their implications. Putting this together with Theorem 30 allows us to achieve Objectives 2–4.

6. Weakly Ramsey Ultrafilters

An ultrafilter $\mathcal{U}$ is *weakly Ramsey* if for each $l \geq 2$ and each coloring $c : [\omega]^2 \to l$, there is a member $U \in \mathcal{U}$ such that the restriction of c to $[U]^2$ takes on at most two colors. This is denoted symbolically as

$$\mathcal{U} \to (\mathcal{U})^2_{l,2}. \tag{14}$$

Blass showed in [4] that weakly Ramsey ultrafilters have exactly one Rudin–Keisler predecessor, and that is a Ramsey ultrafilter. Thus, the initial Rudin–Keisler structure below an ultrafilter forced by $\mathbb{P}_1$ is simply a chain of length two (disregarding the principal ultrafilters).

In [23], Laflamme constructed a partial ordering $(\mathbb{P}_1, \leq_1)$ which forces a weakly Ramsey ultrafilter. This partial ordering has conditions which are simply infinite subsets of ω, but the partial order $\leq_1$ is stronger than inclusion.

Definition 32 $((\mathbb{P}_1, \leq_1)$, **[23]**$)$. $\mathbb{P}_1 = [\omega]^\omega$. Let $X, Y \in [\omega]^\omega$. Enumerate them in increasing order and in blocks of increasing size as $X = \langle x_1^1, x_1^2, x_2^2, \ldots, x_1^n, \ldots, x_n^n, \ldots \rangle$, and $Y = \langle y_1^1, y_1^2, y_2^2, \ldots, y_1^n, \ldots, y_n^n, \ldots \rangle$. We call $\{x_1^n, \ldots, x_n^n\}$ the nth block of X and similarly for Y. Define $Y \leq_1 X$ if and only if

$$\forall n \, \exists m \, \{y_1^n, \ldots, y_n^n\} \subseteq \{x_1^m, \ldots, x_m^m\}. \tag{15}$$

Note that $Y \leq_1 X$ implies $Y \subseteq X$ but not vice versa. It is this stronger ordering which is responsible for producing an ultrafilter which is weakly Ramsey but not Ramsey.

Laflamme gave a combinatorial characterization of ultrafilters forced by $\mathbb{P}_1$, showing that these ultrafilters have complete combinatorics in the original sense of Blass. An ultrafilter $\mathcal{V}$ *satisfies the Ramsey partition relation* $\mathrm{RP}(k)$ if for all functions $f : [\omega]^k \to 2$ and all partitions $\langle A_m : m < \omega \rangle$ of ω with each $A_m \notin \mathcal{V}$, there is a set $X \in \mathcal{V}$ such that

(1) $|X \cap A_m| < \omega$ for all $m < \omega$.
(2) $|f''[X \cap A_m]^k| \leq 1$ for all $m < \omega$.

Baumgartner and Taylor showed in [1] that $\mathrm{RP}(2)$ is equivalent to weakly Ramsey (see also [4]). Laflamme also proved in [23] that for each k, $\mathrm{RP}(k)$ is equivalent to $\mathcal{U} \to (\mathcal{U})^k_{l,2^{k-1}}$, a fact which he credits to Blass.

Laflamme proved that the ultrafilter forced by $\mathbb{P}_1$ satisfies $\mathrm{RP}(k)$ for all k. Thus, it would seem that the ultrafilters forced by $\mathbb{P}$ are a strong sort of weakly Ramsey ultrafilter. Indeed, it follows by work of Trujillo in that, assuming CH, there are weakly Ramsey ultrafilters which do not satisfy $\mathrm{RP}(k)$ for some $k > 2$. (See Corollary 4.2.5 in [36] for the exact statement.) The properties $\mathrm{RP}(k)$ for all k are what completely characterize ultrafilters being forced by $\mathbb{P}_1$ over a canonical inner model, as follows. Recall that an ultrafilter $\mathcal{U}$ is *rapid* if for each strictly increasing function $h : \omega \to \omega$, there is a member $X \in \mathcal{U}$ such that for each $n < \omega$, $|X \cap h(n)| \leq n$.

Theorem 33 (Laflamme [23]). *Let κ be a Mahlo cardinal and G be generic for the Lévy collapse of κ. If $\mathcal{U} \in V[G]$ is a rapid ultrafilter satisfying $\mathrm{RP}(k)$ for all k, but $\mathcal{U}$ is not Ramsey, then $\mathcal{U}$ is $\mathbb{P}_1$-generic over $HOD(\mathbb{R})^{V[G]}$.*

The proof uses a key theorem Laflamme proves earlier in [23] which is worth mentioning, as the reader will see the correlation with the topological Ramsey space formulation presented shortly.

Theorem 34 (Laflamme [23]). *Let $\mathcal{U}$ be a nonprincipal ultrafilter. Then the following are equivalent:*

(1) $\mathcal{U}$ *is rapid and satisfies RP(k) for all k.*

(2) *For all Σ_1^1 sets $\mathcal{X} \subseteq [\omega]^\omega$, there is a set $X \in \mathcal{U}$ such that*

$$\{Y \in [\omega]^\omega : Y \leq_1 X\} \subseteq \mathcal{X} \quad \text{or} \quad \{Y \in [\omega]^\omega : Y \leq_1 X\}$$

$$\cap \mathcal{X} = \emptyset. \tag{16}$$

(3) $\mathcal{U}$ *is rapid and satisfies $\mathcal{U} \to (\mathcal{U})_{l,2^k-1}^k$ for all k.*

It is condition (2) that we shall soon see is very closely related to the topological Ramsey space formulation of complete combinatorics.

The following topological Ramsey space was constructed to essentially form a dense subset of $\mathbb{P}_1$, so that the Ramsey space is forcing equivalent to $\mathbb{P}_1$.

Definition 35 $((\mathcal{R}_1, \leq_{\mathcal{R}_1}, r),$ [15]). Let

$$\mathbb{T} = \{\langle\rangle\} \cup \{\langle n\rangle : n < \omega\} \cup \bigcup_{n<\omega}\{\langle n, i\rangle : i \leq n\}. \tag{17}$$

$\mathbb{T}$ is an infinite tree of height two and consists of an infinite sequence of finite trees, where the *nth subtree* of $\mathbb{T}$ is

$$\mathbb{T}(n) = \{\langle\rangle, \langle n\rangle\} \cup \{\langle n, i\rangle : i \leq n\}. \tag{18}$$

Thus $\mathbb{T}(n)$ is a finite tree of height two with one node $\langle\rangle$ on level 0, one node $\langle n\rangle$ on level 1, and $n + 1$ nodes on level 2. An infinite subtree $X \subseteq \mathbb{T}$ is a member of $\mathcal{R}_1$ if and only if X is tree-isomorphic to $\mathbb{T}$. This means X must be an infinite sequence of finite subtrees such that the *nth* subtree $X(n)$ has the node $\langle\rangle$ on level 0, one node $\langle k_n\rangle$ on level 1, and $n + 1$ many nodes $\{\langle k_n, i\rangle : i \in I_n\}$ on level 2, where I_n is some subset of $k_n + 1$ of size $n + 1$. Moreover, we require that for $n < n'$, $X(n)$ and $X(n')$ come from subtrees $\mathbb{T}(k_n)$ and $\mathbb{T}(k_{n'})$ where $k_n < k_{n'}$.

The partial ordering $\leq_{\mathcal{R}_1}$ on $\mathcal{R}_1$ is simply that of subtree. The restriction map r is defined by $r_n(X) = \bigcup_{i<n} X(i)$, for each $n < \omega$ and $X \in \mathcal{R}_1$. $\mathcal{AR}_n$ is the set $\{r_n(X) : X \in \mathcal{R}_1\}$;

$\mathcal{AR} = \{r_n(X) : X \in \mathcal{R}_1,\ n < \omega\}$. For $a, b \in \mathcal{AR}$, $b \leq_{\text{fin}} a$ if and only if b is a subtree of a. The basic open sets are given by $[a, X] = \{Y \in \mathcal{R}_1 : a \sqsubseteq Y \text{ and } Y \leq_{\mathcal{R}_1} X\}$.

$\mathbb{T}$ is a tree which codes Laflamme's blocking structure. Instead of taking all infinite sets and ordering them by a partial ordering stricter than inclusion, the shape of the trees allowed in $\mathcal{R}_1$ transfers the strict partial ordering $\leq_1$ of $\mathbb{P}_1$ to the structure of the trees. By restricting $\mathcal{R}_1$ to contain only those subtrees X of $\mathbb{T}$ which have each nth subtree $X(n)$ coming from within one mth subtree $\mathbb{T}(m)$, the partial ordering $\leq_1$ gets transferred to the structure of the tree. The further restriction that each nth subtree of X, $X(n)$, must come from a different subtree of $\mathbb{T}$ further serves to simplify the set of trees we work with, and more importantly, aids in proving that the Pigeonhole Principle, **Axiom A.4**, holds for this space.

For finite sequences $a \in \mathcal{AR}$, we shall write $a = (a(0), \ldots, a(k-1))$ to denote that a is a sequence of length k of subtrees where the ith subtree has $i + 1$ many maximal nodes. We now show how **Axiom A.4** follows from the finite Ramsey Theorem. Let $X \in \mathcal{R}_1$, $k < \omega$, $Y \leq_{\mathcal{R}_1} X$, and $a = r_k(Y)$. Let $\mathcal{O} \subseteq \mathcal{AR}_{k+1}$ be given. Let m be the least integer such that $a \subseteq r_m(X)$. In other words, $\text{depth}_X(a)$ is finite. To show **A.4** we need to show there is some $Z \in [r_m(X), X]$ such that either $r_{k+1}[a, Z] \subseteq \mathcal{O}$ or $r_{k+1}[a, Z] \cap \mathcal{O} = \emptyset$. Notice that set $r_{k+1}[a, X]$ is the set of all $c \in \mathcal{AR}_{k+1}$ such that $(c(0), \ldots, c(k-1)) = (a(0), \ldots, a(k-1))$.

Let $\mathcal{R}_1(k)$ denote the set of all $c(k)$ where $c \in \mathcal{AR}_{k+1}$, that is the set of all kth trees of some member of $\mathcal{AR}_{k+1}$. The set $\mathcal{O}$ induces a coloring on

$$\mathcal{R}_1(k) \upharpoonright (X/m) := \{c(k) \in \mathcal{R}_1(k) : \exists i \geq m\, (c(k) \subseteq X(i))\},$$

(19)

since

$$r_{k+1}[a, X] = \{(a(0), \ldots, a(k-1), c(k)) :$$
$$c(k) \in \mathcal{R}_1(k) \upharpoonright (X/m)\}. \tag{20}$$

Given $c(k) \in \mathcal{R}_1(k) \upharpoonright (X/m)$, define $f(c(k)) = 0$ if $(a(0), \ldots, a(k-1), c(k)) \in \mathcal{O}$, and $f(c(k)) = 1$ if $(a(0), \ldots, a(k-1), c(k)) \notin \mathcal{O}$. Thus, to construct a $Z \in [r_m(X), X]$ for which either $r_{k+1}[a, Z] \subseteq \mathcal{O}$ or $r_{k+1}[a, Z] \cap \mathcal{O} = \emptyset$, it suffices to construct such a Z with the property that f is constant on the $\mathcal{R}_1(k) \upharpoonright (Z/m)$.

This follows from the finite Ramsey theorem. Let $r_m(Y) = r_m(X)$. Given $r_j(Y)$, to construct $Y(j)$ take some l large enough that coloring the $k+1$ sized subsets of a set of size l, there is a subset of size j on which the $k+1$ sized subsets are homogeneous. Take some subtree $Y(j)$ of $Y(l)$ with j many maximal nodes such that f is homogeneous on the set of $c(k) \in \mathcal{R}_1(k)$ such that $c(k)$ is a subtree of $Y(j)$. This constructs $r_{j+1}(Y)$. Let Y be the infinite sequence of the $Y(i)$, $i < \omega$. This Y is a member of $\mathcal{R}_1$.

Now the color of f might be different for $c(k)$'s coming from within different subtrees of Y. But there must be infinitely many j for which the color of f on $\{c(k) \in \mathcal{R}_1(k) : c(k) \subseteq Y(j)\}$ is the same. Take an increasing sequence $(j_i)_{i \geq m}$ and thin each $Y(j_i)$ to a subtree $Z(i)$ with $i+1$ many maximal nodes. Let $Z = (X(0), \ldots, X(m-1), Z(m), Z(m+1), \ldots)$. Then Z is a member of $[r_m(X), X]$ for which f is constant on $\mathcal{R}_1(k) \upharpoonright (Z/m)$. Hence, $r_{k+1}[a, Z]$ is either contained in or disjoint from $\mathcal{O}$. This proves **A.4**. The other three axioms are routine to prove, and it is suggested that the reader go through those proofs to build intuition.

Theorem 36 (Dobrinen and Todorcevic [15]). $(\mathcal{R}_1, \leq_{\mathcal{R}_1}, r)$ *is a topological Ramsey space.*

The topological Ramsey space constructed has the property that below any member S of $\mathbb{P}_1$, there is a correspondence between $\mathcal{R}_1$ and a dense set below S.

Fact 37. $(\mathcal{R}_1, \subseteq)$ *is forcing equivalent to* $(\mathbb{P}_1, \leq_1)$.

Proof.   Let S be any member of $[\omega]^\omega$. Enumerate S in blocks of increasing length as $\{s_1^1, s_1^2, s_2^2, s_1^3, \dots\}$ as in Laflamme's blocking procedure. Let $\theta : S \to \mathbb{T}$ be the map which takes s_i^n to the node $\langle n - 1, i - 1 \rangle$ in $[\mathbb{T}]$. (Recall that $[\mathbb{T}]$ denotes the set of maximal nodes in $\mathbb{T}$.) Then for each member $S' \in [\omega]^\omega$ such that $S' \leq_1 S$, $\theta(S')$ induces the subtree $\widehat{\theta(S')}$ of $\mathbb{T}$ consisting of the set of all initial segments of members of $\theta(S')$. Now this $\widehat{\theta(S')}$ might not actually be a member of $\mathcal{R}_1$ as there could be two blocks of S' that lie in one block of S; this would translate to two subtrees $\widehat{\theta(S')}(m)$ and $\widehat{\theta(S')}(n)$, for some $m < n$, being subtrees of the same $\mathbb{T}(k)$ for some k. However, we can take a subtree $X \subseteq \widehat{\theta(S')}$ which has each nth subtree of X coming from a different kth subtree of $\mathbb{T}$ so that $X \in \mathcal{R}_1$. Then $S'' := \theta^{-1}(X)$ will be a member of $[\omega]^\omega$ such that $S'' \leq_1 S$, and $\theta(S'') = [X]$, where X is a member of $\mathcal{R}_1$. Thus, given an $S \in \mathbb{P}_1$, the set $\{\theta^{-1}([X]) : X \in \mathcal{R}_1\}$ is dense below S in the partial ordering $\mathbb{P}_1$. $\square$

For $\mathcal{R}_1$, the σ-closed almost reduction ordering presented in Definition 17 is equivalent to the following. Given $X, Y \in \mathcal{R}_1$, $Y \leq^*_{\mathcal{R}_1} X$ if and only if there is some m such that for all $n \geq m$, there is an i_n such that $Y(n) \subseteq X(i_n)$. Notice that for the space $\mathcal{R}_1$, $Y \leq^*_{\mathcal{R}_1} X$ if and only if $[Y] \subseteq^* [X]$.

Let $\mathcal{U}_{\mathcal{R}_1}$ denote a generic ultrafilter forced by $(\mathcal{R}_1, \leq^*_{\mathcal{R}_1})$. The following from [15] completely characterizes the Tukey types of the ultrafilters Tukey reducible to $\mathcal{U}_{\mathcal{R}_1}$. Furthermore, it characterizes the Rudin–Keisler classes inside those Tukey types. The set $\mathcal{Y}_{\mathcal{R}_1}$ denotes a countable set of P-points which will be defined below.

Theorem 38 (Dobrinen and Todorcevic [15]). $(\mathcal{R}_1, \leq_{\mathcal{R}_1})$ *and* $(\mathbb{P}_1, \leq_1)$ *are forcing equivalent, and* $(\mathcal{R}_1, \leq^*_{\mathcal{R}_1})$ *and* $(\mathbb{P}_1, \leq^*_1)$ *are forcing equivalent. Let* $\mathcal{U}_{\mathcal{R}_1}$ *denote the ultrafilter forced by* $(\mathcal{R}_1, \leq^*_{\mathcal{R}_1})$, *and let* $\mathcal{U}_0$ *denote the Ramsey ultrafilter obtained from projecting* $\mathcal{U}_{\mathcal{R}_1}$ *to level 1 on the tree* $\mathbb{T}$.

(1) *For each nonprincipal ultrafilter* $\mathcal{V} \leq_T \mathcal{U}_{\mathcal{R}_1}$, *either* $\mathcal{V} \equiv_T \mathcal{U}_{\mathcal{R}_1}$ *or else* $\mathcal{V} \equiv_T \mathcal{U}_0$.

(2) *If* $\mathcal{V} \equiv_T \mathcal{U}_{\mathcal{R}_1}$, *then* $\mathcal{V}$ *is Rudin–Keisler equivalent to some Fubini iterate of ultrafilters each of which is in a specific countable collection of P-points, denoted* $\mathcal{Y}_{\mathcal{R}_1}$. *If* $\mathcal{V} \equiv_T \mathcal{U}_0$, *then* $\mathcal{V}$ *is Rudin–Keisler equivalent to some* α*th Fubini power of* $\mathcal{U}_0$, *where* $\alpha < \omega_1$.

The proof follows the same five steps as in Sec. 3.9. For the rest of this section, let $\mathcal{U}$ denote $\mathcal{U}_{\mathcal{R}_1}$. Let $\mathcal{V}$ be a nonprincipal ultrafilter on base set ω such that $\mathcal{V} \leq_T \mathcal{U}$. For Step (1), it is shown that for each monotone cofinal map $f : \mathcal{U} \to \mathcal{V}$, there is a member $X \in \mathcal{U}$ such f is continuous when restricted to $\mathcal{U} \upharpoonright X$. With a bit more work, one can show that in fact there is montone cofinal map $f : \mathcal{R}_1 \to [\omega]^{<\omega}$ witnessing this Tukey reduction which is continuous with respect to the metric topology on $\mathcal{R}_1$; that is the topology generated by basic open sets of the form $[a, \mathbb{T}]$, where $a \in \mathcal{AR}$. Thus, there is a finitary map $\hat{f} : \mathcal{AR} \to [\omega]^{<\omega}$ which recovers f as follows: for each $X \in \mathcal{R}_1$,

$$f(X) = \bigcup_{k<\omega} \hat{f}(r_k(X)). \tag{21}$$

For Step (2), let $\mathcal{F}$ be the front on $\mathcal{R}_1$ defined as follows: For each $X \in \mathcal{R}_1$, let $n(X)$ be the least integer such that $\hat{f}(r_{n(X)}(X)) \neq \emptyset$. Define

$$\mathcal{F} = \{r_{n(X)}(X) : X \in \mathcal{R}_1\}. \tag{22}$$

Then $\mathcal{F}$ is a front, since each member of $\mathcal{R}_1$ has a finite initial segment in $\mathcal{F}$, and by minimality of $n(X)$ no member of $\mathcal{F}$ is a proper initial segment of another member of $\mathcal{F}$.

Steps (3) and (4) are the same for each topological Ramsey space. Define

$$\mathcal{F}|X = \{a \in \mathcal{F} : \exists Y \leq X, \ \exists n < \omega \ (a = r_n(Y))\}. \qquad (23)$$

Then let $\mathcal{U} \upharpoonright \mathcal{F}$ be the filter on the countable base set $\mathcal{F}$ generated by $\{\mathcal{F}|X : X \in \mathcal{U}\}$. By genericity of $\mathcal{U}$, $\mathcal{U} \upharpoonright \mathcal{F}$ is an ultrafilter. The set-up to Step (4) is the same as in Sec. 3.9. Define $g : \mathcal{F} \to \omega$ by $g(a) = \min(\hat{f}(a))$, for each $a \in \mathcal{F}$. By Theorem 30, if $\mathcal{V}$ is nonprincipal then the RK image of $\mathcal{U} \upharpoonright \mathcal{F}$ under g, $g(\mathcal{U} \upharpoonright \mathcal{F})$, equals $\mathcal{V}$.

Now we want to understand these maps g so that we can understand the isomorphism class of $\mathcal{V}$ for Step (5). g induces an equivalence relation on the front $\mathcal{F}$. As in the Ellentuck space, there is a notion of canonical equivalence relation for $\mathcal{R}_1$, found in [15] and described now. Since it is dense in $\mathcal{R}_1$ to find a Z such that the equivalence relation $\mathcal{F}|Z$ is canonical, such a Z will be in $\mathcal{U}$. Assume then that $Z \in \mathcal{U}$ and g is canonical on $\mathcal{F}|Z$. We now describe these canonical equivalence relations proved in [15] beginning with some simple examples, building up intuition for the general case.

First we describe the canonical equivalence relations on the set $\mathcal{R}_1(n)$ as these are the building blocks for the canonical equivalence relations on fronts.

Definition 39 (Canonical equivalence relations on $\mathcal{R}_1(n)$ [15]). Let T be any subtree of $\mathbb{T}(n)$. Given $a(n) \in \mathcal{R}_1(n)$, let $\pi_T(a(n))$ be the projection of the tree $a(n)$ to the nodes in the same position as T inside $\mathbb{T}(n)$. That is, if $\iota : \mathbb{T}(n) \to a(n)$ is the tree isomorphism between them, then $\pi_T(a(n)) = \iota''T$. The canonical equivalence relation E_T on $\mathcal{R}_1(n)$ is defined by $a(n) \, E_T \, b(n)$ if and only if $\pi_T(a(n)) = \pi_T(b(n))$.

If $\mathcal{F} = \mathcal{A}\mathcal{R}_1$, then the canonical equivalence relations on $\mathcal{A}\mathcal{R}_1$ are simply those given by subtrees of $\mathbb{T}(0) = \{\langle\rangle, \langle 0 \rangle, \langle 0, 0 \rangle\}$. Thus, there are three canonical equivalence relations on $\mathcal{A}\mathcal{R}_1$.

By density, we may without loss of generality assume that g induces a canonical equivalence relation on all of $\mathcal{AR}_1$; say this is given by π_T. If $T = \{\langle\rangle\}$, then all members of $\mathcal{AR}_1$ are equivalent; this means that g is a constant function. In this case, $\mathcal{V}$ is a principal ultrafilter generated by the singleton $g(a(0))$ for each/every $a(0) \in \mathcal{AR}_1$. If $T = \{\langle\rangle, \langle 0\rangle\}$, then two members $a(0), b(0) \in \mathcal{AR}_1$ are E_T equivalent if and only if their level 1 nodes are equal. For example, if $a(0) = \{\langle\rangle, \langle 3\rangle, \langle 3, 0\rangle\}$ and $b(0) = \{\langle\rangle, \langle 3\rangle, \langle 3, 2\rangle\}$, then they are equivalent, but if $a(0) = \{\langle\rangle, \langle 3\rangle, \langle 3, 0\rangle\}$ and $b(0) = \{\langle\rangle, \langle 4\rangle, \langle 4, 0\rangle\}$, then they are not equivalent. This means that g is (up to permutation) the projection map from the maximal nodes in $\mathbb{T}$ to the the nodes in level 1 of $\mathbb{T}$. This projection map yields the Ramsey ultrafilter $\mathcal{U}_0$. Hence, in this case, $\mathcal{V} = g(\mathcal{U} \upharpoonright \mathcal{AR}_1)$ which is the ultrafilter on base $\{\langle n\rangle : n < \omega\}$ generated by the projections of the members of $\mathcal{U}$ to their first levels. This is RK equivalent to $\mathcal{U}_0$. If $T = \{\langle\rangle, \langle 0\rangle, \langle 0, 0\rangle\}$, then this means that g is one-to-one on $\mathcal{AR}_1$. Notice that $\mathcal{U}$ is isomorphic to the ultrafilter $\mathcal{U} \upharpoonright \mathcal{AR}_1$. Thus, if g is one-to-one, then $\mathcal{V} = g(\mathcal{U} \upharpoonright \mathcal{AR}_1) \equiv_{RK} \mathcal{U}_1$.

If $\mathcal{F} = \mathcal{AR}_2$, then the possible canonical equivalence relations are given by two independent canonical equivalence relations: on the zeroth subtrees $\mathcal{R}_1(0)$ and on the first subtrees $\mathcal{R}_1(1)$. Thus, there are five canonical equivalence relations on $\mathcal{R}_1(1)$. Here, we start to see the more general pattern emerging.

Fact 40. *For each n, the sets $\mathcal{R}_1(n)|X$, $X \in \mathcal{U}$, generate an ultrafilter on base set $\mathcal{R}_1(n)$. Denote these as $\mathcal{U} \upharpoonright \mathcal{R}_1(n)$.*

These are actually P-points, as genericity of $\mathcal{U}$ ensures pseudointersections.

g on $\mathcal{AR}_2$ is canonized by (E_{T_0}, E_{T_1}), where T_0 is a subtree of $\mathbb{T}(0)$ and T_1 is a subtree of $\mathbb{T}(1)$. With some work, one can check that the sets $\{\pi_{T_0}(a(0)) : a(0) \in \mathcal{R}_1|X\}$, $X \in \mathcal{U}_1$, generate an ultrafilter on base set $\{\pi_{T_0}(a(0)) : a(0) \in \mathcal{R}_1(0)\}$; and the sets $\{\pi_{T_1}(a(1)) : a(1) \in \mathcal{R}_1|X\}$, $X \in \mathcal{U}_1$, generate an ultrafilter on base set $\{\pi_{T_1}(a(1)) : a(1) \in \mathcal{R}_1(1)\}$. Denote these by

$\pi_{T_i}(\mathcal{U}_1 \upharpoonright \mathcal{R}_1(i))$, $i \in 2$. For both $i \in 2$, if $T = \{\langle\rangle\}$, then the ultrafilter $\pi_T(\mathcal{U}_1 \upharpoonright \mathcal{R}_1(i))$ is principal. If $T = \{\langle\rangle, \langle i \rangle\}$, then $\pi_T(\mathcal{U}_1 \upharpoonright \mathcal{R}_1(i))$ is isomorphic to the Ramsey ultrafilter $\mathcal{U}_0$. For T_1 equal to $\{\langle\rangle, \langle 1 \rangle, \langle 1, 0 \rangle\}$ or $\{\langle\rangle, \langle 1 \rangle, \langle 1, 1 \rangle\}$, $\pi_{T_1}(\mathcal{U}_1 \upharpoonright \mathcal{R}_1(1))$ is isomorphic to $\mathcal{U}_1$. The new ultrafilter we now see is in the case of $T_1 = \mathbb{T}(1)$, in which case $\pi_T(\mathcal{U}_1 \upharpoonright \mathcal{R}_1(1)) = \mathcal{U}_1 \upharpoonright \mathcal{R}_1(1)$, which is a P-point.

With some more work, one finds that $\mathcal{V}$ is isomorphic to a Fubini product of two ultrafilters which are either P-points, Ramsey, or principal. We write $\mathcal{U} * \mathcal{V}$ to denote the Fubini product $\lim_{n \to \mathcal{U}} \mathcal{V}_n$ where for all n, $\mathcal{V}_n = \mathcal{V}$.

$$\mathcal{V} = g(\mathcal{U}_1 \upharpoonright \mathcal{AR}_2) = \pi_{T_0}(\mathcal{U}_1 \upharpoonright \mathcal{R}_1(0)) * \pi_{T_1}(\mathcal{U}_1 \upharpoonright \mathcal{R}_1(1)). \quad (24)$$

In general, the canonical equivalence relations on $\mathcal{R}_1(n)$ are given by the subtrees of $\mathbb{T}(n)$. There are $2^{n+1} + 1$ many of subtrees of $\mathbb{T}(n)$, hence that many canonical equivalence relations on the nth subtrees. As in the cases for $\mathcal{R}_1(0)$ and $\mathcal{R}_1(1)$, for any $n < \omega$, if the canonical equivalence relation on $\mathcal{R}_1(n)$ is given by $T_n = \{\langle\rangle\}$, then the projection map induces a principal ultrafilter. If $T_n = \{\langle\rangle, \langle n \rangle\}$, then the projection map induces an ultrafilter isomorphic to the Ramsey ultrafilter $\mathcal{U}_0$. If $T_n = \{\langle\rangle, \langle n \rangle\} \cup \{\langle n, i \rangle : i \in I_n\}$ where I_n is some nonempty subset of $n+1$, then letting $k = |I_n|$, the projection map induces an ultrafilter isomorphic to the P-point $\mathcal{U}_1 \upharpoonright \mathcal{R}_1(k - 1)$.

Thus, if $\mathcal{F}$ is equal to $\mathcal{AR}_m$ for some m, then there are trees $T_i \subseteq \mathbb{T}(i)$ for each $i < m$ such that

$$\mathcal{V} \equiv_{RK} \pi_{T_0}(\mathcal{U}_1 \upharpoonright \mathcal{R}_1(0)) * \cdots * \pi_{T_{m-1}}(\mathcal{U}_1 \upharpoonright \mathcal{R}_1(m - 1)). \quad (25)$$

Each of these ultrafilters in the m-iterated Fubini product is either principal, $\mathcal{U}_0$, or $\mathcal{U}_0 \upharpoonright \mathcal{R}_1(k)$ for some k.

Now it may well be that the front $\mathcal{F}$ is not of the form $\mathcal{AR}_m$ for any m, but is more complex. In this case, we still can do an analysis, using the complexity (uniform rank) of the front to conclude that $\mathcal{V}$ is isomorphic to a countable iteration of Fubini products where each ultrafilter in the construction is a member of the collection

$$\mathcal{Y}_{\mathcal{R}_1} = \{1, \mathcal{U}_1\} \cup \{\mathcal{U}_1 \upharpoonright \mathcal{R}_1(k) : k < \omega\}, \qquad (26)$$

where 1 denotes any principal ultrafilter. This completely classifies the isomorphism types within the Tukey types of ultrafilters Tukey reducible to $\mathcal{U}_1$.

7. Ultrafilters of Blass Constructed by n-square Forcing and Extensions to Hypercube Forcings

In the study of the structure of the Rudin–Keisler classes of P-points, Blass showed that not only can there be chains of order type ω_1 and $\mathbb{R}$, but also that there can be Rudin–Keisler incomparable P-points. In [3], Blass proved that assuming Martin's axiom, there is a P-point which has two Rudin–Keisler incomparable P-points RK below it. We will call this forcing n-*square forcing*, $\mathbb{P}_{\text{square}}$, since a subset $X \subseteq \omega \times \omega$ is in the partial ordering $\mathbb{P}_{\text{square}}$ if and only if for each $n < \omega$ there are sets a, b each of size n such that the product $a \times b$ is a subset of X. $\mathbb{P}$ is partially ordered by $\subseteq$. The projections to the first and second coordinates yield the two RK-incomparable P-points. The P-point obtained from this construction satisfies the partition relation

$$\mathcal{U} \to (\mathcal{U})^2_{k,5}. \qquad (27)$$

The forcing $\mathbb{P}_{\text{square}}$ contains a dense subset which forms a topological Ramsey space denoted $\mathcal{H}^2$, which appears in [13, 36]. The members of this space are essentially infinite sequences which are a product of two members of $\mathcal{R}_1$ in the following sense. Let $\mathbb{T}^2$ be the sequence of trees $\langle \mathbb{T}^2(n) : n < \omega \rangle$ such that for each $n < \omega$,

$$\mathbb{T}^2(n) = \{\langle \rangle, \langle n \rangle\} \cup \{\langle n, \langle i, j \rangle \rangle : i, j \in n+1\}. \qquad (28)$$

$\mathbb{T}^2(n)$ should be thought of as a tree with height two where levels 0 and 1 have one node, and level 2 has an $(n+1) \times (n+1)$ square of nodes. A sequence $X = \langle X(n) : n < \omega \rangle$ is a member of $\mathcal{H}^2$ if and only if it is a subtree of $\mathbb{T}^2$ with the same structure as $\mathbb{T}^2$. Specifically, $X \in \mathcal{H}^2$ if and only if there is a strictly increasing sequence (k_n) such that for each $n < \omega$,

$$X(n) = \{\langle\rangle, \langle k_n \rangle\} \cup \{\langle k_n, \langle i, j \rangle\rangle : i \in I_n, \ j \in J_n\}, \qquad (29)$$

where $I_n, J_n \in [k_n + 1]^{n+1}$. We call $X(n)$ the *nth block* of X. For X and Y in $\mathcal{H}^2$, $Y \leq X$ if and only if for each n there is a k_n such that $Y(n) \subseteq X(k_n)$ and moreover, the sequence $(k_n)_{n<\omega}$ is strictly increasing. However, by the structure of the members of $\mathcal{H}^2$, it turns out that $\leq$ is the same as $\subseteq$.

For the space $\mathcal{H}^2$, the almost reduction $\leq^*$ is simply $\subseteq^*$. The forcing $(\mathcal{H}^2, \leq^*)$ is σ-complete and produces a new P-point, $\mathcal{U}_2$. The RK structure below $\mathcal{U}_2$ is a diamond shape. $\mathcal{U}_2$ has two RK incomparable predecessors, namely the projections to the first and second directions. These projected ultrafilters are actually generic for $(\mathcal{R}_1, \leq^*_{\mathcal{R}_1})$. The projection to the nodes of length one produces a Ramsey ultrafilter. Thus, we see that the structure of the RK classes reducible to $\mathcal{U}_2$ includes the structure of the Boolean algebra $\mathcal{P}(2)$. In fact, Ramsey-theory techniques along with the canonical equivalence relations, similar to those in the previous section, allow us to deduce that these are the only RK types of ultrafilters RK reducible to $\mathcal{U}_2$.

Similarly to the space $\mathcal{R}_1$, for each n, there is an ultrafilter $\mathcal{U}_2 \restriction \mathcal{H}^2(n)$ which is the ultrafilter on base set $\mathcal{H}^2(n)$ generated by the sets

$$\mathcal{H}^2(n)|X := \{a(n) : \exists Y \leq X \, (a = r_{n+1}(Y))\}. \qquad (30)$$

Each of these ultrafilters is a P-point. We point out that the ultrafilter $\mathcal{U}_2$ is isomorphic to the ultrafilter $\mathcal{U}_2 \restriction \mathcal{H}^2(0)$.

The canonical equivalence relations on the nth blocks $\mathcal{H}^2(n) = \{X(n) : X \in \mathcal{H}^2\}$ are given by canonical projections of the following forms. Recall the Erdős–Rado canonical projections on finite sets of natural numbers: Given $I \subseteq n+1$, for any

$c = \{c_0, \ldots, c_n\}$, $\pi_I(c) = \{c_i : i \in I\}$. Suppose $a(n)$ is a member of $\mathcal{H}^2(n)$ and $a(n) = \{\langle\rangle, \langle k\rangle\} \cup \{\langle k, \langle i, j\rangle\rangle : i \in I_a, j \in J_a\}$. Given T_0, T_1 subtrees of $\mathbb{T}(n)$ (the nth block of the tree $\mathbb{T}$ from $\mathcal{R}_1$) define

$$\pi_{T_0,T_1}(a(n)) = \{\langle\rangle\} \text{ if } T_0 = T_1 = \{\langle\rangle\}, \tag{31}$$

$$\pi_{T_0,T_1}(a(n)) = \{\langle\rangle, \langle k\rangle\} \text{ if } T_0 = T_1 = \{\langle\rangle, \langle k\rangle\}, \tag{32}$$

$$\pi_{T_0,T_1}(a(n)) = \{\langle\rangle, \langle k\rangle\} \cup \{\langle k, i\rangle : i \in \pi_{I_0}(I_a)\}$$
$$\text{if } T_0 = \{\langle\rangle, \langle k\rangle\} \cup \{\langle k, i\rangle : i \in I_0\} \text{ and}$$
$$T_1 = \{\langle\rangle, \langle k\rangle\}, \tag{33}$$

$$\pi_{T_0,T_1}(a(n)) = \{\langle\rangle, \langle k\rangle\} \cup \{\langle k, j\rangle : j \in \pi_{I_1}(J_a)\}$$
$$\text{if } T_0 = \{\langle\rangle, \langle k\rangle\} \quad \text{and}$$
$$T_1 = \{\langle\rangle, \langle k\rangle, \langle k, i\rangle : i \in I_1\}, \tag{34}$$

$$\pi_{T_0,T_1}(a(n)) = \{\langle\rangle, \langle k\rangle\} \cup \{\langle k, \langle i, j\rangle\rangle : i \in \pi_{I_0}(I_a),$$
$$j \in \pi_{I_1}(J_a)\}. \tag{35}$$

An equivalence relation E on $\mathcal{H}^2(n)$ is *canonical* if for each $i \in \{0, 1\}$, there is some tree $T_i \subseteq \mathbb{T}(n)$ such that for $a(n), b(n) \in \mathcal{H}^2(n)$,

$$a(n) \, \text{E} \, b(n) \iff \pi_{T_0,T_1}(a(n)) = \pi_{T_0,T_1}(b(n)). \tag{36}$$

The *initial Rudin–Keisler structure* below $\mathcal{U}_2$ is the collection of all isomorphism types of nonprincipal ultrafilters RK reducible to $\mathcal{U}_2$. This turns out to be exactly the shape of $(\mathcal{P}(2), \subseteq)$, that is, a diamond shape. Since $\mathcal{U}_2$ is isomorphic to $\mathcal{U}_2 \upharpoonright \mathcal{H}^2(0)$, if $\mathcal{V} \leq_{RK} \mathcal{U}_2$, then there is a map h from $\mathcal{H}^2(0)$ into ω such that $h(\mathcal{U}_2 \upharpoonright \mathcal{H}^2(0)) = \mathcal{V}$. Without loss of generality, we may assume that h is canonical, represented by some projection map π_{T_0,T_1}. If both of T_0, T_1 are $\{\langle\rangle\}$, then the RK image of $\mathcal{U}_2$ is a principal ultrafilter. If both are $\{\langle\rangle, \langle 0\rangle\}$, then the h-image of $\mathcal{U}_2$ is a Ramsey ultrafilter. If $T_0 = \{\langle\rangle, \langle 0\rangle, \langle 0, 0\rangle\}$ and $T_1 = \{\langle\rangle, \langle 0\rangle\}$, then the h-image of $\mathcal{U}_2$ is isomorphic to the ultrafilter forced by $(\mathcal{R}_1, \leq^*_{\mathcal{R}_1})$ and hence is weakly Ramsey; denote this as $\mathcal{V}_0$. Likewise if $T_1 = \{\langle\rangle, \langle 0\rangle, \langle 0, 0\rangle\}$ and $T_0 = \{\langle\rangle, \langle 0\rangle\}$, then the h-image of $\mathcal{U}_2$ $\mathcal{V}_1$. Lastly, if both $T_0 = T_1 = \{\langle\rangle, \langle 0\rangle, \langle 0, 0\rangle\}$, then h-image is isomorphic to $\mathcal{U}_2$. Thus, we find the exact structure of the RK types below $\mathcal{U}_2$. We call this an initial RK structure, since it is downwards closed in the RK classes.

The following ultrafilters form the building blocks for understanding the Tukey types of ultrafilters Tukey reducible to $\mathcal{U}_2$. The canonical projections applied to the P-point $\mathcal{U}_2 \upharpoonright \mathcal{H}^2(n)$ yield the following P-points. Let T_0 and T_1 be subtrees of $\mathbb{T}(n)$. If $T_0 = T_1 = \{\langle\rangle\}$, then $\pi_{T_0, T_1}(\mathcal{U}_2 \upharpoonright \mathcal{H}^2(n))$ is simply a principal ultrafilter. If $T_0 = T_1 = \{\langle\rangle, \langle n\rangle\}$, then $\pi_{T_0, T_1}(\mathcal{U}_2 \upharpoonright \mathcal{H}^2(n))$ is isomorphic to the projected Ramsey ultrafilter $\mathcal{U}_0$, similarly to the space $\mathcal{R}_1$. If T_0 and T_1 are as in Eq. (33), then $\pi_{T_0, T_1}(\mathcal{U}_2 \upharpoonright \mathcal{H}^2(n))$ is isomorphic to $\mathcal{U}_1 \upharpoonright \mathcal{R}_1(l)$, where $l = |I_0|$. Likewise, if T_0 and T_1 are as in Eq. (34), then $\pi_{T_0, T_1}(\mathcal{U}_2 \upharpoonright \mathcal{H}^2(n))$ is isomorphic to $\mathcal{U}_1 \upharpoonright \mathcal{R}_1(l)$, where $l = |I_1|$. If T_0 and T_1 are as in Eq. (35), then $\pi_{T_0, T_1}(\mathcal{U}_2 \upharpoonright \mathcal{H}^2(n))$ is a new type of P-point which has as base set the collection $\{\pi_{T_0, T_1}(a(n)) : a(n) \in \mathcal{H}^2(n)\}$ and is generated by the sets $\pi_{T_0, T_1}(\mathcal{H}^2(n)|X) := \{\pi_{T_0, T_1}(a(n)) : \exists Y \leq X (a = r_{n+1}(Y))\}$, $X \in \mathcal{U}_2$. These are finite trees of height two which have $|I_0| \times |I_1|$ rectangles as their maximal nodes. Let $\mathcal{Y}^2(n)$ denote the collection of all these ultrafilters obtained by canonical projections on $\mathcal{H}^2(n)$. Note that $\mathcal{Y}^2(n)$ is finite.

The Tukey types are handled similarly as for $\mathcal{R}_1$. Each monotone cofinal map from $\mathcal{U}_2$ to an ultrafilter $\mathcal{V}$ on ω is continuous when restricted below some member of $\mathcal{U}_2$. As in the case of $\mathcal{R}_1$, there exist some front $\mathcal{F}$ and a function $g : \mathcal{F} \to \omega$ such that $\mathcal{V} = g(\mathcal{U}_2 \upharpoonright \mathcal{F})$. Again, g may be assumed to be canonical, either by a forcing argument or construction some extra hypothesis like CH or less. For fronts $\mathcal{F}$ of the form $\mathcal{AR}_m$, g is canonized by a sequence $(\pi_n : n \leq m)$ of canonical projection maps so that $g(\mathcal{U}_2 \upharpoonright \mathcal{F})$ is Rudin–Keisler equivalent to the Fubini iteration

$$\pi_0(\mathcal{U}_2 \restriction \mathcal{H}^2(0)) * \cdots * \pi_{m-1}(\mathcal{U}_2 \restriction \mathcal{H}^2(m)), \qquad (37)$$

where π_n is one of the canonical projection maps on $\mathcal{H}^2(n)$.

Theorem 41 (Dobrinen, Trujillo [36]). *Let $\mathcal{U}_2$ be the ultra-filter forced by $(\mathcal{H}^2, \leq^*)$. Then*

(1) *If $\mathcal{W} \leq_{\mathrm{RK}} \mathcal{U}_2$, then $\mathcal{W}$ is isomorphic to one of $\mathcal{U}_2$, $\mathcal{V}_0$, $\mathcal{V}_1$, the projected Ramsey ultrafilter, or a principal ultrafilter. Thus, the initial RK structure of the nonprincipal ultrafilters RK reducible to $\mathcal{U}_2$ is simply the structure of the Boolean algebra $\mathcal{P}(2)$.*

(2) *If $\mathcal{W} \leq_T \mathcal{U}_2$, then $\mathcal{W}$ is Tukey equivalent to one of $\mathcal{U}_2$, $\mathcal{V}_0$, $\mathcal{V}_1$, the projected Ramsey ultrafilter, or a principal ultrafilter. Thus, the initial Tukey structure of the nonprincipal ultra-filters Tukey reducible to $\mathcal{U}_2$ is simply the structure of the Boolean algebra $\mathcal{P}(2)$.*

(3) *If $\mathcal{W} \leq_T \mathcal{U}_2$, then the Rudin–Keisler classes inside of $\mathcal{W}$ are exactly those of the countable iterates of Fubini products of ultrafilters from the countable collection $\bigcup_{n<\omega} \mathcal{Y}^2(n)$.*

Continuing this to higher dimensions, the *hypercube* topological Ramsey spaces $\mathcal{H}^k$ have as elements infinite sequences $X = \langle X(n) : n < \omega \rangle$ such that the nth block $X(n)$ consists of $\{\langle\rangle, \langle k_n \rangle\}$, where (k_n) is a strictly increasing sequence, along with a k-dimensional cube which is the product $I_{k,i}$, where $I_{k,i} \in [k_n + 1]^k$ for each $i < k$. The partial ordering on the spaces $\mathcal{H}^k$ are analogous to $\mathcal{H}^2$. These spaces are shown by Trujillo and the author to be topological Ramsey spaces and appear in [13]. There, a larger countable collection of P-points $\mathcal{Y}^k$ is found, and the analogous results to Theorem 41 for k-dimensions are proved. Thus, the Boolean algebra $\mathcal{P}(k)$ is shown to be both an initial RK structure as well as an initial Tukey structure in the P-points. The analogue of (3) in the theorem also holds, where the iterated Fubini products range over ultrafilters in $\mathcal{Y}^k$.

8.   More Initial Rudin–Keisler and Tukey Structures Obtained from Topological Ramsey Spaces

The previous two sections provided details of how the Rudin–Keisler and Tukey structures below certain P-points can be completely understood if the P-points were forced by some topological Ramsey space in which canonical equivalence relations on fronts are well-understood. This section gives the reader the flavor of a collection of broader results.

8.1.   *k-arrow, not (k + 1)-arrow ultrafilters*

The k-arrow ultrafilters are a class of P-points which satisfy asymmetric partition relations.

Definition 42 ([1]). An ultrafilter $\mathcal{U}$ is *n-arrow* if $2 \leq n < \omega$ and for every function $f : [\omega]^2 \to 2$, either there exists a set $X \in \mathcal{U}$ such that $f([X]^2) = \{0\}$, or else there is a set $Y \in [\omega]^n$ such that $f([Y]^2) = \{1\}$. $\mathcal{U}$ is an *arrow* ultrafilter if $\mathcal{U}$ is n-arrow for each $n \leq 3 < \omega$.

Baumgartner and Taylor showed in [1] that for each $2 \leq n < \omega$, there are P-points which are n-arrow but not $(n + 1)$-arrow. Note that every ultrafilter is 2-arrow. Similarly to the $\mathcal{R}_1$ and $\mathcal{H}^2$ spaces, for each $k \geq 2$, there is a topological Ramsey space $\mathcal{A}_k$ which is dense in the forcing that Baumgartner and Taylor used to construct an k-arrow, not $(k+1)$-arrow ultrafilter. The members of this space are infinite sequences, $X = \langle X(n) : n < \omega \rangle$, such that each $X(n)$ is a certain type of ordered graph omitting $k + 1$-cliques. The fact that these ultrafilters are forced by a topological Ramsey space shows that they have complete combinatorics, by Theorem 23. For details, the reader is referred to [13].

Similarly to $\mathcal{R}_1$, both the initial RK structure and initial Tukey structure for the k-arrow, not $k + 1$-arrow ultrafilter $\mathcal{W}_k$

forced by $(\mathcal{A}_k, \leq^*)$ are of size 2: $\mathcal{W}_k$ and its projection to a Ramsey ultrafilter. However, when we look at the Rudin–Keisler classes inside of the Tukey type of $\mathcal{W}_k$, the picture becomes more complex as we shall now see.

The canonical equivalence relations on the collection of nth blocks (that is, $\{X(n) : X \in \mathcal{A}_n\}$) were obtained by the author and we found to be again given by projections. This depended heavily on the flexibility of the structure of the Fraïssé limit of the class of finite ordered graphs omitting $k+1$-cliques. The following is a specific case of a more general theorem for canonical equivalence relations, attributed to Dobrinen in [13]. For graphs A, B, the notation $\binom{B}{A}$ denotes the set of all subgraphs of B which are isomorphic to A. For an ordered graph A with vertices $\{v_0, \dots, v_j\}$ and $I \subseteq j+1$, $\pi_I(A)$ denotes the subgraph of A induced by the vertices $\{v_i : i \in I\}$.

Theorem 43 (Dobrinen [13]). *Let $k \geq 3$ be given and let A and B be finite ordered graphs omitting k-cliques and such that A embeds into B as a subgraph. Then there is a finite ordered graph C omitting k-cliques which is large enough that the following holds. Given any equivalence relation E on $\binom{C}{A}$, there is an $I \subseteq |A|$ and a $B' \in \binom{C}{B}$ such that E restricted to $\binom{B'}{A}$ is given by E_I.*

The building blocks of the Rudin–Keisler classes inside the Tukey are ultrafilters obtained by the canonical projection maps resulting in the following. Let $\mathcal{K}_{k+1}$ denote the Fraïssé class of all finite ordered graphs omitting $k+1$-cliques. More precisely, we take one finite ordered graph omitting $k+1$-cliques from each isomorphism class of these graphs. This set is partially ordered by graph embedding.

Theorem 44 (Dobrinen, Mijares and Trujillo [13]). *Let $\mathcal{W}_k$ be a k-arrow, not $k+1$-arrow P-point forced by the topological Ramsey space $\mathcal{A}_k$ partially ordered by the σ-closed order $\leq^*$.*

(1) *The initial Rudin–Keisler structure below $\mathcal{W}_k$ is a chain of length 2.*

(2) *The initial Tukey structure below $\mathcal{W}_k$ is a chain of length 2.*
(3) *The isomorphism classes inside the Tukey type of $\mathcal{W}_k$ have the same structure as $\mathcal{K}_{k+1}$ partially ordered by embedding.*

This is a particular example of more general results in [13] handling other Fraïssé classes of ordered relational structures with the Ramsey property, and finite products of such structures, producing quite complex Rudin–Keisler structures inside the Tukey types. For instance, there are topological Ramsey spaces which produce initial Tukey structure of the form $([\omega]^{<\omega}, \subseteq)$. Furthermore, *arrow ultrafilters* (ultrafilters which are k-arrow for all $k < \omega$) are also seen to be forced by a topological Ramsey space, and have similar results for their initial RK and Tukey structures. These and more general results are found as Theorems 60 and 67 in [13].

8.2. *Ultrafilters of Laflamme with increasingly weak partition relations*

The ultrafilter $\mathcal{U}_1$ in Sec. 6 was only the beginning of a hierarchy of P-points satisfying successively weaker partition relations constructed by Laflamme in [23]. These forcings $\mathbb{P}_\alpha$, $1 \le \alpha < \omega_1$, were found to have dense subsets forming topological Ramsey spaces in [16]. The reader interested in more details is referred to that paper. Here, we merely state that this yields rapid P-points $\mathcal{V}_\alpha$ for each $1 \le \alpha < \omega_1$ which have complete combinatorics (proved by Laflamme for the version over $\text{HOD}^{V[G]}$, and obtained over $L(\mathbb{R})$ in the presence of large cardinals by Dobrinen and Todorcevic by virtue of being forced by a topological Ramsey space).

Theorem 45 (Dobrinen and Todorcevic [16]). *For each $1 \le \alpha < \omega_1$, there is a topological Ramsey space $\mathcal{R}_\alpha$ forcing a P-point $\mathcal{V}_\alpha$ such that the initial Rudin–Keisler structure and the initial Tukey structure are both decreasing chains of order-type $(\alpha + 1)^*$.*

For each $1 \leq \alpha < \omega_1$, the Rudin–Keisler types inside the Tukey type of $\mathcal{V}_\alpha$ are the countable Fubini iterates of the P-points obtained by canonical projections on the blocks of the sequences forming members of $\mathcal{R}_\alpha$.

Recent work of Zheng in [39] showed that these ultrafilters are preserved by countable support side-by-side Sacks forcing. Zheng had already shown this to be the case for the ultrafilter on base set FIN $= [\omega]^{<\omega} \setminus \{\emptyset\}$ which is constructed by the Milliken space of infinite increasing block sequences (see [38]).

8.3. *Ultrafilters forced by $\mathcal{P}(\omega \times \omega)/(\mathrm{Fin} \otimes \mathrm{Fin})$*

The forcing $\mathcal{P}(\omega)/\mathrm{Fin}$ which adds a Ramsey ultrafilter has a natural generalization to $\mathcal{P}(\omega \times \omega)/\mathrm{Fin} \otimes \mathrm{Fin}$, where $\mathrm{Fin} \otimes \mathrm{Fin}$ is the ideal of the sets $X \subseteq \omega \times \omega$ such that for all but finitely many $i < \omega$, the set $\{j < \omega : (i,j) \in X\}$ is finite. We let $(X)_i$ denote $\{j < \omega : (i,j) \in X\}$ and call it the *ith fiber* of X. This forcing adds a new ultrafilter $\mathcal{W}_2$ which is not a P-point but satisfies the best partition property that a non-P-point can have, namely, $\mathcal{W}_2 \to (\mathcal{W}_2)^2_{l,4}$. Letting $\pi_0 : \omega \times \omega \to \omega$ by $\pi_0(i,j) = i$, the projection $\pi_0(\mathcal{W}_2)$ to its first coordinates is a Ramsey ultrafilter.

Many properties of the ultrafilter $\mathcal{W}_2$ were investigated by Blass, Dobrinen and Raghavan in [6]. That paper included bounds on Tukey type of $\mathcal{W}_2$ showing that it is neither minimal nor maximal in the Tukey types of ultrafilters, but the question of the exact structure of the Tukey types below it remained open.

In [10], the author proved that $\mathcal{P}(\omega \times \omega)/\mathrm{Fin} \otimes \mathrm{Fin}$ is forcing equivalent to a topological Ramsey space when partially ordered by its almost reduction relation. The coideal $(\mathrm{Fin} \otimes \mathrm{Fin})^+$ is the collection of all $X \subseteq \omega \times \omega$ such that for all but finitely many $i < \omega$, the ith fiber of X is infinite. For $X, Y \subseteq \omega \times \omega$, write $Y \subseteq^{*2} X$ if and only if $Y \setminus X \in \mathrm{Fin} \otimes \mathrm{Fin}$. It is routine to check that $\mathcal{P}(\omega \times \omega)/\mathrm{Fin} \otimes \mathrm{Fin}$ is forcing equivalent to $((\mathrm{Fin} \otimes \mathrm{Fin})^+, \subseteq^{*2})$. The forcing $((\mathrm{Fin} \otimes \mathrm{Fin})^+, \subseteq^{*2})$ contains a dense subset which forms a topological Ramsey space. We denote this space $\mathcal{E}_2$, since it is the two-dimensional Ellentuck space. Here, we will only present an overview of this work, referring the interested reader to [10].

In order to find the initial Rudin–Keisler and Tukey structures below $\mathcal{W}_2$, a new kind of canonical Ramsey theorem for equivalence relations on fronts had to be proved. The canonical equivalence relations are again given by canonical projection functions, projecting to subtrees. However, they have a quite different structure than the previous examples in that they are not sequences of finitary projections, since the structure of the members of $\mathcal{E}_2$ are isomorphic to the ordinal ω^2.

Similarly to how $\text{Fin}^{\otimes 2} := \text{Fin} \otimes \text{Fin}$ was defined given Fin, the process can be recursively continued to define ideals $\text{Fin}^{\otimes k+1}$ for each $k < \omega$, where $X \subseteq \omega^{k+1}$ is a member of $\text{Fin}^{\otimes k+1}$ if and only if for all but finitely many $i_0 < \omega$, $\{(i_1, \ldots, i_k) \in \omega^k : (i_0, i_1, \ldots, i_k) \in X\}$ is a member of $\text{Fin}^{\otimes k}$. Then $\mathcal{P}(\omega^k)/\text{Fin}^{\otimes k}$ forces an ultrafilter $\mathcal{W}_k$ which is not a P-point and projects to $\mathcal{W}_j$ for each $1 \leq j < k$, where $\mathcal{W}_1$ is Ramsey. The initial Rudin–Keisler and Tukey structures of these ultafilters are as follows.

Theorem 46 (Dobrinen [10]). *For each $2 \leq k < \omega$, there is a k-dimensional Ellentuck space, $\mathcal{E}_k$, such that $(\mathcal{E}_k, \subseteq^{*k})$ is forcing equivalent to $\mathcal{P}(\omega^k)/\text{Fin}^{\otimes k}$. The forced ultrafilter $\mathcal{U}_k$ satisfies complete combinatorics, and its initial Rudin–Keisler and its initial Tukey structures are both chains of length k.*

Although these ultrafilters $\mathcal{W}_k$ are not P-points, the high-dimensional Ellentuck spaces which force them treat them as P-points, in the sense that every $\subseteq^{*k}$-decreasing sequence has a $\subseteq^{*k}$-pseudointersection in $\mathcal{W}_k$. This is the sense in which these ultrafilters are similar to P-points; they satisfy diagonalization with respect to some σ-closed ideal. It is efficacious to think of these as P-points with respect to topological Ramsey spaces with respect to almost reduction.

9. Further Directions

The construction of topological Ramsey spaces to has served to fine-tune our understanding of several classes of ultrafilters satisfying some partition relations. It seems to us that these are just a few examples of a broader scheme. Listed below are some guiding themes for further investigation in which topological Ramsey spaces will likely play a vital role. Recall our Conjecture 1: Every ultrafilter satisfying some partition relation and forced by some σ-closed forcing is actually forced by some topological Ramsey space. If this turns out to be true, then topological Ramsey spaces will be exactly the correct spaces in which to investigate such ultrafilters, and moreover, all such ultrafilters will have complete combinatorics.

Finding the initial Tukey structures is a way of approximating the exact structure of *all* the Tukey types of ultrafilters starting from the bottom of the hierarchy and going as high up as possible. We have shown that $(\alpha + 1)^*$ for each $1 \leq \alpha < \omega_1$, $\mathcal{P}(k)$ for each $1 \leq k < \omega$, and $([\omega]^{<\omega}, \subseteq)$ all appear as initial Tukey structures of P-points. We have also shown that each finite chain of length two or more appears as an initial Tukey structure of a non-P-point. Furthermore, [11] and a forthcoming paper obtain uncountable linear orders as initial Tukey structures of non-P-points. In [14], Dobrinen and Todorcevic constructed $2^{\mathfrak{c}}$ many Tukey incomparable Ramsey ultrafilters assuming $\operatorname{cov}(\mathcal{M}) = \mathfrak{c}$, showing that this large antichain appears as an initial Tukey structure. This is in contrast to other work in [14, 28, 30, 31] showing that certain structures embed into the Tukey types of ultrafilters. We would like to know the structure of downward closed Tukey structures which are as large as possible as a means of gaining information about the exact structure of all Tukey types of ultrafilters.

Problem 47. Given an ultrafilter $\mathcal{U}$ satisfying some partition property, what is the structure of the Tukey types of all ultrafilters Tukey reducible to $\mathcal{U}$?

For the examples analyzed in previous sections, knowing that the generating partial ordering is essentially a topological Ramsey space aided greatly in solving this problem. The Ramsey theory available also enabled us to find initial Rudin-Keisler structures and precisely, the structure of the RK classes inside the Tukey types of an initial structure of Tukey types. Similar questions can be asked for these two foci.

A related but more challenging problem is the following.

Problem 48. What are the most complex structures initial Tukey structures that can be found?

If one can find the initial Tukey structure below the maximal Tukey type, then one has completely found the full Tukey structure of all ultrafilters. It should be pointed out that it may be consistent that there is only one Tukey type. This is what remains of Isbell's Problem in [21] which is one of the most important questions on Tukey types of ultrafilters. Such a model would have to contain no P-points and hence no Ramsey ultrafilters, so it was not the focus of this paper.

As we briefly saw in Sec. 8.3, forcing with the σ-closed forcings $\mathcal{P}(\omega^k)/\mathrm{Fin}^{\otimes k}$, and more generally $\mathcal{P}(\omega^\alpha)/\mathrm{Fin}^{\otimes \alpha}$ for each countable ordinal α (see [11]), and for the other examples covered in previous sections, forcing with some partial ordering modulo a σ-ideal produces an ultrafilter which has complete combinatorics, since, for these examples, they are forced by topological Ramsey spaces. This leads to the following question which we find quite compelling.

Problem 49. Given a countable set X and a σ-closed ideal $\mathcal{I}$ on X, if the forcing $\mathcal{P}(X)/\mathcal{I}$ adds an ultrafilter which satisfies some weak partition properties, is there some topological Ramsey space $\mathcal{R}$ such that $(\mathcal{R}, \leq^*)$ is forcing equivalent to $\mathcal{P}(X)/\mathcal{I}$?

A related question is the following.

Problem 50. For which σ-closed ideals $\mathcal{I}$ on a countable set X, such that the forcing $\mathcal{P}(X)/\mathcal{I}$ adds an ultrafilter which has complete combinatorics, is there some topological Ramsey space $\mathcal{R}$ such that $(\mathcal{R}, \leq^*)$ is forcing equivalent to $\mathcal{P}(X)/\mathcal{I}$?

In [20], Hrušak and Verner proved that if $\mathcal{I}$ is a tall F_σ P-ideal, then $\mathcal{P}(\omega)/\mathcal{I}$ adds a P-point which has no rapid RK-predecessor and which is not Canjar. Thus, there is no Ramsey ultrafilter RK below this forced ultrafilter, but the Mathias forcing with tails in this ultrafilter does add a dominating real. It seems unlikely that such ideals give an affirmative answer to Problem 50 since all know topological Ramsey spaces have ultrafilters with Ramsey ultrafilters RK below them, but this remains open.

Lastly, we would like to have a more user-friendly characterization of complete combinatorics for topological Ramsey spaces. The characterization of complete combinatorics given by Di Prisco, Mijares, and Nieto in Theorem 23 requires one to understand selectivity of an ultrafilter in the sense of diagonalizations of certain dense open sets with respect to the ultrafilter. The complete combinatorics of Blass and Laflamme, on the other hand, characterize complete combinatorics in terms of Ramsey properties. We conjecture that a similar characterization can be given for the topological Ramsey space setting. We say that a filter $\mathcal{U}$ on base set $\mathcal{R}$ is *Ramsey* with respect to $(\mathcal{R}, \leq, r)$ if and only if for each $2 \leq n < \omega$, for each coloring $c : \mathcal{AR}_n \to 2$, there is an $X \in \mathcal{U}$ such that c has one color on $\mathcal{AR}_n | X$.

Conjecture 2. *Let $(\mathcal{R}, \leq, r)$ be a topological Ramsey space, and let $\mathcal{U}$ be a filter on base set $\mathcal{R}$. Suppose that there is a supercompact cardinal in V. If $\mathcal{U}$ is Ramsey, then $\mathcal{U}$ is generic for the forcing $(\mathcal{R}, \leq^*)$ over the Solovay model $L(\mathbb{R})$.*

It would suffice to prove that $\mathcal{U}$ is selective (in the sense of Definition 21) if and only if $\mathcal{U}$ is Ramsey. This seems likely, as similar (but not exactly the same) results were obtained for the ultrafilters in [13] and for a class of topological Ramsey spaces of trees in [36]. Such a representation of complete combinatorics over $L(\mathbb{R})$ for topological Ramsey spaces would be the ideal analogue of the complete combinatorics of Blass and Laflamme.

Acknowledgments

This work was partially supported by National Science Foundation Grant DMS-1600781.

References

[1]   J. E. Baumgartner and A. D. Taylor, Partition theorems and ultra-filters, *Tans. Amer. Math. Soc.* **241** (1978) 283–309.

[2]   A. Blass, Orderings of ultrafilters, Ph.D. thesis, Harvard University (1970).

[3]   A. Blass, The Rudin–Keisler ordering of p-points, *Trans. Amer. Math. Soc.* **179** (1973) 145–166.

[4]   A. Blass, Ultrafilter mappings and their Dedekind cuts, *Trans. Amer. Math. Soc.* **188** (1974) 327–340.

[5]   A. Blass, Selective ultrafilters and homogeneity, *Ann. Pure Appl. Logic* **38** (1988) 215–255.

[6]   A. Blass, N. Dobrinen and D. Raghavan, The next best thing to a p-point, *J. Symbolic Logic* **80**(3) (2015) 866–900.

[7]   D. Booth, Ultrafilters on a countable set, *Ann. Math. Logic* **2**(1) (1970/1971), 1–24.

[8]   T. J. Carlson and S. G. Simpson, Topological Ramsey theory, in *Mathematics of Ramsey Theory*, Algorithms and Combinatorics, Vol. 5 (Springer, 1990), pp. 172–183.

[9]   C. DiPrisco, J. G. Mijares and J. Nieto, Local Ramsey theory: an abstract approach, *Math. Logic Quart.* **63**(5) (2017) 384–396.

[10] N. Dobrinen, High dimensional Ellentuck spaces and initial chains in the Tukey structure of non-p-points, *J. Symbolic Logic* **81**(1) (2016) 237–263.

[11] N. Dobrinen, Infinite dimensional Ellentuck spaces and Ramsey-classification theorems, *J. Math. Logic* **16**(1) (2016) 37.

[12] N. Dobrinen, Continuous and other finitely generated canonical cofinal maps on ultrafilters, *Fund. Math.* (2020) 37 pp, Online version appeared December 6, 2019.

[13] N. Dobrinen, J. G. Mijares and T. Trujillo, Topological Ramsey spaces from Fraïssé classes, Ramsey-classification theorems, and initial structures in the Tukey types of p-points, *Arch. Math. Logic*, special issue in honor of James E. Baumgartner **56**(7–8) (2017) 733–782 (Invited submission).

[14] N. Dobrinen and S. Todorcevic, Tukey types of ultrafilters, *Illinois J. Math.* **55**(3) (2011) 907–951.

[15] N. Dobrinen and S. Todorcevic, A new class of Ramsey-classification: Theorems and their applications in the Tukey theory of ultrafilters, Part 1, *Trans. Amer. Math. Soc.* **366**(3) (2014) 1659–1684.

[16] N. Dobrinen and S. Todorcevic, A new class of Ramsey-classification: Theorems and their applications in the Tukey theory of ultrafilters, Part 2, *Trans. Amer. Math. Soc.* **367**(7) (2015) 4627–4659.

[17] E. Ellentuck, A new proof that analytic sets are Ramsey, *J. Symbolic Logic* **39**(1) (1974) 163–165.

[18] I. Farah, Semiselective coideals, *Mathematika* **45**(1) (1998) 79–103.

[19] F. Galvin and K. Prikry, Borel sets and Ramsey's Theorem, *J. Symbolic Logic* **38** (1973).

[20] M. Hrušák and J. L. Verner, Adding ultrafilters by definable quotients, *Rend. Circ. Mat. Palermo* **60** (2011) 445–454.

[21] J. Isbell, The category of cofinal types. II, *Trans. Amer. Math. Soc.* **116** (1965) 394–416.

[22] O. Klein and O. Spinas, Canonical forms of Borel functions on the Milliken space, *Trans. Amer. Math. Soc.* **357**(12) (2005) 4739–4769.

[23] C. Laflamme, Forcing with filters and complete combinatorics, *Ann. Pure Appl. Logic* **42** (1989) 125–163.

[24] H. Lefmann, Canonical partition relations for ascending families of finite sets, *Studia Sci. Math. Hungarica* **31**(4) (1996) 361–374.

[25] A. Louveau, Une démonstration topologique de théorèmes de Silver et Mathias, *Bull. Sci. Math.* **98**(2) (1974) 97–102.

[26] A. R. D. Mathias, Happy families, *Ann. Math. Logic* **12**(1) (1977) 59–111.

[27] J. G. Mijares, A notion of selective ultrafilter corresponding to topological Ramsey spaces, *Math. Logic Quart.* **53**(3) (2007) 255–267.

[28] D. Milovich, Tukey classes of ultrafilters on ω, *Topology Proc.* **32** (2008) 351–362.

[29] C. St. J. A. Nash-Williams, On well-quasi-ordering transfinite sequences, *Proc. Cambridge Philos. Soc.* **61** (1965) 33–39.

[30] D. Raghavan and S. Shelah, Embedding partial orders into the P-points under Rudin-Keisler and Tukey reducibility, *Trans. Amer. Math. Soc.* **369**(6) 4433–4455.

[31] D. Raghavan and S. Todorcevic, Cofinal types of ultrafilters, *Ann. Pure Appl. Logic* **163**(3) (2012) 185–199.

[32] S. Shelah and H. Woodin, Large cardinals imply that every reasonably definable set of reals is Lebesgue measurable, *Israel J. Math.* **70** (1990) 381–384.

[33] J. Silver, Every analytic set is Ramsey, *J. Symbolic Logic* **35** (1970) 60–64.

[34] S. Solecki and S. Todorcevic, Cofinal types of topological directed orders, *Ann. Inst. Fourier* **54**(6) (2004) 1877–1911.

[35] S. Todorcevic, *Introduction to Ramsey Spaces* (Princeton University Press, 2010).

[36] T. Trujillo, Topological Ramsey spaces, associated ultrafilters, and their applications to the Tukey theory of ultrafilters and Dedekind cuts of nonstandard arithmetic, Ph.D. thesis, University of Denver (2014).

[37] T. Trujillo, Selective but not Ramsey, *Topology Appl.* **202** (2016) 61–69.

[38] Y. Y. Zheng, Selective ultrafilters on FIN, *Proc. Amer. Math. Soc.* **145**(12) (2017) 5071–5086.

[39] Y. Y. Zheng, Preserved under Sacks forcing again? *Acta Math. Hungarica* **154**(1) (2018) 1–28.

Chapter 2

Infinitary Partition Properties of Sums of Selective Ultrafilters

Andreas Blass

Department of Mathematics, University of Michigan,
Ann Arbor, MI 48109–1043, USA

ablass@umich.edu

We consider two sorts of ultrafilters on ω^2. The first are the sums of nonisomorphic selective ultrafilters, indexed by another selective ultrafilter. The second are ultrafilters generic with respect to the forcing whose conditions are subsets of ω^2 that have infinite intersection with infinitely many of the columns $\{n\} \times \omega$. It is known that the two sorts share many properties, for example, finitary square-bracket partition relations, being Q-points but not P-points, and not being at the top of the Tukey ordering. They differ, however, in other respects: Only the first are basically generated and only the second are weak P-points. The first main result of the present paper improves the known infinitary partition property of the first sort to match what is known about the second sort. This leads to an application of "complete combinatorics" and some additional forcing arguments to show that, in certain situations, the two sorts are actually the same ultrafilters, but viewed in different models of set theory. We use this to give alternative proofs of the similarities between the two sorts of ultrafilters, and we explain how the differences between the two sorts arise.

1.   Introduction

This paper is a sequel to [2, 3]. As in these earlier papers, we deal with two sorts of special ultrafilters on ω^2. One sort is obtained as sums of nonisomorphic selective ultrafilters on ω, indexed by another such selective ultrafilter. The other sort is obtained by forcing with the Boolean algebra $\mathcal{P}(\omega^2)/\mathfrak{Fr}^{\otimes 2}$, where $\mathfrak{Fr}$ is the filter of cofinite subsets of ω and $\mathfrak{Fr}^{\otimes 2}$ is its tensor square. (For definitions of unfamiliar terms and notations used in this introduction, see Sec. 2.)

These two sorts of ultrafilters are similar in many respects. The following properties were proved for both sorts in [3], except that "not Tukey top" was already known earlier in the case of selective-indexed sums of selective ultrafilters.

- The ultrafilter is a Q-point but not a P-point.
- Every function on ω^2 becomes, when restricted to a suitable set in the ultrafilter, either one-to-one, or constant, or a composition $g \circ \pi_1$ where g is one-to-one and $\pi_1 : \omega^2 \to \omega$ is the projection to the first factor.
- The ultrafilter is $(n, T(n))$-weakly Ramsey.
- The second projection $\pi_2 : \omega^2 \to \omega$ is one-to-one on a set in the ultrafilter.
- The ultrafilter is not at the top of the Tukey ordering of ultrafilters on countable sets.

On the other hand, there are some significant differences, also proved in [3] for $\mathcal{P}(\omega^2)/\mathfrak{Fr}^{\otimes 2}$-generic ultrafilters; the contrasting information for selective-indexed sums of selective ultrafilters was known earlier.

- $\mathcal{P}(\omega^2)/\mathfrak{Fr}^{\otimes 2}$-generic ultrafilters are weak P-points; sums are not weak P-points.
- Selective-indexed sums of selective ultrafilters are basically generated; $\mathcal{P}(\omega^2)/\mathfrak{Fr}^{\otimes 2}$-generic ultrafilters are not.

In [2], infinitary partition properties were established for both sorts of ultrafilters. These properties were very similar but not identical; the version for $\mathcal{P}(\omega^2)/\mathfrak{Fr}^{\otimes 2}$-generic ultrafilters was slightly stronger.

- Let τ be any ω-type, and let the set of all type-τ subsets of ω^2 be partitioned into an analytic piece and a co-analytic piece. Then the ultrafilter contains a set H all of whose type-τ subsets lie in the same piece. In the case of a $\mathcal{P}(\omega^2)/\mathfrak{Fr}^{\otimes 2}$-generic ultrafilter, H can be taken to have type τ as well.

The first main result of this paper is a combinatorial property of selective-indexed sums of selective ultrafilters that allows us to upgrade the infinitary partition property for these ultrafilters to exactly match what was proved in [2] for $\mathcal{P}(\omega^2)/\mathfrak{Fr}^{\otimes 2}$-generic ultrafilters. That is, the homogeneous set H in the infinitary partition property for such sums can be found with the prescribed type τ.

After proving this result and the analogous results for HOD$\mathbb{R}$ partitions in the Lévy–Mahlo model, we shall investigate the consequences in the light of the complete combinatorics for $\mathcal{P}(\omega^2)/\mathfrak{Fr}^{\otimes 2}$-generic ultrafilters established in [2]. This investigation will show that $\mathcal{P}(\omega^2)/\mathfrak{Fr}^{\otimes 2}$-generic ultrafilters become sums of selective ultrafilters when one suitably enlarges the set-theoretic universe.

This result accounts for the similarities noted above between the two sorts of ultrafilters, and it leads to simplifications of some of the proofs in [3]. We also explain how the dissimilarities noted above are compatible with our results.

2. Notation and Preliminaries

In this section, we review some definitions and results, including those presupposed in the introduction. We shall usually be concerned with filters on ω and ω^2, but everything we do in this context can be transferred to arbitrary countably infinite sets via an arbitrary bijection. Unless the contrary is explicitly stated, filters are assumed to contain all cofinite sets; in particular ultrafilters are assumed to be nonprincipal.

Notation 1. $\mathfrak{Fr}$ is the filter of cofinite subsets of ω, often called the *Fréchet filter*.

Definition 2. If $f : X \to Y$ and $\mathcal{F}$ is a filter on X, then

$$f(\mathcal{F}) = \{A \subseteq Y : f^{-1}(Y) \in \mathcal{F}\}.$$

This is a filter on Y except that it might not contain all cofinite subsets of Y. If f and g are equal on a set in $\mathcal{F}$ then $f(\mathcal{F}) = g(\mathcal{F})$. The converse can fail, even if $\mathcal{F}$ is an ultrafilter, but it holds if $\mathcal{F}$ is an ultrafilter and f is one-to-one on some set in $\mathcal{F}$.

Definition 3. Ultrafilters $\mathcal{U}$ on X and $\mathcal{V}$ on Y are *isomorphic* if $\mathcal{V} = f(\mathcal{U})$ for some $f : X \to Y$ that is one-to-one on some set in $\mathcal{U}$.

Isomorphism is an equivalence relation. If X and Y have the same cardinality, then ultrafilters on X and Y are isomorphic if and only if they are related by a bijection between X and Y. Indeed, any f as in the definition of isomorphism agrees, on some set in $\mathcal{U}$, with some bijection.

The definition of isomorphism also makes sense for filters that are not ultrafilters, but we shall not need this additional generality.

Definition 4. If $\mathcal{F}$ is a filter on X, then we say that a subset A of X is *positive* with respect to $\mathcal{F}$ if its complement $X - A$ is not in $\mathcal{F}$. (Equivalently: A has nonempty intersection with every set in $\mathcal{F}$.)

We write $A \subseteq^* B$ to mean that $A - B$ is finite; in this case we say that A is *almost included* in B.

When we mention topological notions, like Borel or analytic sets, in the context of a power set $\mathcal{P}(X)$, we mean the topology obtained by identifying $\mathcal{P}(X)$ with $\{0,1\}^X$ and giving this the product topology induced from the discrete topology on $\{0,1\}$.

2.1. *Special ultrafilters*

Definition 5. An ultrafilter $\mathcal{U}$ on a set X is

- *selective* if every function on X becomes constant or one-to-one when restricted to a suitable set in $\mathcal{U}$,
- *a P-point* if every function on X becomes constant or finite-to-one when restricted to a suitable set in $\mathcal{U}$, and
- *a Q-point* if every finite-to-one function on X becomes one-to-one when restricted to a suitable set in $\mathcal{U}$.

An ultrafilter is selective if and only if it is both a P-point and a Q-point.

An equivalent characterization of P-points, among ultrafilters on countable sets, is that, for any countably many sets $A_n \in \mathcal{U}$, there is a set $B \in \mathcal{U}$ that is almost included in each A_n.

If $\mathcal{U}$ is selective, then any nonprincipal ultrafilter of the form $f(\mathcal{U})$ is isomorphic to $\mathcal{U}$. Indeed, because $f(\mathcal{U})$ is nonprincipal, f cannot be constant on any set in $\mathcal{U}$, so, by selectivity, it is one-to-one on some set in $\mathcal{U}$. But then, as mentioned above, it coincides, on some set in $\mathcal{U}$, with a bijection from ω to ω, and this witnesses the desired isomorphism.

Neither P-points nor Q-points can be proved to exist on the basis of the usual ZFC axioms of set theory. But under additional hypotheses, such as the continuum hypothesis (CH) or Martin's axiom (MA), many (in fact $2^{2^{\aleph_0}}$) selective ultrafilters on ω exist, and there are also nonselective P-points and nonselective Q-points on ω. It is easily provable in ZFC that there exist ultrafilters that are neither P-points nor Q-points. It is not known whether it is consistent that neither P-points nor Q-points exist.

Definition 6. An ultrafilter $\mathcal{U}$ on a set X is a *weak P-point* if, for any countably many (nonprincipal) ultrafilters $\mathcal{V}_n$ distinct from $\mathcal{U}$, there is a set $B \in \mathcal{U}$ that is in none of the $\mathcal{V}_n$'s.

An equivalent characterization is that a weak P-point is not in the closure of any countable set of other ultrafilters in the Stone–Čech remainder $\beta(X) - X$.

The existence of weak P-points on ω is provable in ZFC [6]. As the terminology suggests, every P-point is a weak P-point.

2.2.   *Sums and tensor products*

Definition 7. Let $\mathcal{F}$ and $\mathcal{G}_n$ for $n \in \omega$ be filters on ω. The $\mathcal{F}$-*indexed sum* of the $\mathcal{G}_n$'s is the filter

$$\mathcal{F}\text{-}\sum_n \mathcal{G}_n = \{A \subseteq \omega^2 : \{x \in \omega : \{y \in \omega : (x,y) \in A\} \in \mathcal{G}_x\} \in \mathcal{F}\}$$

on ω^2.

That is, a set A is in the sum if and only if, for $\mathcal{F}$-almost all x, the vertical section

$$A_x = \{y : (x,y) \in A\}$$

is large in the sense of $\mathcal{G}_x$.

Convention 8. Throughout this paper, "section" means "vertical section", and we use the notation A_x as above to mean the section of $A \subseteq \omega^2$ at horizontal coordinate $x \in \omega$.

When $\mathcal{F}$ and all of the $\mathcal{G}_n$'s are ultrafilters, then so is $\mathcal{F}\text{-}\sum_n \mathcal{G}_n$. This sum is never a P-point or even a weak P-point. Indeed, the images $\mathcal{V}_n = i_n(\mathcal{G}_n)$ of the summands $\mathcal{G}_n$ under the injection maps $i_n : \omega \to \omega^2 : y \mapsto (n,y)$ constitute a counterexample for the definition of weak P-point.

The sum of ultrafilters $\mathcal{G}_n$ with respect to an ultrafilter $\mathcal{F}$ is a special case of the general topological notion of limit along a filter. Specifically, $\mathcal{F}\text{-}\sum_n \mathcal{G}_n$ is the limit, with respect to $\mathcal{F}$, of the sequence of points $i_n(\mathcal{G}_n)$ in the Stone–Čech compactification $\beta(\omega^2)$.

When all the $\mathcal{G}_n$ are the same filter $\mathcal{G}$, we write $\mathcal{F} \otimes \mathcal{G}$ for the sum $\mathcal{F}\text{-}\sum_n \mathcal{G}$, and we call it the *tensor product* of $\mathcal{F}$ and $\mathcal{G}$. (Other authors sometimes call it the Fubini product.) In the even more special case where $\mathcal{G} = \mathcal{F}$, we write $\mathcal{F}^{\otimes 2}$.

2.3. *The square of the Fréchet filter*

We shall be particularly interested in $\mathfrak{Fr}^{\otimes 2}$, the tensor product of the Fréchet filter with itself. According to the definitions above, the sets in this filter are those subsets $A \subseteq \omega^2$ such that, for all but finitely many $x \in \omega$ the (vertical) section A_x contains all but finitely many $y \in \omega$. Therefore, a set $B \subseteq \omega$ is positive with respect to $\mathfrak{Fr}^{\otimes 2}$ if and only if, for infinitely many x, the section A_x is infinite.

The positive sets with respect to $\mathfrak{Fr}^{\otimes 2}$ constitute the most natural notion of forcing to adjoin a generic ultrafilter extending $\mathfrak{Fr}^{\otimes 2}$. As shown in [3, Lemma 3.1], the separative quotient of this forcing is countably closed, so no new subsets of ω are added; this makes it easy to check that the generic object is indeed an ultrafilter.

It will be convenient to clean up the notion of forcing a bit, by passing to a dense subset. Specifically, we work with the dense subset $\mathbb{P}$ defined, as in [2, Definition 6], to consist of those $A \subseteq \omega^2$ that satisfy

(1) A has infinitely many infinite sections and no nonempty finite sections,
(2) the sections of A are pairwise disjoint,
(3) all elements (x, y) of A have $x < y$, and
(4) for any (x, y) and (x', y') in A, we have $x \neq y'$.

See [2] for the verification that this $\mathbb{P}$ is dense in the poset of positive sets with respect to $\mathfrak{Fr}^{\otimes 2}$.

2.4.   *Types*

In this subsection, we recall the notion of ω-types, which will play a crucial role in this paper, and the related notion of n-types for finite n, which was implicitly used in the introduction above.

Consider any $A \in \mathbb{P}$. By clause (2) in the definition of $\mathbb{P}$, all elements of A have distinct second components, so we can list A as an ω-sequence $\langle (x_n, y_n) : n \in \omega \rangle$ in order of increasing second components, $y_0 < y_1 < \dots$. In what follows, we shall assume this order of listing elements of any set in $\mathbb{P}$, and we shall sometimes refer to "x-coordinates" and "y-coordinates" rather than first and second components of elements of ω^2.

By the ω-type of a set $A \in \mathbb{P}$, we mean, intuitively, all the information about the relative order (as natural numbers) of the coordinates of members of A, but without any information about the magnitude of those components. Thus, for example, $y_7 < x_4$ could be part of an ω-type, but $x_4 = 17$ could not. Formally, we use the following definition from [2, Definitions 32 and 33].

Definition 9. An *ω-type* is a linear pre-order of the infinite set of formal symbols $x_1, x_2, \dots$ and $y_1, y_2, \dots$ such that

- each equivalence class of the pre-order consists of either infinitely many x's or a single y,
- there are infinitely many equivalence classes of x's,
- the induced linear order of the equivalence classes has order-type ω,
- $y_1 < y_2 < \dots$, and
- each x_i strictly precedes the corresponding y_i.

The ω-type *realized* by a set $A \in \mathbb{P}$ consists of exactly those inequalities that are satisfied when the x_i and y_i are interpreted as the x- and y-coordinates of all the elements of A, listed in order of increasing y-coordinates.

These ω-types will be important here because of their connection to infinitary partition relations for $\mathbb{P}$. They provide a strong counterexample to any attempted partition theorem of the form "when $\mathbb{P}$ is partitioned into nice (e.g., Borel) pieces, there is some $H \in \mathbb{P}$ such that all its subsets in $\mathbb{P}$ lie in the same piece". The set $\mathbb{P}_\tau$ of members of $\mathbb{P}$ realizing a specific ω-type τ is a (low-level) Borel set and is dense in the partial order $\mathbb{P}$. These sets $\mathbb{P}_\tau$ constitute a Borel partition of $\mathbb{P}$ into $2^{\aleph_0}$ pieces such that every $H \in \mathbb{P}$ has subsets in all of the pieces.

On the other hand, this is essentially the only counterexample to this sort of partition theorem. Theorem 35(1) of [2] says that, for any ω-type τ, when $\mathbb{P}_\tau$ is partitioned into nice (e.g., Borel) pieces, there is some $H \in \mathbb{P}_\tau$ such that all its subsets in $\mathbb{P}_\tau$ lie in the same piece. As noted in [2, Theorem 35(3)], it follows that such a homogeneous H can be found in any $\mathcal{P}(\omega^2)/\mathfrak{Fr}^{\otimes 2}$-generic ultrafilter.

For finite n, we can define n-*types* to be simply the restrictions of ω-types to $x_1, x_2, \ldots, x_n$ and $y_1, y_2, \ldots, y_n$. The only use of these n-types in the present paper is that the number of them, called $T(n)$, occurs in the finitary weak Ramsey properties quoted in the introduction. An ultrafilter $\mathcal{U}$ on a set X is (n, h)-*weakly Ramsey* if, whenever the set $[X]^n$ of n-element subsets of X is partitioned into finitely many pieces, there is an $H \in \mathcal{U}$ such that $[H]^n$ meets at most h of the pieces.

2.5. *The Lévy–Mahlo model*

It is well known that infinitary partition relations contradict the axiom of choice unless one requires the partitions to be well-behaved. In the case where the homogeneous sets are to be found in a given ultrafilter, two notions of well-behaved have played a prominent role since the fundamental work of Mathias [7].

One is topological, namely that the partition consists of an analytic and a coanalytic piece. Mathias proved, in ZFC, that such partitions have homogeneous sets in any selective ultrafilter. That is, if $\mathcal{U}$ is a selective ultrafilter on ω and if $[\omega]^\omega$ is partitioned into an analytic piece and its complement, then there is a set $H \in \mathcal{U}$ such that all its infinite subsets lie in the same piece of the partition.

The other main notion of well-behaved in such contexts is ordinal definability from reals, but this notion "works" only in certain particular set-theoretic universes. Specifically, suppose κ is a Mahlo cardinal, and consider the model obtained by adjoining a Lévy-generic collapsing of all cardinals below κ to ω. Thus, in this model, which we call the *Lévy–Mahlo* model, κ has become $\aleph_1$.

Mathias proved that, in the Lévy–Mahlo model, if $[\omega]^\omega$ is partitioned into two pieces ordinal definable from reals, then any selective ultrafilter $\mathcal{U}$ on ω contains a set H all of whose infinite subsets lie in the same piece. It follows that, in the Lévy–Mahlo model, every selective ultrafilter is $\mathcal{P}(\omega)/\mathfrak{Fr}$-generic over the submodel HODℝ of sets hereditarily ordinal definable from reals. It should be noted that this result is not vacuous; the Lévy–Mahlo model satisfies CH, so it has many selective ultrafilters.

These two notions of well-behaved, analyticity in the context of arbitrary ZFC universes and ordinal definability from reals in the context of the Lévy–Mahlo model, have continued to serve as standard hypotheses in infinitary partition theorems for other sorts of ultrafilters. Typical examples are Theorems 35 and 37 of [2].

2.6. *Tukey reduction*

The original motivation for the study in [3] of $\mathcal{P}(\omega^2)/\mathfrak{Fr}^{\otimes 2}$-generic ultrafilters came from the Tukey ordering. We shall make some comments in Sec. 5 about this ordering, so we recall here its definition and some related information.

The general context for the Tukey ordering is that of directed partial orders, i.e., orders in which every finite subset has an upper bound. A subset X of a partially ordered set D is said to be *cofinal* if every element of D is $\leq$ some element of X.

Definition 10. Let D and E be directed partial orders. A function $f : D \to E$ is said to be *convergent* if, whenever X is a cofinal subset of D, then $f(X)$ is cofinal in E. When such a function exists, one says that D is *Tukey-above* E and one writes $D \geq_T E$.

Among all directed partial orders of cardinality κ, there is a largest one with respect to the Tukey (pre)order, namely $[\kappa]^{<\omega}$ ordered by inclusion.

Ultrafilters can be naturally viewed as directed sets with respect to the ordering of reverse inclusion, $\supseteq$, so a cofinal subset of an ultrafilter is what is usually called a base for the ultrafilter. There are ultrafilters on ω of maximum Tukey type, i.e., Tukey-equivalent to $[2^{\aleph_0}]^{<\omega}$. It is not known whether ZFC proves the existence of ultrafilters that are not at the top of the Tukey ordering, although this can be proved under additional hypotheses like CH. In particular, P-points are never Tukey-equivalent to $[2^{\aleph_0}]^{<\omega}$.

A widely used tool for showing that certain ultrafilters on ω (in particular all P-points) are not at the top of the Tukey ordering is to show that they are *basically generated*. This means that the ultrafilter has a base $\mathcal{B}$ such that, whenever a sequence $\langle B_n : n \in \omega \rangle$ of members of $\mathcal{B}$ converges (in the usual topology of $\mathcal{P}(\omega)$) to an element of $\mathcal{B}$, then there is an infinite $X \subseteq \omega$ such that $\bigcap_{n \in X} B_n \in \mathcal{U}$. A basically generated ultrafilter is never Tukey-equivalent to $[2^{\aleph_0}]^{<\omega}$; in fact, it cannot be $\geq_T [\omega_1]^{<\omega}$. Furthermore, not only all P-points [4] but all ultrafilters obtainable from them by (transfinitely) iterated sums [8] are basically generated (by means of a base that is closed under finite intersections). So, once one has some P-points, this provides many Tukey nonmaximum ultrafilters.

Part of the motivation for [3] was to produce, in some model of ZFC, an ultrafilter that is not Tukey-maximum but also not basically generated; it was shown in that paper (Theorems 4.10

and 6.1) that $\mathcal{P}(\omega)/\mathfrak{Fr}^{\otimes 2}$-generic ultrafilters have this combination of properties.

3.   All Types Are Realized

In this section, we shall establish a partition theorem for certain sums of selective ultrafilters, improving Theorem 35(2) of [2] to match Theorem 35(3). The sums under consideration are those where the summands and the indexing ultrafilter are selective and pairwise non-isomorphic. As in [2] we find it convenient to abbreviate selective-indexed sum of nonisomorphic selectives as follows.

Definition 11. A *sisnis* ultrafilter is an ultrafilter on ω^2 of the form $\mathcal{U}\text{-}\sum_n \mathcal{V}_n$ where $\mathcal{U}$ and all of the $\mathcal{V}_n$'s are pairwise nonisomorphic, selective ultrafilters on ω.

We shall prove the following partition theorem for these ultrafilters.

Theorem 12. *Let τ be any ω-type, and let $\mathbb{P}_\tau$ be partitioned into an analytic set and its complement. Then any sisnis ultrafilter contains a set $H \in \mathbb{P}_\tau$ such that all its subsets in $\mathbb{P}_\tau$ lie in the same piece of the partition. Furthermore, in the Lévy–Mahlo model, the hypothesis of analyticity can be weakened to ordinal definability from reals and the conclusion remains unchanged.*

Notice that this theorem is very close to Theorems 35(2) and 37 of [2]; the only difference is that we now assert that the homogeneous set H is in $\mathbb{P}_\tau$, i.e., that it realizes the specified ω-type τ. So the only new information that we need to prove here is the following.

Theorem 13. *Let τ be any ω-type, let $\mathcal{W}$ be any sisnis ultrafilter on ω^2 and let A be any set in $\mathcal{W}$. Then A has a subset $B \in \mathcal{W} \cap \mathbb{P}_\tau$.*

The proof of Theorem 13 will occupy the rest of this section. Throughout the section, fix τ, $\mathcal{W}$, and A as in the theorem. The overall strategy for the proof will be to shrink A in several steps so as to obtain a set $C \in \mathcal{W}$ where we have good control over its structure, and then to enlarge C again, adding suitable elements from the original A, so as to finally obtain the desired B.

Since $\mathcal{W}$ is sisnis, we have $\mathcal{W} = \mathcal{U}\text{-}\sum_n \mathcal{V}_n$ for some pairwise nonisomorphic, selective ultrafilters $\mathcal{U}$ and $\mathcal{V}_n$. Since $A \in \mathcal{W}$, there are sets $X \in \mathcal{U}$ and $Y_x \in \mathcal{V}_x$ for each $x \in X$, such that

$$\{(x, y) \in \omega^2 : x \in X \text{ and } y \in Y_x\},$$

which we abbreviate as $X\text{-}\bigsqcup_n xY_x$ and which is clearly in $\mathcal{W}$, is a subset of A. Our overall strategy of shrinking and then re-expanding A will be carried out by shrinking and re-expanding the sets X and Y_x. The first shrinking will use the following lemma to make these sets pairwise disjoint.

Lemma 14. *Let $\mathcal{D}_n$ for $n \in \omega$ be countably many pairwise distinct ultrafilters on ω. If they are all selective (or merely weak P-points), then there are sets $D_n \in \mathcal{D}_n$ such that all these D_n's are pairwise disjoint.*

Proof. Because the $\mathcal{D}_n$'s are weak P-points, each $\mathcal{D}_n$ contains a set E_n that is in none of the other $\mathcal{D}_k$'s. Let

$$D_n = E_n \cap \bigcap_{k<n} (\omega - E_k).$$

Since the $\mathcal{D}_n$ are ultrafilters, and in particular closed under finite intersections, we have $D_n \in \mathcal{D}_n$. The sets D_n are obviously pairwise disjoint, so they fulfill the requirements of the lemma.   □

Apply the lemma to get pairwise disjoint sets, one in $\mathcal{U}$ and one in each $\mathcal{V}_x$ for $x \in X$. Intersecting X and the Y_x's (for $nx \in X$) with these sets, respectively, we obtain a new X and new Y_x's that have all the properties required of the old ones and, in addition, are pairwise disjoint. Fix this X and these Y_x's for the remainder of the proof. We shall refer to them as "the disjointified X and Y_x's".

Lemma 15. *Let $\mathcal{D}_n$ for $n \in \omega$ be countably many pairwise non-isomorphic selective ultrafilters on ω. Also, let ω be partitioned into finite intervals*

$$I_0 = [0, i_1), \quad I_1 = [i_1, i_2), \quad \ldots, \quad I_n = [i_n, i_{n+1}), \quad \ldots.$$

There exist sets $D_n \in \mathcal{D}_n$ such that

(1) *each interval I_k meets D_n for at most one n, and*
(2) *no two consecutive intervals I_k and I_{k+1} both meet D_n's.*

The conclusion of the lemma can be visualized as saying that all of the D_n's occupy distinct intervals I_k and, furthermore, between two occupied intervals, there is always a buffer interval occupied by no D_n.

Proof.   Let $f : \omega \to \omega$ be the function that is constant on each interval I_n with value n. Since the ultrafilters $f(\mathcal{D}_n)$ are isomorphic to the corresponding $\mathcal{D}_n$'s, they are selective and pairwise nonisomorphic. So the preceding lemma applies to them and produces pairwise disjoint sets $D'_n \in f(\mathcal{D}_n)$. Then the sets $f^{-1}(D'_n)$ are in $\mathcal{D}_n$ and can serve as D_n for conclusion (1) of the lemma.

To get conclusion (2), use, instead of f, the functions obtained by dividing by 2 and rounding down or up. The function $\lfloor f(x)/2 \rfloor$ is constant on unions $I_k \cup I_{k+1}$ for even k, and the function $\lceil f(x)/2 \rceil$ is constant on unions $I_k \cup I_{k+1}$ for odd k. Apply the argument of the preceding paragraph to both of these functions and intersect the resulting two sets in each $\mathcal{D}_n$. The result satisfies a slightly weakened version of conclusion (2): No two consecutive intervals meet two distinct D_n's. To finish the proof, we need to ensure that no two consecutive intervals meet the same D_n. But this is easy; since $\mathcal{D}_n$ is an ultrafilter, we can shrink each D_n to a set, still in $\mathcal{D}_n$, that meets I_k only for even k or only for odd k.   □

We shall apply this lemma to $\mathcal{U}$ and the $\mathcal{V}_n$'s, for a carefully chosen sequence of intervals I_k. Intuitively, we want the intervals to be very long, so that we have large buffers, because we shall eventually need to find enough useful elements in these buffer intervals.

We work with the ω-type τ, which was fixed at the start of the proof, and with the disjointified X and Y_x's that we obtained from Lemma 14.

Definition 16. A *queue* is a function, q, defined on an initial segment of the pre-order τ, taking values in ω, with the following properties.

- q is strictly monotone, i.e., $q(a) \leq q(b)$ in ω if and only if $a \leq b$ in the pre-order τ. In particular, q is constant on each equivalence class of τ and takes distinct values on distinct equivalence classes.
- $q(x_n) \in X$ for all n.
- $q(y_n) \in Y_{q(x_n)}$ for all n.

In this definition, "initial segment of the pre-order τ" means a subset S of τ such that whenever $p \leq q \in S$ then $p \in S$. Thus, S is a union of equivalence classes of τ. Since the induced linear ordering of equivalence classes has order-type ω, it follows that S is either all of τ or the union of finitely many equivalence classes. According to which of these possibilities applies to the domain of a queue, we shall refer to the queue as "infinitely long" or "finitely long".

In the definition of queue, the $Y_{q(x_n)}$ in the last property is defined, because the subscript $q(x_n)$ is known to be in X.

The importance of queues for our purposes arises from the following observation. If q is a queue defined on all of τ, if

$$\{q(x_n) : n \in \omega\} \in \mathcal{U},$$

and if, for each v of the form $q(x_n)$,

$$\{q(y_m) : m \in \omega \text{ and } q(x_m) = v\} \in \mathcal{V}_v,$$

then the set

$$B = \{(q(x_n), q(y_n)) : n \in \omega\}$$

is as required in Theorem 13. Indeed, the y-coordinates of elements of B are the $q(y_n)$ in increasing order (because q is monotone) and the last two requirements in the definition of queues imply that the ω-type realized by B is τ and $B \subseteq X\text{-}\bigsqcup_n Y_n \subseteq A$. Furthermore, $B = X'\text{-}\bigsqcup_v Y_v'$, where $X' = \{q(x_n) : n \in \omega\} \in \mathcal{U}$, and $Y_v' = \{q(y_m) : m \in \omega \text{ and } q(x_m) = v\} \in \mathcal{V}_v$ for $v \in X'$, so $B \in \mathcal{U}\text{-}\sum \mathcal{V}_n = \mathcal{W}$.

Thus, to complete the proof, we need only produce an infinitely long queue with the appropriate sets X' and Y_v' in the corresponding ultrafilters. We shall eventually do this by means of longer and longer finite queues, approaching the desired infinite queue.

If we didn't need to get X' and the Y_v''s into the appropriate ultrafilters, it would be easy to construct an infinitely long queue. Indeed, any finitely long queue can be extended to include one more equivalence class of τ in its domain, simply because X and all the Y_v's are infinite. Just map the next equivalence class to an element of the required X or Y_v that is larger than everything that is already in the range of the queue.

We now proceed to the definition of the intervals I_k, by induction on k. At the stage of the induction where we want to define I_k, with the earlier intervals already defined, we proceed as follows. We know what the left endpoint i_k of I_k must be, since it's the first natural number not in any of the previous intervals. Consider all (finitely long) queues q whose range is included in the union $[0, i_k)$ of the previous intervals. Notice that there are only finitely many such queues.

For each such q find an equivalence class of x's in τ that is not in the domain of q. This is possible because the definition of ω-type requires infinitely many equivalence classes of x's. Also, for each $v < i_k$ that occurs as $q(x_i)$ for some i (and thus for a whole τ-equivalence class of x_i's), find a y_i, for one of these i's, that is not already in the domain of q. This is possible, for each such v, because there are infinitely many x_i's in the relevant τ-equivalence class and all the corresponding y_i's are inequivalent in τ. We have found here a finite number of τ-equivalence classes for our q — one equivalence class of x's and, for each relevant v, one equivalence class consisting of a single y. Choose an extension q' of q to a finitely long queue long enough to have all these finitely many equivalence classes in its domain.

Doing this for all queues q with range included in $[0, i_k)$, we obtain a finite number of new, finitely long queues q' altogether. Choose i_{k+1} so large that all these queues have their ranges included in $[0, i_{k+1})$ and (to avoid some trivialities later) $i_{k+1} > k + 1$. This completes the inductive definition of our intervals I_k.

Apply Lemma 15 to the ultrafilters $\mathcal{U}$ and $\mathcal{V}_n$, using the sequence of intervals I_k that we have just defined. We get sets $X^* \in \mathcal{U}$ and $Y_n^* \in \mathcal{V}_n$ satisfying the conclusions of that lemma. By intersecting these sets with our old disjointified X and Y_n's (for $n \in X$), we can assume, without loss of generality, that $X^* \subseteq X$ and $Y_n^* \subseteq Y_n$ for all $n \in X$. (The new Y_n^*'s are defined only for $n \in X$, because Y_n was defined only for these n. This situation will cause no difficulties because we shall need Y_n^* only for $n \in X$.) Furthermore, because our selective ultrafilters are Q-points, we can also assume without loss of generality that X^* and Y_n^* meet each I_k in at most one point. Finally, since a set in a nonprincipal ultrafilter remains in the ultrafilter if we remove finitely many elements, we may assume that Y_n^* has no elements in any of the finitely many intervals $I_0, \ldots, I_n$.

We shall complete the proof by constructing an infinitely long queue q such that $X' = \{q(x_n) : n \in \omega\}$ is a superset of X^* (and a subset of X, by definition of "queue"), and $Y_v' = \{q(y_m) : q(x_m) = v\}$ is a superset of Y_v^* (and a subset of Y_v, by definition of "queue") for all $v \in X'$. Thus, we shall have $X' \in \mathcal{U}$ and $Y_v' \in \mathcal{V}_v$ for all relevant v. As explained above, this will complete the proof of the theorem, with $B = \{(q(x_n), q(y_n)) : n \in \omega\}$.

The required infinitely long queue q will be constructed inductively, working through the intervals I_k, but not in their natural order. Rather, we concentrate on those intervals I_k that contain a point from X^* or from one of the Y_n^*'s. By construction, each of these intervals, which we call *critical intervals*, meets only one of X^* and the Y_n^*'s, and meets it in only a single point.

We shall treat the critical intervals in their natural order; while treating a critical interval, we shall also treat the non-critical intervals that precede it but come after the last previous critical interval. Notice that there is always at least one such interval, namely a buffer interval resulting from our use of Lemma 15 in defining the sets X^* and Y_n^*.

Suppose, therefore, that I_k is a critical interval, that I_j is the last critical interval before I_k, and that we have already defined the part $\bar{q}$ of q that maps into $I_0 \cup \cdots \cup I_j$. (If I_k is the first critical interval, so I_j doesn't exist, then we take $\bar{q}$ to be the empty queue.)

In the induction step associated to the critical interval I_k, we shall produce a queue $\tilde{q}$ with the following properties:

(1) $\tilde{q}$ is a finitely long queue with $\text{Range}(\tilde{q}) \subseteq I_0 \cup \cdots \cup I_k$.
(2) $\bar{q}$ is an initial segment of $\tilde{q}$ and $\text{Range}(\tilde{q}) - \text{Range}(\bar{q})$ has no elements $< i_{j+1}$.
(3) If I_k contains an element p of X^*, then $p \in \text{Range}(\tilde{q})$.
(4) If I_k contains an element of $Y_{\bar{q}(x_i)}$ for some x_i, then $p \in \text{Range}(\tilde{q})$.

Properties (1) and (2), holding at every stage of the construction, ensure that the union of all the finitely long queues that we construct is a queue q. Property (2) further ensures that $\text{Range}(\bar{q})$ contains all the elements of $\text{Range}(q)$ that are $< i_{j+1}$. Property (3) ensures that q is infinitely long and that $X' = \{q(x_n) : n \in \omega\} \supseteq X^*$, so $X' \in \mathcal{U}$. Finally, property (4) ensures that, for each $v = q(x_i) \in X'$, we have $Y_v' = \{q(y_m) : q(x_m) = v\} \supseteq Y_v^*$, so $Y_v' \in \mathcal{V}_v$. As explained above, the existence of such a queue

q will complete the proof of the theorem, so it only remains to carry our the construction of $\tilde{q}$ satisfying properties (1)–(4).

We know, because of buffer intervals, that $j \leq k - 2$. If this inequality is strict, then we decide to put no elements of $I_{j+1} \cup \cdots \cup I_{k-2}$ into the range of $\tilde{q}$. So $\bar{q}$ will be the part of q that maps into $I_0 \cup \cdots \cup I_{k-2} = [0, i_{k-1})$.

Now we look at the point $p \in I_k$ that belongs to X^* or to some Y_n^*. Consider first the case that $p \in X^*$. We shall put p into the range of $\tilde{q}$, along with some elements of I_{k-1}. Those elements will be chosen carefully to avoid violating the requirements for a queue. When we defined the interval I_{k-1} we considered all queues whose range is included in $[0, i_{k-1})$; our present $\bar{q}$ is one of those. So part of our construction was to find an extension q' of $\bar{q}$ that is long enough to have, in its domain, some equivalence class C of x's that is not already in the domain of $\bar{q}$. We then chose i_k large enough so that the range of q' is included in $[0, i_k)$. We modify q' as follows. Leave it unchanged at all arguments strictly before C, but change it to map C to p, and remove from its domain all elements (if any) greater than C. It is easy to check, using the fact that $p \in X^* \subseteq X$, that this modification $\tilde{q}$ of q' is still a queue. So we have extended $\bar{q}$ to a queue $\tilde{q}$ whose range contains p and is included in $I_0 \cup \cdots \cup I_k$. This completes the inductive step for the critical interval I_k in the case that $p \in X^*$.

It remains to consider the case that $p \in Y_n^*$ for some $n \in X^*$. Recall that, when constructing Y_n^*, we arranged for it to have no elements in $I_0 \cup \cdots \cup I_n$, so we must have $n < k$. When we defined the intervals I_k, we ensured (to avoid some trivialities) that $k < i_k$ for all k, so from $n \leq k - 1$ we infer that $n < i_{k-1}$. Since $\bar{q}$ is already known to be the part of our ultimate q with range below i_{k-1}, we already know whether n will be in the range of q and, if so, which τ-equivalence class will map to it. If n is not of the form $\bar{q}(x_i)$ for some i (equivalently, if n will not be the image, in q, of an equivalence class of x's) then put no elements of I_{k-1} and I_k into the range of q, i.e., set $\tilde{q} = \bar{q}$, and proceed to the next critical interval.

The only case that remains is that $p \in Y_n^*$, where $n = \bar{q}(C)$ for an equivalence class C of x's. In this case, as part of our construction of I_{k-1}, we chose some $x_i \in C$ whose corresponding y_i was not in the domain of $\bar{q}$, we chose a queue q' extending $\bar{q}$ and having y_i in its domain, and we ensured that i_k is large enough so that this q' maps into $[0, i_k)$. Now we can proceed in analogy with the case where p was in X^*. Modify q' by redefining its value at y_i to be p, and delete from its domain all elements beyond y_i. The result is still a queue, because $p \in Y_n^* \subseteq Y_n$. So we have extended $\bar{q}$ to a queue $\tilde{q}$ whose range contains p and is included in $I_0 \cup \cdots \cup I_k$. This completes the inductive step for the critical interval I_k, the proof of Theorem 13, and thus also the proof of Theorem 12.

4.   Almost a Contradiction

There seems to be a clash between the last part of Theorem 12 above and Proposition 38 of [2]. The former says that, in the Lévy–Mahlo model, all sisnis ultrafilters satisfy the HODℝ partition property for all ω-types τ. The latter says that, in that same model, this partition property for even one ω-type implies that the ultrafilter is ℙ-generic over HODℝ. (The latter also assumes that the ultrafilter is a non-P-point in standard position, but this assumption is satisfied by all sisnis ultrafilters.) So the sisnis ultrafilters in the Lévy–Mahlo model are ℙ-generic over the submodel HODℝ. Yet, as pointed out in the introduction, sisnis ultrafilters differ from ℙ-generic ones in at least two respects: sisnis ultrafilters are basically generated, and they are not weak P-points. Since sisnis ultrafilters certainly exist in the Lévy–Mahlo model, because CH holds there and provides plenty of selective ultrafilters, this appears to be a contradiction.

On closer inspection, fortunately, the contradiction dissolves, leaving behind some useful information. Working in the

Lévy–Mahlo model, let $\mathcal{W} = \mathcal{U}\text{-}\sum_n \mathcal{V}_n$ be a sisnis ultrafilter. Then the argument in the preceding paragraph shows that $\mathcal{W}$ is $\mathbb{P}$-generic over HOD$\mathbb{R}$. From this, it follows, by Theorems 3.11 and 6.1 of [3], that $\mathcal{W}$ is a weak P-point and is not basically generated in the forcing extension that it generates, HOD$\mathbb{R}[\mathcal{W}]$. But this extension is not the whole Lévy–Mahlo model. It does not contain the ultrafilters $\mathcal{V}_n$ that witness that $\mathcal{W}$ is sisnis, and whose images in the vertical slices of ω^2 witness that it is not a weak P-point; nor does it contain the basis that witnesses that $\mathcal{W}$ is basically generated. (It does contain $\mathcal{U}$, since this ultrafilter is $\pi_1(\mathcal{W})$, where π_1 is the projection of ω^2 to its first factor ω.)

The situation can be summarized by saying that the $\mathbb{P}$-generic ultrafilter $\mathcal{W}$ in the model HOD$\mathbb{R}[\mathcal{W}]$ becomes a sisnis ultrafilter when one extends that model by adjoining the sequence of ultrafilters $\mathcal{V}_n$.

That extension, HOD$\mathbb{R}[\mathcal{W}, \langle \mathcal{V}_n \rangle_{n \in \omega}]$, can equivalently be described as HOD$\mathbb{R}[\mathcal{U}, \langle \mathcal{V}_n \rangle_{n \in \omega}]$, because, in the presence of the sequence of $\mathcal{V}$'s, each of $\mathcal{U}$ and $\mathcal{W}$ can be obtained from the other. In this form, the extension is just the result of adjoining to HOD$\mathbb{R}$ an ω-sequence of pairwise nonisomorphic selective ultrafilters. It is known from [1, Theorem 12] that any such sequence is generic over HOD$\mathbb{R}$ with respect to the full-support product forcing $(\mathcal{P}(\omega)/\mathfrak{Fr})^\omega$.

So, when forcing over the HOD$\mathbb{R}$ of a Lévy–Mahlo model, we can regard the forcing by $\mathbb{P}$ (or equivalently by $\mathcal{P}(\omega^2)/\mathfrak{Fr}^{\otimes 2}$) as a complete subforcing of $(\mathcal{P}(\omega)/\mathfrak{Fr})^\omega$. By general results from [5], the larger forcing $(\mathcal{P}(\omega)/\mathfrak{Fr})^\omega$, which adjoins $\mathcal{U}$ and the sequence of $\mathcal{V}$'s, can be regarded as a two-step iteration, in which the first step adjoins the sum $\mathcal{W} = \mathcal{U}\text{-}\sum_n \mathcal{V}_n$, and the second step adjoins the sequence of $\mathcal{V}$'s witnessing that $\mathcal{W}$ is sisnis. Note that none of these forcings adjoin reals; everything takes place in the Lévy–Mahlo model, all of whose reals are already in the ground model HOD$\mathbb{R}$. In this situation, therefore, an ultrafilter that is $\mathbb{P}$-generic in the model that it generates can be regarded as a sisnis ultrafilter in a further forcing extension.

5.   Generic Ultrafilters Become Sums

The discussion in the preceding section took place in the Lévy–Mahlo model, but the final conclusion, that $\mathbb{P}$-generic ultrafilters become sisnis in a forcing extension that adds no reals, is true more generally. In this section, we prove this result and indicate some applications of it.

Our results will be formulated in terms of forcing over the universe V, but they can easily be reformulated in any of the other familiar frameworks for forcing, e.g., countable transitive models.

We begin by considering the forcing notion $([\omega]^\omega)^\omega$, the full-support countable product of copies of the forcing $[\omega]^\omega$ consisting of the infinite subsets of ω ordered by inclusion. Its separative quotient $(\mathcal{P}(\omega)/\mathfrak{Fr})^\omega$ is countably closed, so it adjoins no new reals. The generic object is an ω-sequence of mutually $[\omega]^\omega$-generic ultrafilters, which we shall write as $(\mathcal{U}, \mathcal{V}_0, \mathcal{V}_1, \dots)$, singling out one component $\mathcal{U}$ to play a special role in what follows. It follows easily from genericity that all the ultrafilters $\mathcal{U}$ and $\mathcal{V}_n$ are selective and pairwise nonisomorphic. So $\mathcal{W} = \mathcal{U}\text{-}\sum_n \mathcal{V}_n$ is a sisnis ultrafilter in this forcing extension.

Theorem 17. *The ultrafilter $\mathcal{W}$ defined above is a $\mathbb{P}$-generic filter over the ground model.*

Proof.   It suffices to check that $\mathcal{W}$ meets every dense subset $\mathcal{D}$ of $\mathbb{P}$. So let some dense $\mathcal{D}$ be given; we need only show that a dense set of conditions $(A, B_0, B_1, \dots)$ in $([\omega]^\omega)^\omega$ force $\mathcal{W}$ to meet $\mathcal{D}$. So let $(A, B_0, B_1, \dots)$ be any forcing condition in $([\omega]^\omega)^\omega$; we shall find an extension of it that forces $\mathcal{W}$ to meet $\mathcal{D}$.

The disjoint union

$$A\text{-}\bigsqcup_n B_n = \{(x,y) \in \omega^2 : x \in A \text{ and } y \in B_x\}$$

is a subset of ω^2 with infinitely many points in each of infinitely many vertical slices. As such, it has a subset in $\mathbb{P}$, and, since $\mathcal{D}$ is dense, we can arrange for this subset to be in $\mathcal{D}$. The subset in question has the form $A'\text{-}\bigsqcup_n B'_n$ for some infinite $A' \subseteq A$ and some infinite sets $B'_n \subseteq B_n$. (Note that the inclusion $B'_n \subseteq B_n$ follows from $A'\text{-}\bigsqcup_n B'_n \subseteq A\text{-}\bigsqcup_n B_n$ only when $n \in A'$, but we can arrange that $B'_n \subseteq B_n$ for all n by choosing $B'_n = B_n$ for $n \notin A'$, since these B'_n have no effect on $A'\text{-}\bigsqcup_n B'_n$.)

Then $(A', B'_0, B'_1, \dots)$ is an extension of $(A, B_0, B_1, \dots)$ in $([\omega]^\omega)^\omega$, and it forces $A' \in \mathcal{U}$ and $B'_n \in \mathcal{V}_n$, and therefore $A'\text{-}\bigsqcup_n B'_n \in \mathcal{W}$. In particular, it forces that $\mathcal{W}$ meets $\mathcal{D}$, as required. $\qquad\square$

Corollary 18. $\mathcal{P}(\omega^2)/\mathfrak{Fr}^{\otimes 2}$ *is equivalent as a forcing notion to a complete sub-ordering of* $(\mathcal{P}(\omega)/\mathfrak{Fr})^\omega$. *Forcing by* $(\mathcal{P}(\omega)/\mathfrak{Fr})^\omega$ *is equivalent to a two-step iteration* $\mathcal{P}(\omega^2)/\mathfrak{Fr}^{\otimes 2} * \dot{\mathbb{Q}}$, *where* $\dot{\mathbb{Q}}$ *names a certain forcing that adds no reals. The* $(\mathcal{P}(\omega)/\mathfrak{Fr})^\omega$- *generic ultrafilter* $\mathcal{W}$ *added by the first step in this iteration is made into a sisnis ultrafilter by the second step.*

Proof. All the assertions in the corollary follow by standard forcing theory from the theorem plus the facts that (1) the $\mathcal{W}$ of the theorem was introduced as a sisnis ultrafilter in the $(\mathcal{P}(\omega)/\mathfrak{Fr})^\omega$ forcing extension and (2) the forcing $(\mathcal{P}(\omega)/\mathfrak{Fr})^\omega$ adds no reals and therefore neither does $\dot{\mathbb{Q}}$. $\qquad\square$

The fact that any $\mathcal{P}(\omega^2)/\mathfrak{Fr}^{\otimes 2}$-generic ultrafilter becomes sisnis in a forcing extension that adds no reals can be used to give alternative proofs for several results from [3]. In the introduction above, we listed five properties that $\mathcal{P}(\omega^2)/\mathfrak{Fr}^{\otimes 2}$-generic ultrafilters share with sisnis ultrafilters. Four of those are clearly absolute for forcing extensions that add no new reals, namely

- The ultrafilter is a Q-point but not a P-point.
- Every function on ω^2 becomes, when restricted to a suitable set in the ultrafilter, either one-to-one, or constant, or a composition $g \circ \pi_1$ where g is one-to-one and $\pi_1 : \omega^2 \to \omega$ is the projection to the first factor.
- The ultrafilter is $(n, T(n))$-weakly Ramsey.
- The second projection $\pi_2 : \omega^2 \to \omega$ is one-to-one on a set in the ultrafilter.

Thus, the fact that $\mathcal{P}(\omega^2)/\mathfrak{Fr}^{\otimes 2}$-generic ultrafilters have these properties follows immediately from the corresponding fact for sisnis ultrafilters.

The fifth item on the list, not being at the top of the Tukey ordering, is not immediately absolute, but it can be obtained, even in a stronger form, with a little more work as follows.

The stronger form in question, Theorem 4.12 of [3], is that a $\mathcal{P}(\omega^2)/\mathfrak{Fr}^{\otimes 2}$-generic ultrafilter $\mathcal{W}$ is not Tukey-above the directed set $[\omega_1]^{<\omega}$ (ordered by inclusion); this is stronger because $[\omega_1]^{<\omega} \leq_T [2^{\aleph_0}]^{<\omega}$ and the inequality is consistently strict (if CH fails).

We need a general fact from the theory of the Tukey order, namely that $D \geq_T E$ if and only if there is a function $f : E \to D$ that maps unbounded sets in E to unbounded sets in D. ("Unbounded" means having no upper bound in the poset.) In the present situation, if we had $\mathcal{W} \geq_T [\omega_1]^{<\omega}$, this would give us a family of $\aleph_1$ sets (namely f of singletons) in $\mathcal{W}$ such that every infinite subfamily has intersection $\notin \mathcal{W}$.

Such an uncountable family would remain in the forcing extension where $\mathcal{W}$ becomes sisnis, and it would still be the case that every infinite subfamily has intersection $\notin \mathcal{W}$. Indeed, it suffices to check this for countably infinite subfamilies, and these are all in the ground model because the forcing adds no new reals.

But sisnis ultrafilters, being basically generated, cannot be $\geq_T [\omega_1]^{<\omega}$. This contradiction shows that $\mathcal{P}(\omega^2)/\mathfrak{Fr}^{\otimes 2}$-generic ultrafilters cannot be $\geq_T [\omega_1]^{<\omega}$ either.

These relatively easy proofs of several properties of $\mathcal{P}(\omega^2)/\mathfrak{Fr}^{\otimes 2}$-generic ultrafilters, using the same properties of sisnis ultrafilters, help to explain the prominent role of sisnis ultrafilters in some of the forcing proofs in [3].

Finally, let us consider the properties of being a weak P-point or being basically generated, properties where, as we mentioned in the introduction, $\mathcal{P}(\omega^2)/\mathfrak{Fr}^{\otimes 2}$-generic ultrafilters differ from sisnis ultrafilters. In the light of our results above, what must happen is that these properties are not sufficiently absolute in the $\dot{\mathbb{Q}}$ forcing extension. Indeed, the extension adjoins the ultra-filters $\mathcal{V}_n$ whose isomorphic copies in vertical slices of ω^2 show that $\mathcal{W}$ is no longer a weak P-point, and the extension adjoins a basis witnessing that $\mathcal{W}$ has become basically generated.

Acknowledgments

A part of the research for this paper was done while the author was a visiting fellow in the "Mathematical, Foundational and Computational Aspects of the Higher Infinite" program at the Isaac Newton Institute for Mathematical Sciences.

References

[1] A. Blass, Selective ultrafilters and homogeneity, *Ann. Pure Appl. Logic* **38** (1988) 215–255.

[2] A. Blass, Partitions and conservativity, *Topology Appl.* **213** (2016) 167–189.

[3] A. Blass, N. Dobrinen and D. Raghavan, The next best thing to a P-point, *J. Symbolic Logic* **80** (2015) 866–900.

[4] N. Dobrinen and S. Todorcevic, Tukey types of ultrafilters, *Illinois J. Math.* **55** (2011) 907–951.

[5] S. Grigorieff, Intermediate extensions and generic models in set theory, *Ann. of Math.* **101** (2) (1975) 447–490.

[6] K. Kunen, Weak P-points in $\mathbf{N}^*$, in *Topology, Vol. II, Proc. Fourth Colloq., Budapest, 1978,* Colloq. Math. Soc. János Bolyai **23** (North-Holland, 1980) 741–749.

[7] A. R. D. Mathias, Happy families, *Ann. Math. Logic* **12** (1977) 59–111.

[8] D. Raghavan and S. Todorcevic, Cofinal types of ultrafilters, *Ann. Pure Appl. Logic* **163** (2012) 185–199.

Part II
Algorithmic Randomness and Information

Chapter 3

Limits of the Kučera–Gács Coding Method

George Barmpalias[*,†,§,‖] and Andrew Lewis-Pye[‡,¶,**]

*State Key Lab of Computer Science, Institute of Software,
Chinese Academy of Sciences, Beijing, China
†School of Mathematics, Statistics and Operations Research,
Victoria University of Wellington, New Zealand
‡Department of Mathematics, Columbia House,
London School of Economics, Houghton St.,
London, WC2A 2AE, UK
§barmpalias@gmail.com
¶A.Lewis7@lse.ac.uk
‖http://barmpalias.net
**http://aemlewis.co.uk

Every real is computable from a Martin-Löf random real. This well-known result in algorithmic randomness was proved by Kučera [Measure, Π_1^0-classes and complete extensions of PA, in *Recursion Theory Week (Oberwolfach, 1984), Lecture Notes in Mathematics,* Vol. 1141 (Springer, Berlin, 1985), pp. 245–259.] and Gács [Every sequence is reducible to a random one, *Inform. Control* **70**(2–3) (1986) 186–192]. In this chapter, we discuss various approaches to the problem of coding an arbitrary real into a Martin-Löf random real, and also describe new results concerning optimal methods of coding. We start with a simple presentation of the original methods of Kučera and Gács and then rigorously demonstrate their limitations in terms of the size of the redundancy in the codes that they produce. Armed with a deeper understanding of these methods, we then proceed to motivate and illustrate aspects of the new coding method that was recently introduced by Barmpalias and Lewis-Pye in [Optimal redundancy in computations from random oracles, preprint (2016), arXiv:1606.07910.] and which achieves optimal logarithmic redundancy, an exponential improvement over the original redundancy bounds.

1. Introduction

Information means structure and regularity, while randomness means the lack of structure and regularity. One can formalize and even quantify this intuition in the context of algorithmic randomness and Kolmogorov complexity, where the interplay between information and randomness has been a principal driving force for much of the research.

$$\text{How much information can be coded into a random} \atop \text{binary sequence?} \tag{1.0.1}$$

This question has various answers, depending on how it is formalized, but as we are going to see in the following discussion, for sufficiently strong randomness the answer is "not much".

1.1. *Finite information*

In the case of a finite binary sequence (string) σ, let $K(\sigma)$ denote the prefix-free complexity of σ. Then σ is c-incompressible if $K(\sigma) \geq |\sigma| - c$. Here we view the underlying optimal universal prefix-free machine U as a *decompressor* or *decoder*, which takes a string/program τ and may output another string σ, in which case τ is regarded as a description of σ. Then $K(\sigma)$ is the length of the shortest description of σ and the random strings are the c-incompressible strings for some c, which is known as the *randomness deficiency*. It is well known that the shortest description of a string is random, i.e., there exists a constant c such that each shortest description is c-incompressible. In other words,

> every string σ can be coded into a random string
> (its shortest description), of length the Kolmogorov
> complexity of σ (1.1.1)

which may seem as a strong positive answer to Question (1.0.1), in the sense that every string σ can be coded into a random string. The following proposition, however, points in the opposite direction:

Proposition 1.1 (Folklore). *If U is an optimal universal prefix-free machine, then there exists a constant c such that $U(\sigma) \uparrow$ for all strings σ such that $K(\sigma) \geq |\sigma| + c$.*[a]

Viewing U as a universal decompressor, Proposition 1.1 says that a sufficiently random string cannot be decoded into anything, which means that in that sense it does not effectively code any information. According to this fact, Question (1.0.1) has a strong negative answer.

1.2. *Bennett's analogy for infinite information*

The notions and issues discussed in the previous section have infinitary analogues which concern coding infinite binary sequences (*reals*) into random reals. For sufficiently strong (yet still moderate) notions of randomness for reals (such as the randomness corresponding to statistical tests or predictions that are definable in arithmetic with two quantifiers), the answer to Question (1.0.1) is *not much*; such random reals cannot solve the halting problem or even compute a complete extension of

---

[a]The proof of this fact is based on the idea that each string in the domain of U is a prefix-free description of itself (modulo some fixed overhead). In other words, if $U(\sigma) \downarrow$ then σ can be used to describe itself, with respect to some prefix-free machine that is then simulated by U, producing a U-description of σ of length $|\sigma| + c$ for some constant c.

Peano Arithmetic. Charles Bennett [2] asked if Question (1.0.1) can have a strongly positive answer, just as in the finite case, for a standard notion of algorithmic randomness such as Martin-Löf randomness. Remarkably, Kučera [12] and Gács [10] gave a positive answer to Bennett's question.

Theorem 1.2 (Kučera–Gács theorem). *Every real is computable from a Martin-Löf random real.*

Bennett [2] commented:

> *This is the infinite analog of the far more obvious fact that every finite string is computable from an algorithmically random string (e.g., its minimal program).*

Here we argue that Bennett's suggested analogy between (1.1.1) and Theorem 1.2 is not precise, in the sense that it misses the quantitative aspect of (1.1.1) — namely that the random code can be chosen short (of length the complexity of the string). It is much easier to code σ into a random string which is much longer than σ, than code it into a random string of length at most $|\sigma|$. The analogue of *length of code* for infinite codes, is the *use-function* in a purported Turing reduction underlying the computation of a real X from a random real Y. The use-function for the reduction is a function f such that for each n, the first n bits of X can be uniformly computed from the first $f(n)$ bits of Y.

1.3.   *A quantitative version of the Kučera–Gács theorem?*

The more precise version of Bennett's suggested analogy that we have just discussed can be summarized in Table 1, where σ^* denotes the shortest program for σ.[b] So what is the analogue

---

[b]If there are several shortest strings τ such that $U(\tau) = \sigma$ then σ^* denotes the one that converges the fastest.

Table 1. Quantitative analogy between finite and infinite codes; here $n \mapsto f(n)$ refers to an "optimal" nondecreasing upper bound on the use-function in the computation of X from Y.

Notion	Finite	Infinite
Source	σ	X
Code	σ^*	Y
Code-length	$\|\sigma^*\|$	$n \mapsto f(n)$
Optimal code	$K(\sigma)$	?

of the code length in the Kučera–Gács theorem? If we code a real X into a Martin-Löf random real Y, how many bits of Y do we need in order to compute the first n bits of X? This question has been discussed in the literature (see below) but, until recently, only very incomplete answers were known. Kučera [12] did not provide tight calculations and various textbook presentations of the theorem (e.g., Nies [16, Sec. 3.3]) estimate the use-function in this reduction of X to a Martin-Löf random Y to be of the order n^2. In fact, the actual bound that can be obtained by Kučera's method is $n \log n$. Gács used a more elaborate argument and obtained the upper bound $n + \sqrt{n} \cdot \log n$, which is $n + o(n)$, and the same bound was also obtained later by Merkle and Mihailović [14] who used an argument in terms of supermartingales.

1.4. *Coding into random reals, since Kučera and Gács*

The Kučera–Gács coding method has been combined with various arguments in order to produce Martin-Löf random reals with specific computational properties. The first application already appeared in [13], where a high incomplete Martin-Löf random real computable from the halting problem was constructed. Downey and Miller [7] and later Barmpalias, Downey and Ng [1] presented a variety of different versions of this method, which allow some control over the degree of the random real which is coded into. Doty [9] revisited the Kučera–Gács theorem from the viewpoint of constructive dimension. He characterized the asymptotics of the redundancy in computations of an infinite sequence X from a random oracle in terms of the constructive dimension of X. We should also mention that this is not the only method for coding into members of a positive measure Π_1^0 class (or into the class of Martin-Löf random reals). Barmpalias, Lewis-Pye and Ng [4] used a different method in order to show that every degree that computes a complete extension of Peano Arithmetic is the supremum of two Martin-Löf random degrees.

It is fair to say that all of these methods rely heavily on the density of reals inside a nonempty Π_1^0 class that consists entirely of Martin-Löf reals. This is also true of more recent works such as Bienvenu, Greenberg, Kučera, Nies and Turetsky [3], Day and Miller [8] and Miyabe, Nies and Zhang [15]. Khan [11] explicitly studies the properties of density inside Π_1^0 classes, not necessarily consisting entirely of Martin-Löf random reals. Much of this work is concerned with lower bounds on the density that a Martin-Löf real has inside every Π_1^0 class that contains it. In our analysis of the Kučera–Gács theorem, we isolate the role of density in the argument and show that, in a sense, tighter oracle-use in computations from Martin-Löf random oracles is only possible through methods that do not rely on such density requirements.

2. Coding into an Effectively Closed Set Subject to Density Requirements

The arguments of Kučera and Gács both provide a method for coding an arbitrary real X into a member of an effectively closed set P (a Π_1^0 class), and rely on certain density requirements for the set of reals $\mathcal{P}$. The connection to Theorem 1.2 is that the class of Martin-Löf random reals is the effective union of countably many Π_1^0 classes of positive measure. The only difference in the two methods is that Kučera codes X one-bit-at-a-time (with each bit of X coded into a specified segment of Y) while Gács codes X block-by-block into Y, with respect to a specified segmentation of X.

2.1. *Overview of the Kučera–Gács argument*

In general, we can code m_i many bits of X at the ith coding step, using a block in Y of length ℓ_i, as Table 2 indicates. We leave the parameters $(m_i), (\ell_i)$ unspecified for now, while in the following it will become clear what the growth of this sequence needs to be in order for the argument to work. Note that the bit-by-bit version of the coding is the special case where $m_i = 1$ for all i. The basic form of the coding process (which we shall elaborate on later) can be outlined as follows.

(1) Start with a Π_1^0 class $\mathcal{P} \neq \emptyset$ which only contains (Martin-Löf) randoms.
(2) Choose the *length m_i of the block* coded at step i.
(3) Choose the length $\ell_i = m_i + g(i)$ used for coding the ith block.
(4) The *oracle-use* for the first $M_n = \sum_{i<n} m_i$ bits is $L_n = \sum_{i<n} \ell_i$.
(5) Form a subclass $\mathcal{P}^*$ of $\mathcal{P}$ with the property that for all but finitely many n and for each $X \in \mathcal{P}^*$, there are at least

Table 2. Parameters of the Kučera–Gács coding of X into Y.

m_i	Length of the ith block of X
ℓ_i	Length of the ith block of Y
M_n	Number of bits of X coded after n-many coding steps: $M_n := \sum_{i<n} m_i$
L_n	Length of Y used in the computation of $X \restriction_{M_n}$: $L_n := \sum_{i<n} \ell_i$

2^{m_n} extensions of $X \restriction_{L_n}$ of length L_{n+1} which have infinite extensions in $\mathcal{P}^*$.

(6) Argue that $\mathcal{P}^* \neq \emptyset$ (due to the growth of (ℓ_i), relative to (m_i)).

A crucial fact here is that if $\mathcal{P}$ is a Π_1^0 class then $\mathcal{P}^*$ is also a Π_1^0 class. In Sec. 2.2, we turn this outline into a modular proof, which makes the required properties of the parameters (m_i), (ℓ_i) transparent. We will show that apart from the computability of (m_i), (ℓ_i), the following facts characterize the necessary and sufficient constraints on the two sequences for the coding to work.

(i) If $\sum_i 2^{m_i - \ell_i} < \infty$, then there exists a Π_1^0 class of positive measure that consists entirely of Martin-Löf random reals such that $\mathcal{P}^* \neq \emptyset$.

(ii) If $\sum_i 2^{m_i - \ell_i} = \infty$ and $\mathcal{P}$ is a Π_1^0 class such that $\mathcal{P}^* \neq \emptyset$, then $\mathcal{P}$ contains a real which is not Martin-Löf random.

2.2. *The general Kučera–Gács argument*

We give a modular argument in terms of Π_1^0 classes, showing that every real is computable from a Martin-Löf random real, and consisting of a few simple lemmas. We use Martin-Löf's paradigm of algorithmic randomness, much like in the original argument of Kučera and Gács.[c] In the following definition, recall that for finite σ, $[\sigma]$ is the set of all infinite extensions of σ.

Definition 2.1 (Extension property). Given a Π_1^0 class P and sequences (m_i), (ℓ_i) of positive integers, let $M_n := \sum_{i<n} m_i$, $L_n := \sum_{i<n} \ell_i$ and say that P has the extension property with respect to (m_i), (ℓ_i) if for each i, every string σ of length L_i with $[\sigma] \cap P \neq \emptyset$ has at least 2^{m_i} extensions τ of length L_{i+1} such that $P \cap [\tau] \neq \emptyset$.

---

[c]However, our presentation has been significantly assisted by Merkle and Mihailović [14], who phrased the argument in terms of martingales.

Table 3.   Overheads in the Kučera–Gács coding of X into Y.

$\ell_i - m_i$	Overhead at the ith coding step
$\sum_{i<n}(\ell_i - m_i)$	Accumulated overhead after n coding steps
$\sum_i 2^{m_i-\ell_i} < \infty$	Necessary and sufficient condition for successful coding

The first lemma says that subject to certain density conditions on a Π_1^0 class P, every real is computable from a member of P.

Lemma 2.2 (General block coding). *Suppose that P is a Π_1^0 class, and (m_i), (ℓ_i) are computable sequences of positive integers. If P has the extension property with respect to (m_i), (ℓ_i), then every sequence is computable from a real in P with use L_{s+1} for bits in $[M_s, M_{s+1})$.*

Proof. For any string σ of length L_i consider the variables $w_0(\sigma)[s], \ldots, w_{2^{m_i}-1}(\sigma)[s]$ for strings, which are defined dynamically according to the approximation (P_s) to P as follows. At stage 0 let $w_j(\sigma)[0] \uparrow$ for all $j < 2^{m_i}$. At stage $s+1$ find the least $t < 2^{m_i}$ such that one of the following cases holds:

(a) $w_t(\sigma)[s] \uparrow$;
(b) $w_t(\sigma)[s] \downarrow$ and $[w_t(\sigma)[s]] \cap P_{s+1} = \emptyset$.

In case (a), look for the lexicographically least ℓ_i-bit extension τ of σ such that $[\tau] \cap P_{s+1} \neq \emptyset$ and $w_j(\sigma)[s] \not\succeq \tau$ for all $j < 2^{m_i}$. If no such exists, terminate the process (hence let $w_j(\sigma)[n] \simeq w_j(\sigma)[s]$ for all $j < 2^{m_i}$ and all $n > s$). Otherwise define $w_t(\sigma)[s+1] = \tau$ and go to the next stage. In case (b) let $w_t(\sigma)[s + 1] \uparrow$ and go to the next stage.

By the hypothesis of the lemma, for every i and every string σ of length L_i such that $[\sigma] \cap P \neq \emptyset$, the words $w_j(\sigma)[s]$, $j < 2^{m_i}$ reach limits $w_j(\sigma)$ after finitely many stages, such that:

- $j \neq k \Rightarrow w_j(\sigma) \neq w_k(\sigma)$ for all $j, k < 2^{m_i}$;
- $[w_j(\sigma)] \cap P \neq \emptyset$.

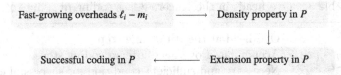

Fig. 1.   Diagrammatic representation of the Kučera–Gács coding argument.

Consider the Turing functional Φ which, given oracle Y, works inductively as follows. Suppose that $\Phi(Y \restriction_{L_i}) \restriction_{M_i}$ has been calculated. The functional then searches for the least pair (j, s) (under a fixed effective ordering of all pairs, of order type ω) such that $j < 2^{m_i}$, $w_j(Y \restriction_{L_i})[s] \downarrow$ and is a prefix of Y. For τ which is the jth string of length m_i (under the lexicographical ordering) the functional then defines $\Phi(Y \restriction_{L_i+\ell_i}) = \Phi(Y \restriction_{L_i}) * \tau$. By construction Φ is consistent, and if $\Phi(Y \restriction_{L_i})$ is defined it has length M_i. Finally we show that Φ is onto the Cantor space. Given X we can inductively construct Y such that $\Phi(Y) = X$. Suppose that we have constructed $Y \restriction_{L_i}$ such that $\Phi(Y \restriction_{L_i}) = X \restriction_{M_i}$ and $Y \restriction_{L_i}$ is extendible in P. Let σ be the unique string of length m_i such that $X \restriction_{M_i} *\sigma$ is a prefix of X. Then $w_j(Y \restriction_{L_i})$ is defined for all $j < 2^{m_i}$ and takes distinct values for different j. Let t be the index of σ in the lexicographical ordering of strings of length m_i. Then let $Y \restriction_{L_{i+1}} = w_t(Y \restriction_{L_i})$. Clearly $Y \restriction_{L_{i+1}}$ is extendible in P and moreover $\Phi(Y \restriction_{L_{i+1}}) = X \restriction_{M_{i+1}}$. This completes the induction step in the construction of Y and shows that $\Phi(Y) = X$. $\square$

Recall that for σ of length n, the P-density of σ is defined to be $2^n \cdot \mu([\sigma] \cap P)$, where μ denotes Lebesgue measure on Cantor space (see diagram of coding in Fig. 1).

Definition 2.3 (Density property). Given P, (m_i), (ℓ_i) as in Definition 2.1 we say that P has the density property with respect to (m_i), (ℓ_i) if for each n, every string of length L_n with $[\sigma] \cap P \neq \emptyset$ has P-density at least $2^{m_n - \ell_n}$.

Lemma 2.4 (Density and extensions). *Given P, (m_i), (ℓ_i) as in Definition 2.1, if P has the density property with respect to (m_i), (ℓ_i) then it also has the extension property with respect to (m_i), (ℓ_i).*

Proof. This follows from the general fact that if the P-density of σ is at least 2^{-t} for some t, then given any m, there are at least 2^m extensions τ of σ of length $|\sigma| + t + m$ such that $[\tau] \cap P \neq \emptyset$. In order to prove the latter fact, suppose for a contradiction that it is not true. Then the P-density of σ would be at most $(2^m - 1) \cdot 2^{-m-t} = 2^{-t} - 2^{-m-t} < 2^{-t}$ which contradicts the hypothesis. □

Lemma 2.5 (Lower bounds on the density). *Let P be a Π_1^0 class and let $(m_i), (\ell_i)$ be computable sequences of positive integers such that $\sum_i 2^{m_i - \ell_i} < \mu(P)$. Then there exists a Π_1^0 class $P^* \subseteq P$ which has the extension property with respect to $(m_i), (\ell_i)$.*

Proof. We construct a Σ_1^0 class Q in stages and let (P_s) be a Π_1^0 approximation to P, where each P_s is a clopen set. A string σ is *active* at stage $s + 1$ if it is of length L_n for some n and $[\sigma] \cap (P_s - Q_s) \neq \emptyset$. Moreover σ of length L_n *requires attention* at stage $s + 1$ if it is active at this stage and the $(P_s - Q_s)$-density of σ is at most $2^{m_n - \ell_n}$. At stage $s + 1$, we pick the least string of length $< s$ which requires attention (if such exists) and enumerate $[\sigma] \cap (P_s - Q_s)$ into Q. If this enumeration occurred, we say that the construction *acted* on string σ at stage $s + 1$. This concludes the construction.

First we establish an upper bound on the measure of $Q = \bigcup_s Q_s$. Clearly, the construction can act on a string at most once. The measure that is added to Q at stage $s + 1$ if the construction acts on σ of length L_n at this stage, is at most $2^{-L_n + m_n - \ell_n}$. Therefore, the total measure enumerated into Q

throughout the construction is bounded above by

$$\sum_n \sum_{\sigma \in 2^{L_n}} 2^{-L_n + m_n - \ell_n} = \sum_n 2^{L_n} \cdot 2^{-L_n + m_n - \ell_n}$$

$$= \sum_n 2^{m_n - \ell_n} < \mu(P).$$

It follows that $P^* := P - Q$ is a nonempty Π_1^0 class, and by the construction we have that for every n and every string σ of length L_n, if $[\sigma] \cap P^* \neq \emptyset$ then the P^*-density of σ is at least $2^{m_n - \ell_n}$. By Lemma 2.4 this means that every P^*-extendible string of length L_n for some n has at least 2^{m_n} many P^*-extendible extensions of length $L_n + m_n - (m_n - \ell_n) = L_{n+1}$. Hence P^* has the extension property with respect to (m_i), (ℓ_i).   □

Corollary 2.6 (General block coding). *Suppose that P is a Π_1^0 class, and (m_i), (ℓ_i) are computable sequences of positive integers. If $\sum_i 2^{m_i - \ell_i} < \mu(P)$ then every sequence is computable from a real in P with use L_{s+1} for bits in $[M_s, M_{s+1})$.*

Proof. By Lemma 2.5 we can consider a Π_1^0 class $P^* \subseteq P$ which has the extension property with respect to (m_i), (ℓ_i). The statement then follows by Lemma 2.2 and the fact that $P^* \subseteq P$.   □

Note that, while Corollary 2.6 seems to require (a) $\sum_i 2^{m_i - \ell_i} < \mu(P)$, if P is of positive measure then the condition (b) $\sum_i 2^{m_i - \ell_i} < \infty$ suffices to ensure that $\sum_{i \geq d} 2^{m_i - \ell_i} < \mu(P)$ for some d — meaning that (c) is sufficient to give the existence of the required functional (albeit with some added nonuniformity required in specifying the index of the reduction).

2.3. *The oracle-use in the general Kučera–Gács coding argument*

Recall that if X can be computed from Y with the use function on argument n bounded by $n + g(n)$, then we say that X

can be computed from Y with *redundancy* $g(n)$. Note that in the following corollary we do not need to require that h, h_r are computable.

Corollary 2.7. *Suppose* (m_i), (ℓ_i) *are computable sequences of positive integers with* $\sum_i 2^{m_i - \ell_i} < 1$ *and suppose* h, h_r *are non-decreasing functions such that:*

$$\sum_{i \leq s} \ell_i \leq h \left(1 + \sum_{i < s} m_i \right) \quad and \quad m_s + \sum_{i \leq s} (\ell_i - m_i)$$

$$\leq h_r \left(\sum_{i < s} m_i \right).$$

Then if P is a Π_1^0 class of positive measure, any sequence is computable from a real in P with oracle-use h and redundancy h_r.

Proof. The first claim follows directly from Corollary 2.6 and for the second, recall that in the same corollary, for each s and each $n \in [M_s, M_{s+1})$, the length of the initial segment of Y that is used for the computation of $X \restriction_n$ is at most

$$L_{s+1} = M_s + m_s + \sum_{i \leq s} (\ell_i - m_i) \leq n + m_s$$

$$+ \sum_{i \leq s} (\ell_i - m_i) \leq n + h_r(M_s) \leq n + h_r(n),$$

where the second inequality was obtained from the main property assumed for h_r, and the last inequality follows from the monotonicity of h_r. □

Without yet specifying the sequences (m_i), (ℓ_i), the condition $\sum_i 2^{m_i - \ell_i} < 1$ means that a near-optimal choice for the sequence $(\ell_i - m_i)$ is $\lceil 2 \log(i + 2) \rceil$. This means that $\sum_i (\ell_i - m_i)$ will be of the order $\log(n!)$ or $n \log n$. We may now consider an appropriate choice for the sequence (m_i), which roughly minimizes the redundancy established in Corollary 2.7. For Kučera's coding we have $m_i = 1$ for all i which means that the redundancy in this type of bit-by-bit coding is $n \log n$ (modulo a constant). Gács chose the sequence $m_i = i + 1$, and the reader may verify that this growth-rate of the blocks of the coded stream gives a near-optimal redundancy in Corollary 2.7.[d] In this case, the function $h_r(n) = \sqrt{n} \cdot \log n$ satisfies the second displayed inequality of Corollary 2.7 (for almost all n), since $n + 1 + n \log n \leq \sqrt{(n+1)n/2} \cdot \log((n+1)n/2)$ for almost n. Hence every real is computable from a Martin-Löf random real with this redundancy, much like Gács had observed.

We can now intuitively understand how the redundancy upper bounds $n \log n$ and $\sqrt{n} \cdot \log n$, of Kučera and Gács respectively, are produced. The argument of Sec. 2.2 describes a coding process where in n coding steps we code M_n many bits of X into L_n many bits of Y. The parameter $g(i) := \ell_i - m_i$ can be seen as an *overhead* of the ith coding step, i.e., the number of additional bits we use in Y in order to code the next m_i bits of X. Moreover, Corollary 2.7 says that these overheads are accumulated along the coding steps and push the redundancy of the computation to become larger over time. In particular, the number $\sum_{i<n} g(i)$ is the redundancy (total overhead accumulated) corresponding to n coding steps. (See Table 3 for summary.) Due to the condition $\sum_i 2^{-g(i)} < 1$ in Corollary 2.7 a representative choice for g is $2 \log(n+1)$, which means that $\sum_{i<n} g(i)$ needs to be of the order $\log(n!)$ or (by Stirling's formula) $n \log n$.

In the case of Kučera's argument, n bits of X are coded in n coding steps, so the redundancy for the computation of n bits of X from Y following Kučera's argument is of the order $n \log n$. If we are free to choose (m_i), note that a fast-growing choice will make the accumulated overhead smaller (since the coding steps for any initial segment of X become less) but a different type

---

[d]For example, the choices $m_i = (i+1)^2$ or $m_i = \sqrt{i+1}$ produce redundancy considerably above Gács' $\sqrt{n} \cdot \log n$ upper bound.

of overhead, namely the parameter m_s in the second inequality of Corollary 2.7, pushes the redundancy higher. Gács' choice of $m_i = i + 1$ means that in n coding steps there are $\sum_{i \le n} m_i \approx n^2$ many bits of X coded into Y. Hence the coding of $X \upharpoonright_n$ requires roughly $\sqrt{n}$ coding steps, which accumulate a total of $\sqrt{n} \cdot \log \sqrt{n} \approx \sqrt{n} \cdot \log n$ in overheads according to the previous discussion. For this reason, Gács' redundancy is of the order $\sqrt{n} \cdot \log n$. We may observe that in Gács' coding, the length of the next coding block m_{n+1} is both:

(a) the number of coding steps performed so far;
(b) roughly equal to the accumulated overhead from the coding steps performed so far.

2.4. Some limits of the Kučera–Gács method

In this section, we will frequently identify a set of finite strings V with the Σ_1^0 class specified by V, i.e., the set of infinite sequences extending elements of V. In the following proof, we use the notation $\mu_\sigma(C)$ for a string σ and a set of reals C, which is the measure of C relative to $[\sigma]$. More precisely $\mu_\sigma(C) = \mu(C \cup [\sigma]) \cdot 2^{|\sigma|}$.

Lemma 2.8. *Let P be a Π_1^0 class, and g be a computable function taking positive values, such that $\sum_i 2^{-g(i)} = \infty$. Let (n_i) be a computable sequence such that $n_{i+1} > n_i + g(i)$ for all i. If*

(U_i) *is a uniformly c.e. sequence with $U_i \subseteq 2^i$ and*
$\mu(P \cap U_i) < 2^{-i}$ *for all i,*

then every Martin-Löf random real $X \in \bigcap_i (P \cap U_i)$ has a prefix in some in U_{n_t} with P-density at most $2^{-g(t)}$.

Proof. We define a uniform sequence (V_i) of Σ_1^0 classes such that $V_t \supseteq V_{t+1}$ for all t, inductively as follows. Let V_0 (as a set of finite strings) consist of all the strings of length n_0. Assuming that V_t has been defined, for each $\sigma \in V_t$ define

$$V_{t+1} \cap [\sigma] = (U_{n_{t+1}} \cap [\sigma])^{[\leq 2^{-|\sigma|} \cdot (1 - 2^{-g(t)-1})]},$$

where for any real r and any Σ_1^0 class C with an underlying computable enumeration $C[s]$ the expression $C^{[\leq r]}$ denotes the class $C[s_*]$ where s_* is the largest stage s such that $\mu(C[s]) \leq r$ if such a stage exists, and $s_* = \infty$ otherwise (in which case we let $C[\infty] = C$). Clearly, for each t the set V_t consists of strings of length n_t. Then for each t we have $\mu(V_{t+1}) \leq (1 - 2^{-g(t)-1}) \cdot \mu(V_t)$ so

$$\mu(V_{t+1}) \leq \prod_{i=0}^{t} (1 - 2^{-g(i)-1}).$$

By hypothesis, $\sum_i 2^{-g(i)} = \infty$ so $\prod_{i=0}^{\infty}(1 - 2^{-g(i)-1}) = 0$. Since g is computable, there exists a computable increasing sequence (k_i) such that $\prod_{i=0}^{k_t}(1 - 2^{-g(i)-1}) < 2^{-t}$ for all $t > 0$. Hence (V_{k_i}) is a Martin-Löf test. Now let X be a Martin-Löf random real with $X \in \bigcap_i (P \cap U_i)$, as in the statement of the lemma. Since X is Martin-Löf random, $X \notin \bigcap_i V_{k_i} = \bigcap_i V_i$ and there exists a maximum t such X has a prefix σ in V_t. By the maximality of t we have $X \notin V_{t+1}$ and since $X \in U_{n_{t+1}}$ we must have $\mu_\sigma(U_{n_{t+1}}) > 1 - 2^{-g(t)-1}$, because otherwise a prefix of X would enter V_{t+1}. Also $\mu_\sigma(P \cap U_{n_{t+1}}) \leq 2^{|\sigma|} \cdot \mu(P \cap U_{n_{t+1}}) \leq 2^{|\sigma|-n_{t+1}}$. Since $\sigma \in V_t$, the length of σ is n_t. Since $n_{t+1} > n_t + g(t)$ we have $\mu_\sigma(P \cap U_{n_{t+1}}) \leq 2^{-g(t)-1}$. From the fact that

$$\mu_\sigma(P) + \mu_\sigma(U_{n_{t+1}}) - \mu_\sigma(P \cap U_{n_{t+1}}) \leq 1$$

we can deduce that $\mu_\sigma(P) \leq 2^{-g(t)}$. Since σ is a prefix of X of length n_t, this concludes the proof. $\square$

Corollary 2.9. *Suppose that $(m_i), (\ell_i)$ are computable sequences of positive integers with $\sum_i 2^{m_i - \ell_i} = \infty$. Then every Π_1^0 class consisting entirely of Martin-Löf random reals, which has the density property with respect to $(m_i), (\ell_i)$, is empty.*

Proof. We apply Lemma 2.8 with $n_k = L_k = \sum_{i<k} \ell_i$ and $g(i) = \ell_i - m_i$. First note that $n_{k+1} = n_k + \ell_k > n_k + g(k)$ because $g(k) < \ell_k$, so the hypothesis of Lemma 2.8 for (n_i) holds. Second, for each i let σ_i^* be the leftmost P-extendible string of length i and let U_i consist of σ_i^* as well as the strings of length i which are lexicographically to the left of σ_i^*. Then (U_i) is uniformly c.e. and $\mu(P \cap U_i) = \mu(P \cap [\sigma_i^*]) \leq 2^{-i}$ for all i. Now suppose that P is nonempty and consider the leftmost path X through P. By our assumptions regarding P, the real X is Martin-Löf random, so by Lemma 2.8 there exists some k such that the P-density of $X \upharpoonright_{L_k}$ is less than $2^{m_k - \ell_k}$. This means that there is a P-extendible string of length L_k with P-density below $2^{m_k - \ell_k}$, so P does not have the density property with respect to $(m_i), (\ell_i)$. □

Corollary 2.10 (Lower bounds on the density inside a Π_1^0 class of random reals). *Let P be a nonempty Π_1^0 class consisting entirely of Martin-Löf random reals, let g be a computable function, and let (L_i) be an increasing sequence of positive integers such that $L_{t+1} > L_t + g(t)$ for all t. Then the following conditions are equivalent:*

(a) *For every i the P-density of any P-extendible string of length L_i is $\Omega(2^{-g(i)})$.*
(b) $\sum_i 2^{-g(i)} < \infty$,

where the asymptotic notation $\Omega(2^{-g(i)})$ means $\geq 2^{-g(i)-c}$ for some constant c.

3. Coding into Randoms Without Density Assumptions

In [5] a new coding method was introduced which allows for coding every real into a Martin-Löf random real with optimal, logarithmic redundancy. We call this method *density-free coding* as it does not rely on density assumptions inside Π_1^0 classes, which is also the reason why it gives an exponentially better redundancy upper bound.

Lemma 3.1 (Density-free coding, from [5]). *Let (u_i) be a nondecreasing computable sequence and let $\mathcal{P}$ be a Π^0_1 class. If $\sum_i 2^{i-u_i} < \mu(\mathcal{P})$ then every binary stream is uniformly computable from some member of $\mathcal{P}$ with oracle-use (u_i).*

Note that by letting P be a Π^0_1 class of Martin-Löf random reals of sufficiently large measure, Lemma 3.1 shows that every real is computable from a Martin-Löf random real with use $n + 2\log n$, i.e., with logarithmic redundancy. On the other hand in [6] it was shown that this is optimal, in the sense that if $\sum_i 2^{i-u_i} = \infty$ then there is a real which is not computable from any Martin-Löf random real with use $n \mapsto u_n$. In particular, given a real ϵ, redundancy $\epsilon \cdot \log n$ in a computation from a random oracle is possible for every real if and only if $\epsilon > 1$.

We shall not give a proof of Lemma 3.1. Instead, we will discuss some aspects of this more general coding method, which contrasts the more restricted Kučera–Gács coding whose limitations we have already explored.

3.1. *Coding as a labeling task*

Coding every real into a path through a tree $\mathcal{T}$ in the Cantor space involves constructing a Turing functional Φ which is onto the Cantor space, even when it is restricted to $\mathcal{T}$. In fact, this is normally done in such a way that there is a subtree $\mathcal{T}^*$ of $\mathcal{T}$ such that Φ is a bijection between $[\mathcal{T}^*]$ and 2^ω. In this case, we refer to $\mathcal{T}^*$ as the *code-tree*. Suppose we fix $\mathcal{T}$ and consider constructing a functional for which the use u_n on argument n does not depend upon the oracle. Then the functional Φ can be constructed as a partial computable "labeling" of the finite branches of $\mathcal{T}$. If the label x_σ is placed on τ, this means that Φ outputs σ when τ is the oracle. If we also suppose that the use function is strictly increasing, then the labeling might be assumed to satisfy the following conditions:

(1) only strings of lengths $u_i, i \in \mathbb{N}$ of $\mathcal{T}$ can have a label;
(2) the labels placed on strings of length u_i of $\mathcal{T}$ are of the type x_σ where $|\sigma| = i$;
(3) if label x_σ exists in $\mathcal{T}$ then all labels x_ρ, $\rho \in 2^{\leq|\sigma|}$ exist in $\mathcal{T}$;
(4) each string in $\mathcal{T}$ can have at most one label;
(5) if ρ of length u_k in $\mathcal{T}$ has label x_σ then for each $i < k$, $\rho \restriction_{u_i}$ has label $x_{\sigma\restriction_i}$.

The reader may take a minute to view the Kučera coding as detailed in Sec. 2.2 as a labeling satisfying the properties (1)–(5) above. It is clear that:

> The code-tree $\mathcal{T}^*$ in the Kučera coding is isomorphic to the full binary tree.

The new coding behind Lemma 3.1 is also a labeling process, but in this case the code-tree can be much more complex.

3.2. *Fully labelable trees*

If (u_i) is an increasing sequence of positive integers, a (u_i)-tree T is a subset of $\{\lambda\} \cup (\bigcup_i 2^{u_i})$ which contains the empty string and is downward closed, in the sense that for each $\sigma \in 2^{u_{i+1}} \cap T$, the string $\sigma \restriction_{u_i}$ belongs to T. The elements of a (u_i)-tree T are called nodes and the t-level of T consists of the nodes of T of length u_t. The full binary tree of height k is $2^{\leq k}$ ordered by the prefix relation. Note that a (u_i)-tree is a tree, in the sense that it is a partially ordered set (with respect to the prefix relation) in which the predecessors of each member are linearly ordered. Hence given any $k \in \mathbb{N}$, we may talk about a (u_i)-tree being isomorphic to the full binary tree of height k. When we talk about two trees being isomorphic, it is in this sense that we shall mean it — as partially ordered sets. A labeling of a

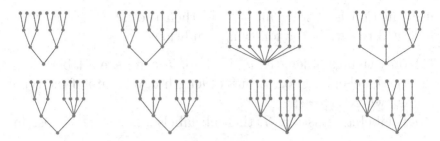

Fig. 2.   Some fully labelable (u_i)-trees of height 3.

(u_i)-tree is a partial map from the nodes of the tree to the set of labels $\{x_\sigma \mid \sigma \in 2^{<\omega}\}$ which satisfies properties (1)–(5) of the previous section. A full labeling of a (u_i)-tree is a labeling $\{x_\sigma \mid \sigma \in 2^{<\omega}\}$ with the property that for every σ there exists a node on the (u_i)-tree which has label x_σ.

A (u_i)-tree is called *fully labelable* if there exists a full labeling of it. Figure 2 illustrates some examples of fully labelable trees of height 3. Note that here the nodes are binary strings (hence nodes of the full binary tree) but since they are nodes of a (u_i)-tree, they can have more than two branches. Clearly, if $T_0 \subseteq T_1$ are (u_i)-trees and T_0 is fully labelable, then T_1 is also fully labelable. These definitions can be easily adapted to finite (u_i)-trees (where the height is the length of its longest leaf). Figure 2 shows some examples of fully labelable finite (u_i)-trees, while Fig. 3 shows some examples of finite (u_i)-trees which are not fully labelable.

Clearly any (u_i)-tree which is isomorphic to the full binary tree, is fully labelable. The success of the Kučera coding was based on this fact, along with the fact that a Π_1^0 class of sufficient measure contains such a canonical tree (subject to the growth of (u_i)). A similar remark can be made about the slightly more general Gács coding. We have already demonstrated that the density property that guarantees the extension property cannot be expected to hold if the growth of (u_i) is significantly less than $n + \sqrt{n} \cdot \log n$. Hence more efficient coding methods, such as the

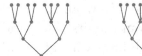

Fig. 3. Some (u_i)-trees of length 3 which are not fully labelable.

one behind Lemma 3.1, need to rely on a wider class of labelable trees.

Given two trees T_0, T_1 (thought of as partially ordered sets), we say that T_0 is *splice-reducible* to T_1 if we can obtain T_1 from T_0 via a series of operations on the nodes of T_0, each consisting of splicing two sibling nodes into one — i.e., the two sibling nodes u_0 and u_1 are replaced by a single node u, for which the set of elements $> u$ is isomorphic to the set of nodes strictly greater than u_0 union the set of nodes strictly greater than u_1. The following result points to a concrete difference between Kučera coding and the general coding from [5]: in Kučera coding the code-tree is an isomorphic copy of the full binary tree, while in [5] the code-tree is only splice-reducible to an isomorphic copy of the full binary tree.[e]

Theorem 3.2. *Given a (u_i)-tree T, the following conditions are equivalent:*

(a) *T is a fully labelable (u_i)-tree.*
(b) *T is splice-reducible to an isomorphic copy of the full binary tree.*

Proof. Suppose that T is fully labelable. We describe how to produce the full binary tree by a repeated application of the splice operation between siblings of T. Fix a full labeling of T and obtain the minimal fully labeled tree T' from T by splicing

---

[e]A similar remark can be made with respect to the Gács coding, only that instead of binary trees we need to consider a homogeneous trees, in the sense that for each level, every node of that level has the same number of successors.

the unlabeled nodes of T onto labeled ones. Now all nodes of T' are labeled. Then gradually, starting from the first level and moving toward the last level of T', splice siblings with identical labels. Inductively, by the properties of the assumed labeling, the resulting (u_i)-tree is isomorphic to the full binary tree.

Conversely, assume that T is splice-reducible to a (u_i)-tree which is isomorphic to the full binary tree. Then reversing the splice operations behind this reduction, we get a sequence of node splitting operations that transform an isomorphic (u_i)-copy of the full binary tree into T. Since this (u_i)-copy of the full binary tree has a full labeling, by making these labels persistent during the series of splitting operations that lead to T, we get a full labeling of T. □

The work in [5] shows that if (u_i) is an increasing computable sequence, then any tree of measure more than $\sum_i 2^{i-u_i} < \mu(\mathcal{P})$ has a full labeling. Moreover, such a labeling has a Π_1^0 approximation, given any Π_1^0 approximation of $\mathcal{P}$.

Acknowledgment

Barmpalias was supported by the 1000 Talents Program for Young Scholars from the Chinese Government, and the Chinese Academy of Sciences (CAS) President's International Fellowship Initiative No. 2010Y2GB03. Additional support was received by the CAS and the Institute of Software of the CAS. Partial support was also received from a Marsden grant of New Zealand and the China Basic Research Program (973) grant No. 2014CB340302.

References

[1]   G. Barmpalias, R. Downey and K. Meng Ng, Jump inversions inside effectively closed sets and applications to randomness, *J. Symbolic Logic* **76**(2) (2011) 491–518.

[2]   C. H. Bennett, Logical depth and physical complexity, in R. Herken (ed.), *The Universal Turing Machine, A Half Century Survey* (Oxford University Press, 1988) 227–257.

[3]   L. Bienvenu, N. Greenberg, A. Kučera, A. Nies and D. Turetsky, Coherent randomness tests and computing the K-trivial sets, *J. European Math. Soc.* **18** (2016) 773–812.

[4]   G. Barmpalias, A. E. M. Lewis and K. M. Ng, The importance of Π_1^0 classes in effective randomness, *J. Symbolic Logic* **75**(1) (2010) 387–400.

[5]   G. Barmpalias and A. Lewis-Pye, Optimal redundancy in computations from random oracles, preprint (2016), arXiv:1606.07910.

[6]   G. Barmpalias, A. Lewis-Pye and J. Teutsch, Lower bounds on the redundancy in computations from random oracles via betting strategies with restricted wagers, *Inform. Comput.* **251** (2016) 287–300.

[7]   R. G. Downey and J. S. Miller, A basis theorem for Π_1^0 classes of positive measure and jump inversion for random reals, *Proc. Amer. Math. Soc.* **134**(1) (2006) 283–288 (electronic).

[8]   A. Day and J. S. Miller, Density, forcing, and the covering problem, *Math. Res. Lett.* **22** (2015) 719–727.

[9]   D. Doty, Every sequence is decompressible from a random one, in *Logical Approaches to Computational Barriers, Second Conf. Computability in Europe, CiE 2006*, Swansea, UK, June 30-July 5, 2006, (2006), pp. 153–162.

[10] P. Gács, Every sequence is reducible to a random one, *Inform. Control* **70**(2–3) (1986) 186–192.

[11] M. Khan, Lebesgue density and Π_1^0-classes, *J. Symbolic Logic* **81** (2016) 80–95.

[12] A. Kučera, Measure, Π_1^0-classes and complete extensions of PA, in *Recursion theory Week (Oberwolfach, 1984)*, Lecture Notes in Mathematics, Vol. 1141 (Springer, Berlin, 1985), pp. 245–259.

[13] A. Kučera, On the use of diagonally nonrecursive functions, in *Logic Colloquium '87 (Granada, 1987)*, Stud. Logic Found. Math. Vol. 129 (North-Holland, Amsterdam, 1989), pp. 219–239.

[14] W. Merkle and N. Mihailović, On the construction of effectively random sets, *J. Symbolic Logic* **69**(3) (2004) 862–878.

[15] K. Miyabe, A. Nies and J. Zhang, Using almost-everywhere theorems from analysis to study randomness, preprint (2015).

[16] A. Nies, *Computability and Randomness* (Oxford University Press, 2009).

https://doi.org/10.1142/9789813228238_0004

Chapter 4

Information vs. Dimension: An Algorithmic Perspective

Jan Reimann

Pennsylvania State University, USA

jan.reimann@psu.edu

This paper surveys work on the relation between fractal dimensions and algorithmic information theory over the past thirty years. It covers the basic development of prefix-free Kolmogorov complexity from an information theoretic point of view, before introducing Hausdorff measures and dimension along with some important examples. The main goal of the paper is to motivate and develop the informal identity "entropy = complexity = dimension" from first principles. The last section of the paper presents some new observations on multifractal measures from an algorithmic viewpoint.

1. Introduction and Preliminaries

Starting with the work by Ryabko [41, 41] and Staiger [47–49] in the 1980s, over the past 30 years researchers have investigated a strong relation between fractal dimension and algorithmic information theory. At the center of this relation is a *pointwise* version of Hausdorff dimension (due to Lutz [29, 30]), that is, a notion of dimension that is defined for individual points in a space instead of subsets. This is made possible by effectivizing the notion of measure (in the sense of Martin-Löf [33]), which restricts the collection of nullsets to a countable family, thereby allowing for singleton sets not to be null. Such singletons are considered *random* in this framework. The correspondence between randomness and the various flavors of Kolmogorov complexity (a cornerstone of the theory of algorithmic randomness) then re-emerges in form of an asymptotic information density that bears close resemblance to the entropy of a stationary process. Effective dimension makes the connections between entropy and fractal dimension that have been unearthed by many authors (arguably starting with Billingsley [2] and Furstenberg [17]) very transparent. Moreover, effective dimension points to new ways to compute fractal dimensions, by investigating the algorithmic complexities of single points in a set (e.g., Lutz and Lutz [31]), in particular when studied in relation to other *pointwise* measures of complexity, such as irrationality exponents in diophantine approximation [1].

This survey tries to present these developments in one coherent narrative. Throughout, we mostly focus on Hausdorff dimension, and on a small fraction of results on effective dimension and information, in order to streamline the presentation. For other notions such as packing or box counting dimension, refer to [10, 15] for the classical theory, and to [13, 37, 52, 53] for many results concerning effective dimension.

The outline of the paper is as follows. In Sec. 2, we cover the basic theory of information measures, introducing entropy and Kolmogorov complexity, highlighting the common foundations of both. In Sec. 3, we briefly introduce Hausdorff measures and dimension, along with some fundamental examples. Arguably the central section of the paper, Sec. 4 develops the theory of effective dimension, establishes the connection with asymptotic Kolmogorov complexity, and describes the fundamental instance of entropy vs dimension — the Shannon–McMillan–Breiman theorem. Finally, in Sec. 5, we present some new observations concerning multifractal measures. It turns out that the basic idea of pointwise dimension emerges here, too.

It is the goal of this paper to be accessible for anyone with a basic background in computability theory and measure theory. Where the proofs are easy to give and of reasonable length, we include them. In all other cases we refer to the literature.

2. Information Measures and Entropy

Suppose X is a finite, nonempty set. We choose an element $a \in X$ and want to communicate our choice to another person, through a binary channel, i.e., we can only transmit bits 0 and 1. How many bits are needed to transmit our choice, if nothing else is known about X but its cardinality? To make this possible at all, we assume that we have agreed on a numbering of $X = \{a_1, \ldots, a_n\}$ that is known to both parties. We can then transmit the *binary code* for the index of a, which requires at most $\log_2 n = \log_2 |X|$ bits. Of course, if we had an N-ary channel available, we would need $\log_N |A|$ bits. We will mostly work with binary codes, so log will denote the binary logarithm $\log_2$.

Often our choice is guided by a probabilistic process, i.e., we have X given as a *discrete random variable*. X has countably many outcomes, and we denote the range of X by $\{a_0, a_1, a_2, \ldots\}$. Suppose that we repeatedly choose from X at random and want to communicate our choice to the other party. Arguably the central question of information theory is:

> *How do we minimize the expected number of bits we need to transmit to communicate our choice?*

Depending on the distribution of the random variable, we can tweak our coding system to optimize the length of the code words.

Definition 2.1. A *binary code* for a countable set X is a one-to-one function $c : \{x_i : i \in \mathbb{N}\} \to 2^{<\mathbb{N}}$.

Here, $2^{<\mathbb{N}}$ denotes the set of all finite binary strings. In the nonprobabilistic setting, when X is a finite set, we essentially use a *fixed-length codec* $: X \to \{0,1\}^n$, by mapping

$$x_i \mapsto \text{binary representation of } i.$$

In the probabilistic setting, if the distribution of the random variable X does not distinguish between the outcomes, i.e., if X is equidistributed, this will also be the best we can do.

However, if the possible choices have very different probabilities, we could hope to save on the *expected* code word length

$$\sum_i \text{len}(c(a_i)) \cdot \mathbb{P}(X = a_i),$$

where $\text{len}(c(a_i))$ is the length of the string $c(a_i)$, by assigning shorts code words to a of high probability. The question then becomes how small the expected code word length can become, and how we would design a code to minimize it.

Another way to approach the problem of information transmission is by trying to measure the *information gained* by communicating only partial information. Suppose our random variable X is represented by a partition

$$[0,1] = \bigcup_i X_i,$$

where each X_i is an interval such that $\mathbb{P}(X = a_i) = |X_i|$, i.e., the lengths of the intervals mirror the distribution of X. Pick, randomly and uniformly, an $x \in [0,1]$. If we know which X_i x falls in, we gain knowledge of about the first $-\log|X_i|$ bits of the binary expansion of x. Therefore, on average, we gain

$$H(X) = -\sum \mathbb{P}(X = a_i) \cdot \log \mathbb{P}(X = a_i) = -\sum |X_i| \cdot \log|X_i|$$

bits of information. We put $0 \cdot \log 0 =: 0$ to deal with $\mathbb{P}(X = a_i) = 0$. $H(X)$ is called the *entropy* of X. We will apply

H not only to random variables, but also to measures in general. If $\vec{p} = (p_1, \ldots, p_n)$ is a finite probability vector, i.e., $p_i \geq 0$ and $\sum_i p_i = 1$, then we write $H(\vec{p})$ to denote $\sum_i p_i \cdot \log(p_i)$. Similarly, for $p \in [0, 1]$, $H(p)$ denotes $p \cdot \log p + (1 - p) \cdot \log(1 - p)$.

The emergence of the term $-\log \mathbb{P}$ is no coincidence. Another, more axiomatic way to derive it is as follows. Suppose we want to measure the information gain of an event by a function $I :$ $[0, 1] \to \mathbb{R}^{\geq 0}$, that is, I depends only on the probability of an event. We require the following properties of I:

(I1) $I(1) = 0$. An almost sure event gives us no new information.
(I2) I is decreasing. The lower the probability of an event, the more information we gain from the knowledge that it occurred.
(I3) $I(x \cdot y) = I(x) + I(y)$. If X and Y are independent, then $\mathbb{P}(X \cap Y) = \mathbb{P}(X) \cdot \mathbb{P}(Y)$, and hence $I(\mathbb{P}(X \cap Y)) = I(\mathbb{P}(X)) + I(\mathbb{P}(Y))$. In other words, information gain is additive for independent events.

Proposition 2.2. *If $I : [0, 1] \to \mathbb{R}^{\geq 0}$ is a function satisfying* (I1)–(I3), *then there exists a constant c such that*

$$I(x) = -c \log(x).$$

In this sense, $-\log$ is the only reasonable information function, and the entropy H measures the expected gain in information.

It is also possible to axiomatically characterize entropy directly. Many such characterizations have been found over the years, but the one by Khinchin [23] is arguably still the most popular one.

We will see next how the entropy of X is tied to the design of an optimal code.

2.1. *Optimal prefix codes*

We return to the question of how to find an optimal code, i.e., how to design a binary code for a random variable X such that its average code word length is minimal. While a binary code guarantees that a single code word can be uniquely decoded, we would like to have a similar property for *sequences of code words*.

Definition 2.3. Given a binary code c for a countable set X, its *extension* c^* to $X^{<\mathbb{N}}$ is defined as

$$c^*(x_0 x_1 \ldots x_n) = c(x_0)^\frown c(x_1)^\frown \cdots ^\frown c(x_n),$$

where $\frown$ denotes concatenation of strings. We say a binary code c is *uniquely decodable* if its extension is one-one.

One way to ensure unique decodability is to make c *prefix free*. A set $S \subseteq 2^{<\mathbb{N}}$ is prefix free if no two elements in S are prefixes of one another. A code is prefix free if its range is. It is not hard to see that the extension of a prefix free code is indeed uniquely decodable.

Binary codes live in $2^{<\mathbb{N}}$. The set $2^{<\mathbb{N}}$ is naturally connected to the space of infinite binary sequences $2^{\mathbb{N}}$. We can put a metric on $2^{\mathbb{N}}$ by letting

$$d(x, y) = \begin{cases} 2^{-N} & \text{if } x \neq y \text{ and } N = \min i \colon x(i) \neq y(i) \\ 0 & \text{otherwise.} \end{cases}$$

Given $\varepsilon > 0$, the ε-ball $B(x, \varepsilon)$ around x is the so-called *cylinder set*

$$[\![\sigma]\!] = \{y \in 2^{\mathbb{N}} \colon \sigma \subset y\},$$

where $\subset$ denotes the prefix relation between strings (finite or infinite), and $\sigma = x{\restriction}_n$ is the length-n initial segment of x with $n = \lceil -\log \varepsilon \rceil$.

Hence any string σ corresponds to a basic open cylinder $[\![\sigma]\!]$ with diameter $2^{-\text{len}(\sigma)}$. This induces a Borel measure λ on $2^{\mathbb{N}}$, $\lambda[\![\sigma]\!] = 2^{-\text{len}(\sigma)}$, which is the natural analogue to Lebesgue measure on $[0, 1]$.

Which code lengths are possible for prefix free codes? The *Kraft inequality* gives a fundamental restraint.

Theorem 2.4 (Kraft inequality). *Let* $S \subseteq 2^{<\mathbb{N}}$ *be prefix free. Then*

$$\sum_{\sigma \in S} 2^{-\text{len}(\sigma)} \leq 1.$$

Conversely, given any sequence $l_0, l_1, l_2, \ldots$ *of non-negative integers satisfying*

$$\sum_i 2^{-l_i} \leq 1,$$

there exists a prefix-free set $\{\sigma_0, \sigma_1, \ldots\} \subseteq 2^{<\mathbb{N}}$ *of code words such that* $\text{len}(\sigma_i) = l_i$.

Proof sketch. ($\Rightarrow$) Any prefix free set S corresponds to a disjoint union of open cylinders

$$U = \bigcup_{\sigma \in S} [\![\sigma]\!].$$

The Lebesgue measure λ of $[\![\sigma]\!]$ is $2^{-\text{len}(\sigma)}$. Hence by the additivity of measures,

$$1 \geq \lambda(U) = \sum_{\sigma \in S} 2^{-\text{len}(\sigma)}.$$

($\Leftarrow$) We may assume the l_i are given in nondecreasing order:

$$l_0 \leq l_1 \leq \cdots,$$

which implies the sequence (2^{-l_i}) is nonincreasing. We construct a prefix code by "shaving off" cylinder sets from the left. We choose as σ_0 the leftmost string of length l_0, i.e., 0^{l_0}. Suppose we have chosen $\sigma_0, \ldots, \sigma_k$ such that $\text{len}(\sigma_i) = l_i$ for $i = 0, \ldots, k$. The measure we have left is

$$1 - (2^{-l_0} + \cdots + 2^{-l_k}) \geq 2^{-l_{k+1}},$$

since $\sum_i 2^{-l_i} \leq 1$, and since $2^{-l_0} \geq \cdots \geq 2^{-l_k} \geq 2^{-l_{k+1}}$, the remaining measure is a multiple of $2^{-l_{k+1}}$. Therefore, we can pick σ_{k+1} to be the leftmost string of length l_{k+1} that does not extend any of the $\sigma_0, \ldots, \sigma_k$.

This type of code is known as the *Shannon–Fano code*.   □

Subject to this restraint, what is the minimum expected length of a code for a random variable X? Suppose X is discrete and let $p_i = \mathbb{P}(X = x_i)$. We want to find code lengths l_i such that

$$L = \sum p_i l_i$$

is minimal, where the l_i have to satisfy

$$\sum 2^{-l_i} \le 1,$$

the restriction resulting from the Kraft inequality.

The following result, due to Shannon, is fundamental to information theory.

Theorem 2.5. *Let L be the expected length of a binary prefix code of a discrete random variable X. Then*

$$L \ge H(X).$$

Furthermore, there exists a prefix code for X whose expected code length does not exceed $H(X) + 1$.

To see why no code can do better than entropy, note that

$$L - H(X) = \sum p_i l_i + \sum p_i \log(p_i)$$

$$= -\sum p_i \log 2^{-l_i} + \sum p_i \log(p_i).$$

Put $c = \sum 2^{-l_i}$ and $r_i = 2^{-l_i}/c$. $\vec{r}$ is the probability distribution induced by the code lengths. Then

$$L - H(X) = \sum p_i \log p_i - \sum p_i \log \left(\frac{2^{-l_i}}{c} c \right)$$

$$= \sum p_i \log \frac{p_i}{r_i} - \log c.$$

Note that $c \leq 1$ and hence $\log c \leq 0$. The expression

$$\sum p_i \log \frac{p_i}{r_i}$$

is known as the *relative entropy* or *Kullback–Leibler (KL) distance* $D(\vec{p} \| \vec{r})$ of distributions $\vec{p}$ and $\vec{r}$. It is not actually a distance as it is neither symmetric nor does it satisfy the triangle inequality. But it nevertheless measures the difference between distributions in some important aspects. In particular, $D(\vec{p} \| \vec{r}) \geq 0$ and is equal to zero if and only if $\vec{p} = \vec{r}$. This can be inferred from Jensen's inequality.

Theorem 2.6 (Jensen's inequality). *If f is a convex function and X is a random variable, then*

$$\mathbb{E}f(X) \geq f(\mathbb{E}X).$$

In particular, it follows that $L - H(X) \geq 0$. This can also be deduced directly using the concavity of the logarithm in the above sums, but both KL-distance and Jensen's inequality play a fundamental, almost "axiomatic" role in information theory, so we reduced the problem to these two facts.

To see that there exists a code with average code length close to entropy, let

$$l_i := \lceil \log p_i \rceil.$$

Then $\sum 2^{-l_i} \leq \sum p_i = 1$, and by Theorem 2.4, there exists a prefix code $\vec{\sigma}$ such that $\text{len}(\sigma_i) = l_i$. Hence $x_i \mapsto \sigma_i$ will be a prefix code for X with expected code word length

$$\sum p_i l_i = \sum p_i \lceil \log p_i \rceil \leq \sum p_i (\log p_i + 1) = H(X) + 1.$$

2.2. Effective coding

The transmission of information is an inherently computational act. Given X, we should be able to assign code words to outcomes of X effectively. In other words, if $\{x_i\}$ collects the possible outcomes of X, and $\{\sigma_i\}$ is a prefix code for X, then the mapping $i \mapsto \sigma_i$ should be computable.

The construction of a code is implicit in the proof of the Kraft inequality (Theorem 2.4), and is already of a mostly effective nature. The only noneffective part is the assumption that the word lengths l_i are given in increasing order. However, with careful book keeping, one can assign code words in the order they are requested. This has been observed independently by Levin [25], Schnorr [44], Chaitin [6].

Theorem 2.7 (Effective Kraft inequality). *Let* $L : \mathbb{N} \to \mathbb{N}$ *be a computable mapping such that*

$$\sum_{i \in \mathbb{N}} 2^{-L(i)} \leq 1.$$

Then there exists a computable, one-to-one mapping $c : \mathbb{N} \to 2^{<\mathbb{N}}$ *such that the image of* c *is prefix free, and for all* i, $\text{len}(c(i)) = L(i)$.

A proof can be found, for example, in the book by Downey-Hirschfeldt [13]. Let us call c an effective prefix-code. As c is one-to-one, given $\sigma \in \text{range}(c)$, we can compute i such that $c(i) = \sigma$. Since the image of c is prefix-free, if we are given a sequence of concatenated code words

$$\sigma_1 {}^\frown \sigma_2 {}^\frown \sigma_3 \ldots = c(i_1) {}^\frown c(i_2) {}^\frown c(i_3) {}^\frown \ldots,$$

we can effectively recover the source sequence $i_1, i_2, i_3 \ldots$. Hence, every effective prefix code c comes with an effective *decoder*.

Let us therefore switch perspective now: Instead of mapping events to code words, we will look at functions mapping code words to events. An *effective prefix decoder* is a partial computable function d whose domain is a prefix-free subset of $2^{<\mathbb{N}}$. Since every partial recursive function is computable by a Turing machine, we will just use the term *prefix-free machine* to denote effective prefix decoders. The range of a prefix-free machine will be either $\mathbb{N}$ or $2^{<\mathbb{N}}$. Since we can go effectively from one set to the other (by interpreting a binary string as the dyadic representation of a natural number or by enumerating $2^{<\mathbb{N}}$ effectively), in this section it does not really matter which one we are using. In later sections, however, when we look at infinite binary sequences, it will be convenient to have $2^{<\mathbb{N}}$ as the range.

If M is a prefix-free machine and $M(\sigma) = m$, we can think of σ as an M-code of m. Note that by this definition, m can have multiple codes, which is not provided for in our original definition of a binary code (Definition 2.1). So a certain asymmetry arises when switching perspective. We can eliminate this asymmetry by considering only the *shortest* code. The length of this code will then, with respect to the given machine, tell us how far the information in σ can be *compressed*.

Definition 2.8. Let M be a prefix-free machine. The M-complexity $\mathrm{K}_M(m)$ of a string σ is defined as

$$\mathrm{K}_M(m) = \min\{\mathrm{len}(\sigma)\colon M(\sigma) = m\},$$

where $\mathrm{K}_M(m) = \infty$ if there is no σ with $M(\sigma) = m$.

K_M may not be computable anymore, as we are generally not able to decide whether M will halt on an input σ and output m. But we can *approximate* K_M *from above*: Using a bootstrapping procedure, we run M simultaneously on all inputs. Whenever we see that $M(\sigma) \downarrow = m$, we compare σ with our current approximation of $\mathrm{K}_M(m)$. If σ is shorter, $\mathrm{len}(\sigma)$ is our new approximation of $\mathrm{K}_M(m)$. Eventually, we will have found the shortest M-code for m, but in general, we do not know whether our current approximation is actually the shortest one.

Alternatively, we can combine the multiple codes into a single one probabilistically. Every prefix-free machine M induces a distribution on $\mathbb{N}$ by letting

$$Q_M(m) = \sum_{M(\sigma)\downarrow=m} 2^{-\mathrm{len}(\sigma)}. \tag{2.1}$$

Q_M is not necessarily a probability distribution, however, for it is not required that the domain of M covers all of $2^{<\mathbb{N}}$. Nevertheless, it helps to think of $Q_M(m)$ as the probability that M on a randomly chosen input halts and outputs m.

Any function $Q : \mathbb{N} \to \mathbb{R}^{\geq 0}$ with

$$\sum_m Q(m) \leq 1$$

is called a *discrete semimeasure*. The term "discrete" is added to distinguish these semimeasures from the *continuous semimeasures* which will be introduced in Sec. 5. In this section, all semimeasures are discrete, so we will refer to them just as "semimeasures".

If a semimeasure Q is induced by a prefix-free machine, it moreover has the property that for any m, the set

$$\{q \in \mathbb{Q} : q < Q(m)\}$$

is recursively enumerable, uniformly in m. In general, semimeasures with this property are called *left-enumerable*.

We can extend the effective Kraft Inequality to left-enumerable semimeasures. We approximate a code enumerating better and better dyadic lower bounds for Q. As we do not know the true value of Q, we cannot quite match the code lengths prescribed by Q, but only up to an additive constant.

Theorem 2.9 (Coding Theorem). *Suppose Q is a left-enumerable semimeasure. Then there exists a prefix-free machine M_Q such that for some constant d and for all $m \in \mathbb{N}$,*

$$\mathrm{K}_{M_Q}(m) \leq -\log Q(m) + d.$$

For a proof see Li and Vitányi [26]. Since the shortest M-code contributes $2^{-\mathrm{K}_M(m)}$ to $Q_M(m)$, it always holds that

$$\mathrm{K}_{M_Q}(m) \leq \mathrm{K}_M(m) + d$$

for all m and for some constant d. However, if we consider universal machines, the two processes will essentially yield the same code.

Definition 2.10. A prefix-free machine U is *universal* if for each partial computable prefix-free function $\varphi : 2^{<\mathbb{N}} \to \mathbb{N}$, there is a string σ_φ such that for all τ,

$$U(\sigma_\varphi{}^\frown\tau) = \varphi(\tau).$$

Universal prefix-free machines exist. The proof is similar to the proof that universal (plain) Turing machines exist (see, e.g., [13]).

Fix a universal prefix-free machine U and define

$$\mathrm{K}(m) = \mathrm{K}_U(m).$$

$\mathrm{K}(m)$ is called the *prefix-free Kolmogorov complexity* of m. The universality of U easily implies the following.

Theorem 2.11 (Invariance Theorem). *For any prefix-free machine M there exists a constant d_M such that for all m,*

$$\mathrm{K}(m) \leq \mathrm{K}_M(m) + d_M.$$

Theorems 2.9 and 2.11 together imply that up to an additive constant,

$$\mathrm{K} = \mathrm{K}_U \leq \mathrm{K}_{Q_U} \leq \mathrm{K}_U = \mathrm{K},$$

that is, up to an additive constant, *both coding methods agree* (see Fig. 1).

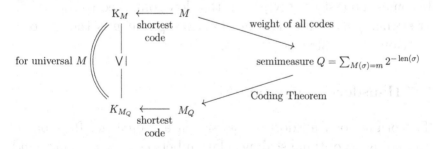

Fig. 1.   Relating complexity and semimeasures.

The Coding Theorem also yields a relation between K and Shannon entropy H. Suppose X is a random variable with a computable distribution p. As K is the length of a prefix code on $2^{<\mathbb{N}}$, Shannon's lower bound (Theorem 2.5) implies that $H(X) \leq \sum_m K(m)p(m)$. On the other hand, it is not hard to infer from the Coding Theorem that, up to a multiplicative constant, $2^{-K(m)}$ dominates every computable probability measure on $\mathbb{N}$. Therefore,

$$H(X) = \sum_m p(m)(-\log p(m)) \geq \sum_m p(m)\,K(m) - d$$

for some constant d depending only on p. In fact, the constant d depends only on the length of a program needed to compute p (which we denote $K(p)$).

Theorem 2.12 (Cover, Gács, and Gray [7]). *There exists a constant d such that for every discrete random variable X with computable probability distribution p on $\mathbb{N}$ with finite Shannon entropy $H(X)$,*

$$H(X) \leq \sum_\sigma K(\sigma)p(\sigma) \leq H(X) + K(p) + d.$$

In other words, probabilistic entropy of a random variable X taking outcomes in N is equal, within an additive constant, to the expected value of complexity of these outcomes. This becomes interesting if $K(p)$ is rather low compared to the length of strings p puts its mass on, for example, if p is the uniform distribution on strings of length n.

3. Hausdorff Dimension

The origins of Hausdorff dimension are geometric. It is based on finding an optimal scaling relation between the diameter and the measure of a set.

Let (Y, d) be a separable metric space, and suppose s is a nonnegative real number. Given a set $U \subseteq Y$, the diameter of U is given as

$$d(U) = \sup\{d(x, y) \colon x, y \in U\}.$$

Given $\delta > 0$, a family $(U_i)_{i \in I}$ of subsets of Y with $d(U_i) \leq \delta$ for all $i \in I$ is called a δ-*family*.

The s-*dimensional outer Hausdorff measure* of a set $A \subseteq Y$ is defined as

$$\mathcal{H}^s(A) = \lim_{\delta \to 0} \left(\inf \left\{ \sum d(U_i)^s \colon A \subseteq \bigcup U_i, \ (U_i)\delta - \text{family} \right\} \right).$$

Since there are fewer δ-families available as $\delta \to 0$, the limit above always exists, but may be infinite. If $s = 0$, $\mathcal{H}$ is a counting measure, returning the number of elements of A if A is finite, or ∞ if A is infinite. If Y is a Euclidean space $\mathbb{R}^n$, then $\mathcal{H}^n$ coincides with Lebesgue measure, up to a multiplicative constant.

It is not hard to see that one can restrict the definition of $\mathcal{H}^s$ to δ-families of open sets and still obtain the same value for $\mathcal{H}^s$.

In $2^{\mathbb{N}}$, it suffices to consider only δ-families of cylinder sets, since for covering purposes any set $U \subseteq 2^{\mathbb{N}}$ can be replaced by $[\![\sigma]\!]$, where σ is the longest string such that $\sigma \sqsubset x$ for all $x \in U$ (provided U has more than one point). Then $d([\![\sigma]\!]) = d(U)$. It follows that δ-covers in $2^{\mathbb{N}}$ can be identified with sets of strings $\{\sigma_i \colon i \in \mathbb{N}\}$, where $\text{len}(\sigma_i) \geq \lceil \log \delta \rceil$ for all i.

In $\mathbb{R}$, one can obtain a similar simplification by considering the *binary net measure* $\mathcal{M}^s$, which is defined similar to $\mathcal{H}^s$, but only coverings by dyadic intervals of the form $[r2^{-k}, (r+1)2^{-k})$ are permitted. For the resulting outer measure $\mathcal{M}^s$, it holds that, for any set $E \subseteq \mathbb{R}$,

$$\mathcal{H}^s E \leq \mathcal{M}^s E \leq 2^{s+1} \mathcal{H}^s E. \qquad (3.2)$$

Instead of dyadic measures, one may use other g-adic net measures and obtain a similar relation as in (3.2), albeit with a different multiplicative constant (see, e.g., [15]).

For each s, $\mathcal{H}$ is a *metric outer measure*, that is, if $A, B \subseteq Y$ have positive distance $(\inf\{d(x, y) \colon x \in A, y \in B\} > 0)$, then

$$\mathcal{H}^s(A \cup B) = \mathcal{H}^s(A) + \mathcal{H}^s(B).$$

This implies that all Borel sets are $\mathcal{H}^s$-measurable [40].

It is not hard to show that

$$\mathcal{H}^s(A) < \infty \;\Rightarrow\; \mathcal{H}^t(A) = 0 \quad \text{for all } t > s, \tag{3.3}$$

and

$$\mathcal{H}^s(A) > 0 \;\Rightarrow\; \mathcal{H}^t(A) = \infty \quad \text{for all } t < s. \tag{3.4}$$

Hence the number

$$\dim_H(A) = \inf\{s \colon \mathcal{H}(A) = 0\} = \sup\{s \colon \mathcal{H}(A) = \infty\}$$

represents the "optimal" scaling exponent (with respect to diameter) for measuring A. $\dim_H(A)$ is called the *Hausdorff dimension* of A. By (3.2), working in $\mathbb{R}$, we can replace $\mathcal{H}^s$ by the binary net measure $\mathcal{M}^s$ in the definition of Hausdorff dimension. One of the theoretical advantages of Hausdorff dimension over other concepts that measure fractal behavior (such as box counting dimension) is its *countable stability*.

Proposition 3.1. *Given a countable family $\{A_i\}$ of sets, it holds that*

$$\dim_H \bigcup A_i = \sup\{\dim_H A_i \colon i \in \mathbb{N}\}.$$

Examples of Hausdorff dimension.

(1) Any finite set has Hausdorff dimension zero.
(2) Any set of nonzero n-dimensional Lebesgue measure in $\mathbb{R}^n$ has Hausdorff dimension n. Hence all "regular" geometric objects of full topological dimension (e.g., squares in $\mathbb{R}^2$, cylinders in $\mathbb{R}^3$) have full Hausdorff dimension, too.

(3) The *middle-third Cantor set*

$$C = \left\{ x \in [0,1] : x = \sum_{i \geq 1} t_i \cdot 3^{-i}, \; t_i \in 0, 2 \right\}$$

has Hausdorff dimension $\ln(2)/\ln(3)$.

(4) Inside Cantor space, we can define a "Cantor-like" set by letting

$$D = \{ x \in 2^{\mathbb{N}} : x(2n) = 0 \text{ for all } n \}.$$

Then $\dim_H(D) = \frac{1}{2}$.

Generalized Cantor sets. Later on, a more general form of Cantor set will come into play. Let $m \geq 2$ be an integer, and suppose $0 < r < 1/m$. Put $I_0^0 = [0,1]$. Given intervals $I_n^0, \ldots, I_n^{m^n-1}$, define the intervals $I_{n+1}^0, \ldots, I_{n+1}^{m^{n+1}-1}$ by replacing each I_n^j by m equally spaced intervals of length $r|I_n^j|$. The Cantor set $C_{m,r}$ is then defined as

$$C_{m,r} = \bigcap_n \bigcup_i I_n^i.$$

It can be shown that

$$\dim_H C_{m,r} = \frac{\ln(m)}{-\ln(r)},$$

see [15, Example 4.5]. In fact, it holds that $0 < \mathcal{H}^{\ln(m)/-\ln(r)}(C_{m,r}) < \infty$. An upper bound on $\mathcal{H}^{\ln(m)/-\ln(r)}$ can be obtained by using the intervals $I_n^0, \ldots, I_n^{m^n-1}$ of the same level as a covering. To show $\mathcal{H}^{\ln(m)/-\ln(r)}(C_{m,r}) > 0$, one can show a mass can be distributed along $C_{m,r}$ that disperses sufficiently fast. A *mass distribution* on a subset E of a metric space Y is a measure on Y with support[a] a contained in Y such that $0 < \mu E < \infty$.

---

[a]Recall that the support of a measure on a metric space Y is the largest closed subset F of Y such that for every $x \in F$, if $U \ni x$ is an open neighborhood of x, $\mu U > 0$.

Lemma 3.2 (Mass Distribution Principle). *Suppose Y is a metric space and μ is a mass distribution on a set $E \subseteq Y$. Suppose further that for some s there are numbers $c > 0$ and $\varepsilon > 0$ such that*

$$\mu U \le c \cdot d(U)^s$$

for all sets U with $d(U) < \varepsilon$. Then $\mathcal{H}^s \ge \mu(E)/c$.

For a proof, refer to [15]. The mass distribution principle has a converse, *Frostman's Lemma*, which we will address in Sec. 4.

4. Hausdorff Dimension and Information

A 1949 paper by Eggleston [14] established a connection between Hausdorff dimension and entropy. Let $\vec{p} = (p_1, \ldots, p_N)$ be a probability vector, i.e., $p_i \ge 0$ and $\sum p_i = 1$. Given $(p_1, \ldots, p_N)$, $i < N$, and $x \in [0,1]$, let

$$\#(x,i,n) = \# \text{ occurrences of digit } i \text{ among the first } n \text{ digits}$$

$$\text{of the } N\text{-ary expansion of } x$$

and

$$D(\vec{p}) = \left\{ x \in [0,1] : \lim_{n \to \infty} \frac{\#(x,i,n)}{n} = p_i \text{ for all } i \le N \right\}.$$

Theorem 4.1 (Eggleston). *For any $\vec{p}$,*

$$\dim_H D(\vec{p}) = -\sum_i p_i \log_N p_i = \log_2 N \cdot H(\vec{p}).$$

Using an effective version of Hausdorff dimension, we can see that Eggleston's result is closely related to a fundamental theorem in information theory.

4.1. *Hausdorff dimension and Kolmogorov complexity*

In the 1980s and 1990s, a series of papers by Ryabko [41, 42] and Staiger [47–50] exhibited a strong connection between Hausdorff dimension, Kolmogorov complexity, and entropy. A main result was that if a set has high Hausdorff dimension, it has to contain elements of high incompressibility ratio. In this section, we focus on the space $2^{\mathbb{N}}$ of infinite binary sequences. It should be clear how the results could be adapted to sequence spaces over arbitrary finite alphabets. We will later comment how these ideas can be adapted to $\mathbb{R}$.

Definition 4.2. Given $x \in 2^{\mathbb{N}}$, we define the *lower incompressibility ratio* of x as

$$\underline{\kappa}(x) = \liminf_{n \to \infty} \frac{K(x{\restriction}n)}{n},$$

respectively. The lower incompressibility ratio of a subset $A \subseteq 2^{\mathbb{N}}$ is defined as

$$\underline{\kappa}(A) = \sup\{\underline{\kappa}(x) \colon x \in A\}.$$

Theorem 4.3 (Ryabko). *For any set $A \subseteq 2^{\mathbb{N}}$,*

$$\dim_H A \leq \underline{\kappa}(A).$$

Staiger was able to identify settings in which this upper bound is also a lower bound for the Hausdorff dimension, in other words, the dimension of a set equals the least upper bound on the incompressibility ratios of its individual members. In particular, he showed it holds for any Σ_2^0 (lightface) subset of $2^{\mathbb{N}}$.

Theorem 4.4 (Staiger). *If $A \subseteq 2^{\mathbb{N}}$ is Σ_2^0, then*

$$\dim_H A = \underline{\kappa}(A).$$

Using the countable stability of Hausdorff dimension, this result can be extended to arbitrary countable unions of Σ_2^0 sets.

The ideas underlying Ryabko's and Staiger's results were fully exposed when Lutz [29, 30] extended Martin-Löf's concept of randomness tests for individual sequences [33] to Hausdorff measures. Lutz used a variant of martingales, so called s-gales.[b] In this paper, we will follow the approach using variants of Martin-Löf tests. We will define a rather general notion of Martin-Löf test which can be used both for probability measures and Hausdorff measures.

4.2. *Effective nullsets*

Definition 4.5. A *premeasure* on $2^\mathbb{N}$ is a function $\rho\colon 2^{<\mathbb{N}} \to [0, \infty)$.

Definition 4.6. Let ρ be a computable premeasure on $2^\mathbb{N}$. A ρ-*test* is a sequence $(W_n)_{n\in\mathbb{N}}$ of sets $W_n \subseteq 2^{<\mathbb{N}}$ such that for all n,

$$\sum_{\sigma\in W_n} \rho(\sigma) \leq 2^{-n}.$$

A set $A \subseteq 2^\mathbb{N}$ is *covered by a ρ-test* (W_n) if

$$A \subseteq \bigcap_n \bigcup_{\sigma\in W_n} [\![\sigma]\!],$$

in other words, A is contained in the G_δ set induced by (W_n).

Of course, whether this is a meaningful notion depends on the nature of ρ. We consider two families of premeasures, *Hausdorff premeasures* of the form $\rho(\sigma) = 2^{-\mathrm{len}(\sigma)s}$, and *probability*

---

[b]A variant of this concept appeared, without explicit reference to Hausdorff dimension, already in Schnorr's fundamental book on algorithmic randomness [43].

premeasures, which satisfy

$$\rho(\varnothing) = 1, \tag{4.5}$$

$$\rho(\sigma) = \rho(\sigma^\frown 0) + \rho(\sigma^\frown 1). \tag{4.6}$$

All premeasures considered henceforth are assumed to be either Hausdorff premeasures or probability premeasures. The latter we will often simply refer to as *measures*.

Hausdorff premeasures are completely determined by the value of s, and hence in this case we will speak of s-tests. By the Carathéodory Extension Theorem, probability premeasures extend to a unique Borel probability measure on $2^{\mathbb{N}}$. We will therefore also identify Borel probability measures μ on $2^{\mathbb{N}}$ with the underlying premeasure $\rho_\mu(\sigma) = \mu[\![\sigma]\!]$.

Even though the definition of Hausdorff measures is more involved than that of probability measures, if we only are interested in Hausdorff nullsets (which is the case if we only worry about Hausdorff dimension), it is sufficient to consider s-tests.

Proposition 4.7. $\mathcal{H}^s(A) = 0$ *if and only if A is covered by an s-test.*

The proof is straightforward, observing that any set $\{W_n\}$ for which $\sum_{\sigma \in W_n} 2^{-\text{len}(\sigma)s}$ is small admits only cylinders with small diameter.

Tests can be made *effective* by requiring the sequence $\{W_n\}$ be uniformly recursively enumerable in a suitable representation [12, 39]. To avoid having to deal with issues arising from representations of noncomputable measures, we restrict ourselves here to computable premeasures.

We call the resulting test a *Martin-Löf ρ-test*.

If A is covered by a Martin-Löf ρ-test, we say that A is *effectively ρ-null*. There are at most countably many Martin-Löf ρ-tests. Since for the premeasures we consider countable unions of null sets are null, for every premeasure ρ there exists a nonempty set of points not covered by any ρ-test. In analogy to the standard definition Martin-Löf random sequences for Lebesgue measure λ, it makes sense to call such sequences ρ-*random*.

Definition 4.8. Given a premeasure ρ, a sequence $x \in 2^{\mathbb{N}}$ is ρ-*random* if $\{x\}$ is not effectively ρ-null.

If ρ is a probability premeasure, we obtain the usual notion of Martin-Löf randomness for probability measures. For Hausdorff measures, similar to (3.3) and (3.4), we have the following (since every s-test is a t-test for all rational $t \geq s$).

Proposition 4.9. *Suppose s is a rational number. If x is not s-random, then it is not t-random for all rational $t > s$, and if x is s-random, then it is t-random for all rational $0 < t < s$.*

We can therefore define the following *pointwise* version of Hausdorff dimension.

Definition 4.10 (Lutz). The *effective* or *constructive* Hausdorff dimension of a sequence $x \in 2^{\mathbb{N}}$ is defined as

$$\dim_H x = \inf\{s > 0 : x \text{ not } s\text{-random}\}.$$

We use the notation $\dim_H x$ to stress the pointwise analogy to (classical) Hausdorff dimension.

4.3. *Effective dimension and Kolmogorov complexity*

Schnorr [13] made the fundamental observation that Martin-Löf randomness for $\lambda(\sigma) = 2^{-\text{len}(\sigma)}$ and incompressibility with respect to prefix-free complexity coincide. Gács [18] extended this to other computable measures.

Theorem 4.11 (Gacs [18]). *Let μ be a computable measure on $2^{\mathbb{N}}$. Then x is μ-random if and only of there exists a constant c such that*

$$\forall n K(x\lceil_n) \geq -\log \mu[\![x\lceil_n]\!].$$

One can prove a similar result for s-randomness.

Theorem 4.12. *A sequence $x \in 2^{\mathbb{N}}$ is s-random if and only if there exists a c such that for all n,*

$$K(x{\upharpoonright}_n) \geq sn - c.$$

In other words, being s-random corresponds to being incompressible only to a factor of s. For this reason s-randomness has also been studied as a notion of *partial randomness* (for example, [5])

Proof. ($\Leftarrow$) Assume x is not s-random. Thus there exists an s-test (W_n) covering x. Define functions $m_n : 2^{<\mathbb{N}} \to \mathbb{R}$ by

$$m_n(\sigma) = \begin{cases} n2^{-\text{len}(\sigma)s-1} & \text{if } \sigma \in W_n, \\ 0 & \text{otherwise,} \end{cases}$$

and let

$$m(\sigma) = \sum_{n=1}^{\infty} m_n(\sigma).$$

All m_n and thus m are enumerable from below. Furthermore,

$$\sum_{\sigma \in 2^{<\mathbb{N}}} m(\sigma) = \sum_{\sigma \in 2^{<\mathbb{N}}} \sum_{n=1}^{\infty} m_n(\sigma) = \sum_{n=1}^{\infty} n \sum_{\sigma \in W_n} 2^{-\text{len}(w)s-1}$$

$$\leq \sum_{n=1}^{\infty} \frac{n}{2^{n+1}} \leq 1.$$

Hence m is a semimeasure and enumerable from below. By the Coding Theorem (Theorem 2.9), there exists a constant d such that for all σ,

$$K(\sigma) \leq -\log m(\sigma) + d.$$

Now let $c > 0$ be any constant. Set $k = \lceil 2^{c+d+1} \rceil$. As x is covered by (W_n) there is some $\sigma \in C_k$ and some $l \in \mathbb{N}$ with $\sigma = x{\upharpoonright}_l$. This implies

$$K(x{\upharpoonright}_l) = K(\sigma) \leq -\log m(\sigma) + d = -\log(k2^{\text{len}(\sigma)s}) + d \leq sl - c.$$

As c was arbitrary, we see that $K(x{\upharpoonright}_n)$ does not have a lower bound of the form $sn - $ constant.

($\Rightarrow$) Assume for every k there exists an l such that $K(x{\restriction}_l) < sl - k$. Let
$$W_k := \{\sigma \colon K(\sigma) \leq \mathrm{slen}(\sigma) - k\}.$$
Then (W_k) is uniformly enumerable, and each W_k covers x. Furthermore,
$$\sum_{\sigma \in W_k} 2^{-\mathrm{len}(\sigma)s} \leq \sum_{\sigma \in W_k} 2^{-K(\sigma)-k} \leq 2^{-k}.$$
Therefore, (W_k) is an s-test witnessing that x is not s-random.

As a corollary, we obtain the "fundamental theorem" of effective dimension [34, 51].

Corollary 4.13. *For any $x \in 2^{\mathbb{N}}$,*

$$\dim_H x = \underline{\kappa}(x).$$

Proof. Suppose first $\dim_H x < s$. This means x is not s-random and by Theorem 4.12, for infinitely many n, $K(x{\restriction}_n) \leq sn$. Thus $\underline{\kappa}(x) \leq s$. On the other hand, if $\underline{\kappa}(x) < s$, there exists $\varepsilon > 0$ such that $K(x{\restriction}_n) < (s-\varepsilon)n$ for infinitely many n. In particular, $sn - K(x{\restriction}_n)$ is not bounded from below by a constant, which means x is not s-random. Therefore, $\dim_H x \leq s$. $\square$

As a further corollary, we can describe the correspondence between Hausdorff dimension and Kolmogorov complexity (Theorem 4.4) as follows.

Corollary 4.14. *For every Σ_2^0 $A \subseteq 2^{\mathbb{N}}$,*

$$\dim_H A = \sup_{x \in A} \dim_H x.$$

The general idea of these results is that if a set of a rather simple nature in terms of definability, then set-wise dimension can be described as the supremum of pointwise dimensions. For more complicated sets, one can use relativization to obtain a suitable correspondence. One can relativize the definition of effective dimension with respect to an oracle $z \in 2^{\mathbb{N}}$. Let us denote the corresponding notion by $\dim_H^z$. If we relativize the notion of Kolmogorov complexity as well, we can prove a relativized version of Corollary 4.13.

Lutz and Lutz [31] were able to prove a most general set-point correspondence principle that is based completely on relativization, in which the Borel complexity does not feature anymore.

Theorem 4.15 (Lutz and Lutz). *For any set* $A \subseteq 2^{\mathbb{N}}$,

$$\dim_H A = \min_{z \in 2^{\mathbb{N}}} \sup_{x \in A} \dim_H^z x.$$

In other words, the Hausdorff dimension of a set is the minimum among its pointwise dimensions, taken over all relativized worlds.

These point-vs.-set principles are quite remarkable in that they have no direct classical counterpart. While Hausdorff dimension is *countably stable* (Proposition 3.1), stability does not extend to arbitrary unions, since singleton sets always have dimension 0.

4.4. *Effective dimension vs. randomness*

Let μ be a computable probability measure on $2^{\mathbb{N}}$. By Theorem 4.11, for a μ-random $x \in 2^{\mathbb{N}}$, the prefix-free complexity of $x{\restriction}n$ is bounded from below by $\log \mu[\![x{\restriction}n]\!]$. If we can bound $\log \mu[\![x{\restriction}n]\!]$ asymptotically from below by a bound that is linear in n, we also obtain a lower bound on the effective dimension of x. This can be seen as an effective/pointwise version of the mass distribution principle: If a point "supports" measure that decays sufficiently fast along it, we can bound its effective dimension from below.

Consider, for example, the *Bernoulli measure* μ_p on $2^{\mathbb{N}}$. Given $p \in [0, 1]$, μ_p is induced by the premeasure

$$\rho_p(\sigma^\frown 1) = \rho_p(\sigma) \cdot p, \quad \rho_p(\sigma^\frown 0) = \rho_p(\sigma) \cdot (1 - p).$$

As we will see later, if x is μ-random,

$$\lim_{n\to\infty} \frac{-\log \mu_p[\![x\lceil n]\!]}{n} = H(p), \qquad (4.7)$$

and therefore, $\dim_H x \geq H(p)$ for any μ-random x. It follows that the "decay" of μ along x is bounded by $2^{-H(p)n}$.

Does the reverse direction hold, too? If x has positive effective dimension, is x random for a suitably fast dispersing measure?

Theorem 4.16 (Reimann [38]). *Suppose $x \in 2^{\mathbb{N}}$ is such that $\dim_H x > s$. Then there exist a probability measure μ on $2^{\mathbb{N}}$ and a constant c such that for all $\sigma \in 2^{<\mathbb{N}}$,*

$$\mu[\![\sigma]\!] \leq c2^{-sn}.$$

This is an effective/pointwise version of *Frostman's lemma* [16], which in Cantor space essentially says that if a Borel set has Hausdorff dimension greater than s, it supports a measures that disperses at least as fast as 2^{-sn} (up to a multiplicative constant).

4.5. *The Shannon–McMillan–Breiman theorem*

In (4.7), we already stated a special case of the *Shannon–McMillan–Breiman Theorem*, also known as the *Asymptotic Equipartition Property* or *Entropy Rate Theorem*. It is one of the central results of information theory.

Let $T : 2^{\mathbb{N}} \to 2^{\mathbb{N}}$ be the *shift map*, defined as

$$T(x)_i = x_{i+1},$$

where y_i is the ith bit of $y \in 2^{\mathbb{N}}$. A measure μ on $2^{\mathbb{N}}$ is *shift-invariant* if for any Borel set A, $\mu T^{-1}(A) = \mu A$. A shift-invariant measure μ is *ergodic* if $T^{-1}(A) = A$ implies $\mu A = 0$ or $\mu A = 1$. Bernoulli measures μ_p are an example of shift-invariant, ergodic measures. For background on ergodic measures on sequence spaces, see [45].

Theorem 4.17. *Let μ be a shift-invariant ergodic measure on $2^{\mathbb{N}}$. Then there exists a nonnegative number h such that almost surely,*

$$\lim_{n\to\infty} -\frac{\log \mu[\![x\lceil n]\!]}{n} = h.$$

For a proof, see, for example, [45]. The number h is called the *entropy rate* of μ and also written as $h(\mu)$. It is also possible to define the entropy of the underlying $\{0, 1\}$-valued process. First, let

$$H(\mu^{(n)}) = -\sum_{\sigma\in A^n} \mu[\![\sigma]\!] \log \mu[\![\sigma]\!].$$

One can show that this is *subadditive* in the sense that

$$H(\mu^{(n+m)}) \leq H(\mu^{(n)}) + H(\mu^{(m)}),$$

which implies that

$$H(\mu) = \lim \frac{1}{n} H(\mu^{(n)})$$

exists. $H(\mu)$ is called the *process entropy* of μ. It is clear that for i.i.d. $\{0, 1\}$-valued processes, $H(\mu)$ agrees with the entropy of the distribution on $\{0, 1\}$ as defined in Sec. 2.

Entropy rate is a local measure, as it follows the behavior of a measure along a typical point, while process entropy captures the entropy over finer and finer partitions of the whole space $2^{\mathbb{N}}$. Process entropy in turn is a special case of a general definition of entropy in measure-theoretic dynamical systems, known as *Kolmogorov–Sinai entropy* (see, for example, [56]).

The Shannon–McMillan–Breiman Theorem states that the pointwise entropy rate

$$\lim_{n\to\infty} -\frac{\log \mu[\![x\lceil n]\!]}{n}$$

not only exists almost surely, but also is constant up to a set of measure zero. Furthermore, it can be shown that this constant entropy rate coincides with the process entropy $H(\mu)$. This fundamental principle has been established in much more general settings, too (e.g., for amenable groups, see [27, 35]).

4.6. *The effective SMB-theorem and dimension*

We have already seen a connection between randomness and complexity in Theorems 4.11 and 4.12.

If μ is a computable ergodic measure, this immediately connects the complexity of infinite sequences to entropy via the Shannon–McMillan–Breiman Theorem. Since μ-almost every sequence is μ-random, we obtain that almost surely,

$$\dim_H x = \liminf_{n \to \infty} \frac{K(x{\upharpoonright}n)}{n} \geq \frac{-\log \mu[\![x{\upharpoonright}n]\!]}{n} = H(\mu).$$

Zvonkin and Levin [58] and independently Brudno [3] showed that for μ-almost every sequence,

$$\lim_{n \to \infty} \frac{K(x{\upharpoonright}n)}{n} \text{ exists,}$$

which implies in turn the asymptotic compression ratio of almost every real equals the metric entropy of μ. Analyzing the proof of the SMB-Theorem by Ornstein and Weiss [35], Výugin [55] was able to establish, moreover, that for *all* μ-random sequences x,

$$\limsup_{n \to \infty} \frac{K(x{\upharpoonright}n)}{n} = H(\mu).$$

Finally, Hoyrup [20] established that the lim inf also equals $H(\mu)$, thereby showing that the SMB-Theorem is effective in the sense of Martin-Löf randomness.

Theorem 4.18 (Effective Shannon–McMillan–Breiman Theorem [20, 55]). *Let μ be a computable, ergodic, shift-invariant measure. Then, for each $x \in 2^{\mathbb{N}}$ that is random with respect to μ,*

$$\dim_H x = \lim_{n \to \infty} -\frac{\log \mu[\![x{\upharpoonright}n]\!]}{n} = H(\mu).$$

Note that effective dimension as a pointwise version of Hausdorff dimension is a natural counterpart to the pointwise nature of entropy rate, the asymptotic compression ratio (in terms of Kolmogorov complexity) forming the bridge between the two, or, to state it somewhat simplified:

$$dimension = complexity = entropy.$$

4.7. *Subshifts*

The principle above turns out to be quite persistent, even when one passes from effective dimension of sequences (points) to (classical) dimension of sets. A *subshift* of $2^{\mathbb{N}}$ is a closed subset $X \subseteq 2^{\mathbb{N}}$ invariant under the shift map on $2^{\mathbb{N}}$.

Subshifts are a *topological dynamical system*, but they can also carry an invariant measure. There is a topological variant of entropy, which for subshifts is given as

$$h_{\text{top}}(X) = \lim_{n \to \infty} \frac{\log |\{\sigma \colon |\sigma| = n \,\&\, \sigma \subset X\}|}{n},$$

that is, h_{top} measures the relative numbers of strings present in X as the length goes to infinity.

Furstenberg [17] showed that for subshifts $X \subseteq 2^{\mathbb{N}}$,

$$h_{\text{top}}(X) = \dim_H X.$$

Simpson [46] extended this result to multidimensional one- and two-sided subshifts. Furthermore, he established that for *any* such subshifts, the coincidence between entropy, complexity, and dimension is complete in the sense that

$$h_{\text{top}}(X) = \dim_H X = \underline{\kappa}(X).$$

Most recently, Day [11] further extended this identity to computable subshifts of A^G, where G is a computable amenable group with computable bi-invariant tempered Følner sequence.

Staiger [48, 49] showed that Furstenberg's result also holds for other families of sets: closed sets definable by finite automata, and for ω-powers of languages definable by finite automata.

4.8.   *Application: Eggleston's theorem*

Using the dimension-complexity-entropy correspondence, we can
also give a short proof of Eggleston's Theorem (Theorem 4.1).
We focus on the binary version. The proof for larger alphabets
is similar.

Let $p \in [0,1]$ be computable, and consider the Cantor space
equivalent of D_p,

$$\overline{D}_p = \left\{ x \in 2^{\mathbb{N}} : \lim \frac{\#(x,0,n)}{n} = p \right\},$$

where $\#(x,i,n)$ now denotes the number of occurrences of digit
i among the first n digits of x.

All μ_p-random sequences satisfy the Law of Large Numbers.
Therefore, there exists a Π^0_1 class $P \subset \overline{D}_p$ consisting of only
μ_p-random sequences. By Theorems 4.4 and 4.18

$$\dim_H D_p \geq \dim_H P = \underline{\kappa}(P) = H(\mu_p) = H(p).$$

On the other hand, for every $x \in D_p$,

$$\underline{\kappa}(x) \leq H(p),$$

as we can always construct a two-part code for $x \in D_p$, by
giving the number of 1's in $x{\restriction}n$, and then the position of $x{\restriction}n$
in a lexicographic enumeration of all binary strings of length n
with that number of 1's. As $\log_2 \binom{n}{k} \approx nH(k/n)$, this gives the
desired upper bound for $\underline{\kappa}(x)$. Therefore, by Theorem 4.3,

$$\dim_H D_p \leq H(p).$$

For noncomputable p, one can use relativized versions of the
results used above.

Cantor space vs the real line.   The reader will note that Eggleston's theorem, as originally stated, applies to the unit interval
$[0,1]$, while the version proved above is for Cantor space $2^{\mathbb{N}}$. It

is in fact possible to develop the theory of effective dimension for real numbers, too. One way to do this is to use net measures $\mathcal{M}^s$ as introduced in Sec. 3. Dyadic intervals correspond directly to cylinder sets in $2^{\mathbb{N}}$ via the finite-to-one mapping

$$x \in 2^{\mathbb{N}} \mapsto \sum_n x_n 2^{-(n+1)}.$$

If we identify a real with its dyadic expansion (which is unique except for dyadic rationals), we can speak of Kolmogorov complexity of (initial segments of) reals, etc. Furthermore, the connection between effective dimension and Kolmogorov complexity carries over to the setting in $[0, 1]$. For a thorough development of effective dimension in Euclidean (and other) spaces, see [28].

5. Multifractal Measures

A measure can have a fractal nature by the way it spreads its mass. To prove a lower bound on the Hausdorff dimension of a Cantor set, we distribute a mass uniformly along it (and appeal to the mass distribution principle, Lemma 3.2). In other words, we define a measure supported on a fractal. But a measure can exhibit fractal properties also through *the way it spreads its mass over its support*, which is not necessarily uniform. Think of earthquake distributions along fault systems. The fault systems themselves are usually of a fractal nature. Moreover, there often seems to be a nonuniformity in the distribution of earthquakes along fault lines. Some "hot spots" produce much more earthquakes than other, more quiet sections of a fault (see, for example, [21, 22, 32]). The underlying mathematical concept has become known as *multifractality of measures*. It is widely used in the sciences today. It would go far beyond the scope of this survey to introduce the various facets of multifractal measures here. Instead, we focus on one aspect of multifractality that appears to be a natural extension of the material presented in Sec. 4. The interested reader may refer to a forthcoming paper [36].

5.1. *The dimension of a measure*

Unless explicitly noted, in the following *measure* will always mean Borel probability measure on $[0, 1]$.

First, we define the *dimension of a measure*, which reflects the fractal nature of its support.

Definition 5.1 (Young [57]). Given a Borel probability measure μ on $\mathbb{R}$, let

$$\dim_H \mu = \inf\{\dim_H E \colon \mu E = 1\}.$$

For example, if we distribute a unit mass uniformly along a Cantor set $C_{m,r}$, we obtain a measure $\mu_{m,r}$ with

$$\dim_H \mu_{m,r} = -\ln(m)/\ln(r).$$

Instead by a single number, one can try to capture the way a measure spreads its mass among fractal sets by means of a distribution.

Definition 5.2 (Cutler [10]). The *dimension distribution* of a measure μ is given by

$$\mu_{\dim}[0, t] = \sup\{\mu D \colon \dim_H D \le t, \ D \subseteq \mathbb{R} \text{ Borel}\}.$$

As $\dim_H$ is countably stable (see Sec. 4.3), $\mu_{\dim}$ extends to a unique Borel probability measure on $[0, 1]$. The dimension of a measure can be expressed in terms of $\mu_{\dim}$:

$$\dim_H \mu = \inf\{\alpha \colon \mu_{\dim}[0, \alpha] = 1\}.$$

For many measures, the dimension distribution does not carry any extra information. For example, if λ is Lebesgue measure on

[0, 1], the dimension distribution is the *Dirac point measure* δ_1, where δ_α is defined as

$$\delta_\alpha A = \begin{cases} 1, & \alpha \in A, \\ 0, & \alpha \notin A. \end{cases}$$

Measures whose dimension distribution is a point measure are called *exact dimensional*. This property also applies to uniform distributions on Cantor sets.

Proposition 5.3. *Let $\mu_{m,r}$ be the uniform distribution along the Cantor set $C_{m,r}$. Then $\mu_{\dim} = \delta_\alpha$, where $\alpha = -\ln(m)/\ln(r)$.*

It is not completely obvious that $\mu_{m,r}$ has no mass on any set of Hausdorff dimension less than $-\ln(m)/\ln(r)$. One way to see this is by connecting the dimension distribution of a measure to its *pointwise dimensions*, which we will introduce next.

5.2. *Pointwise dimensions*

Definition 5.4. Let μ be a probability measure on a separable metric space. The (*lower*) *pointwise dimension* of μ at a point x is defined as

$$\delta_\mu(x) = \liminf_{\varepsilon \to 0} \frac{\log \mu B(x, \varepsilon)}{\log \varepsilon}.$$

(Here $B(x, \varepsilon)$ is the ε-ball around x.)

Of course, one can also define the upper pointwise dimension by considering $\limsup$ in place of $\liminf$, which is connected to packing measures and packing dimension similar to the way lower pointwise dimension is connected to Hausdorff measures and dimension. As this survey is focused on Hausdorff dimension, we will also focus on the lower pointwise dimensions and simply speak of "pointwise dimension" when we mean lower pointwise dimension.

We have already encountered the pointwise dimension of a measure when looking at the Shannon–McMillan–Breiman Theorem (Theorem 4.17), which says that for ergodic measures on $2^{\mathbb{N}}$ μ-almost surely the pointwise dimension is equal to the entropy of μ.

We have also seen (Corollary 4.13) that the effective dimension of a real x is its pointwise dimension with respect to the semimeasure $\widetilde{Q}(\sigma) = 2^{-K(\sigma)}$.

We can make this analogy even more striking by considering a different kind of universal semimeasure that gives rise to the same notion of effective dimension.

Recall that a semimeasure is a function $Q : \mathbb{N} \to \mathbb{R}^{\geq 0}$ such that $\sum Q(m) \leq 1$. Even when seen as a function on $2^{<\mathbb{N}}$, such a semimeasure is of a discrete nature. In particular, it does not take into account the partial ordering of strings with respect to the prefix relation. The notion of a *continuous* semimeasure does exactly that. As the prefix relation corresponds to a partial ordering of basic open sets, and premeasures are defined precisely on those sets, continuous semimeasures respect the structure of $2^{\mathbb{N}}$. Compare the following definition with the properties of a probability premeasure (4.5), (4.6).

Definition 5.5.

(i) A *continuous semimeasure* is a function $M : 2^{<\mathbb{N}} \to \mathbb{R}^{\geq 0}$ such that

$$M(\varnothing) \leq 1,$$
$$M(\sigma) \geq M(\sigma^\frown 0) + M(\sigma^\frown 1).$$

(ii) A continuous semimeasure M is *enumerable* if there exists a computable function $f : \mathbb{N} \times 2^{<\mathbb{N}} \to \mathbb{R}^{\geq 0}$ such that for all n, σ,

$$f(n, \sigma) \leq f(n+1, \sigma) \quad \text{and} \quad \lim_{n \to \infty} f(n, \sigma) = M(\sigma).$$

(iii) An enumerable continuous semimeasure M is *universal* if for every enumerable continuous semimeasure P there exists a constant c such that for all σ,

$$P(\sigma) \leq cM(\sigma).$$

Levin [58] showed that a universal semimeasure exists. Let us fix such a semimeasure $\mathbf{M}$. Similar to K, we can introduce a complexity notion based on $\mathbf{M}$ by letting

$$\mathrm{KM}(\sigma) = -\log \mathbf{M}(\sigma).$$

Some authors denote KM by KA and refer to it as *a priori complexity*. KM is closely related to K.

Theorem 5.6 (Gács [19], Uspensky and Shen [54]). *There exist constants c, d such that for all σ,*

$$\mathrm{KM}(\sigma) - c \leq \mathrm{K}(\sigma) \leq \mathrm{KM}(\sigma) + \mathrm{K}(\mathrm{len}(\sigma)) + d.$$

When writing $K(n)$, we identify n with is binary representation (which is a string of length approximately $\log(n)$). Since $\mathrm{K}(n)/n \to 0$ for $n \to \infty$, we obtain the following alternative characterization of effective dimension.

Corollary 5.7. *For any $x \in 2^{\mathbb{N}}$,*

$$\dim_H x = \liminf_{n \to \infty} \frac{\mathrm{KM}(x \lceil n)}{n} = \liminf_{n \to \infty} \frac{-\log \mathbf{M}(x \lceil n)}{n}.$$

In other words, the effective dimension of a point x in Cantor space (and hence also in $[0, 1]$) is its pointwise dimension with respect to a universal enumerable continuous semimeasure $\mathbf{M}$.

KM "smoothes" out a lot of the complexity oscillations K has. This has a nice effect concerning the characterization of random sequences. On the one hand, $\mathbf{M}$ is universal among enumerable semimeasures, hence in particular among computable probability measures. Therefore, for any $x \in 2^{\mathbb{N}}$, and any computable probability measure μ,

$$-\log \mathbf{M}(x\lceil n) - c \leq -\log \mu[\![x\lceil n]\!] \quad \text{for all } n,$$

where c is a constant depending only on μ.

On the other hand, Levin [24] showed that $x \in 2^{\mathbb{N}}$ is random with respect to computable probability measure μ if and only if for some constant d,

$$\mathrm{KM}(x\lceil n) \geq -\log \mu[\![x\lceil n]\!] - d \quad \text{for all } n.$$

Thus, along μ-random sequences, $-\log \mathbf{M}$ and $-\log \mu$ differ by at most a constant. This gives us the following.

Proposition 5.8. *Let μ be a computable probability measure on $2^{\mathbb{N}}$. If $x \in 2^{\mathbb{N}}$ is μ-random, then*

$$\dim_H x = \delta_\mu(x).$$

Given a measure μ and $\alpha \geq 0$, let

$$D_\mu^\alpha = \{x \colon \delta_\mu(x) \leq \alpha\}.$$

If μ is computable, then since $\mathbf{M}$ is a universal semimeasure,

$$D_\mu^\alpha \subseteq \{x \colon \dim_H x \leq \alpha\}.$$

Let us denote the set on the right hand side by $\dim_{\leq \alpha}$.

Theorem 5.9 (Cai and Hartmanis [4], Ryabko [41]). *For any $\alpha \geq 0$,*

$$\dim_H(\dim_{\leq \alpha}) = \alpha.$$

It follows that $\dim_H D_\mu^\alpha \leq \alpha$. This was first shown, for general μ, by Cutler [8, 9]. Cutler also characterized the dimension distribution of a measure through its pointwise dimensions.

Theorem 5.10 (Cutler [8, 9]). *For each $t \geq 0$,*

$$\mu_{\dim}[0, t] = \mu(D_\mu^t).$$

By the above observation, for computable measures, we can replace D_μ^α by $\dim_{\leq t}$. The theorem explains why $\dim_H \mu_{m,r} = \delta_{-\ln(m)/\ln(r)}$: Since the mass is distributed uniformly over the set $C_{m,r}$, $\delta_{\mu_{m,r}}(x)$ is almost surely constant.

5.3. *Multifractal spectra*

We have seen in the previous section that uniform distributions on Cantor sets result in dimension distributions of the form of Dirac point measures δ_α. We can define a more "fractal" measure by biasing the distribution process. Given a probability vector $\vec{p} = (p_1, \ldots, p_m)$, the measure $\mu_{m,r,\vec{p}}$ is obtained by splitting the mass in the interval construction of $C_{m,r}$ not uniformly, but according to $\vec{p}$. The resulting measure is still exact dimensional (see [10] — this is similar to the SMB Theorem). However, the measures $\mu_{m,r,\vec{p}}$ exhibit a rich fractal structure when one considers the *dimension spectrum* of the pointwise dimensions (instead of its dimension distribution).

Definition 5.11. The (*fine*) *multifractal spectrum* of a measure μ is defined as

$$f_\mu(\alpha) = \dim_H \{x \colon \delta_\mu(x) = \alpha\}.$$

For the measures $\mu_{m,r,\vec{p}}$, it is possible to compute the multifractal spectrum using the *Legendre transform* (see [15]). It is a function continuous in α with $f_\mu(\alpha) \leq \alpha$ and maximum $\dim_H \mu$. In a certain sense, the universal continuous semimeasure $\mathbf{M}$ is a "perfect multifractal", as, by the Cai–Hartmanis result, $f_\mathbf{M}(\alpha) = \alpha$ for all $\alpha \in [0, 1]$.

Moreover, the multifractal spectrum of a computable $\mu = \mu_{m,r,\vec{p}}$ can be expressed as a "deficiency" of multifractality against $\mathbf{M}$.

Theorem 5.12 (Reimann [36]). *For a computable measure* $\mu = \mu_{m,r,\vec{p}}$, *it holds that*

$$f_\mu(\alpha) = \dim_H \left\{ x \colon \frac{\dim_H x}{\dim_\mu x} = \alpha \right\}.$$

Here $\dim_\mu$ denotes the *effective Billingsley dimension* of x with respect to μ (see [37]).

References

[1]   V. Becher, J. Reimann and T. A. Slaman, Irrationality exponent, Hausdorff dimension and effectivization, *Monatsh. Math.* **185**(2) (2018) 167–188.

[2]   P. Billingsley, *Ergodic Theory and Information* (Wiley, New York, 1965).

[3]   A. Brudno, Entropy and the complexity of the trajectories of a dynamic system, *Trudy Moskov. Mat. Obshchestva* **44** (1982) 124–149.

[4]   J.-Y. Cai and J. Hartmanis, On Hausdorff and topological dimensions of the Kolmogorov complexity of the real line, *J. Comput. System Sci.* **49**(3) (1994) 605–619.

[5]   C. Calude, L. Staiger and S. A. Terwijn, On partial randomness. *Ann. Pure Appl. Logic* **138**(1–3) (2006) 20–30.

[6]   G. J. Chaitin, A theory of program size formally identical to information theory, *J. ACM* **22**(3) (1975) 329–340.

[7]   T. M. Cover, P. Gàcs and R. M. Gray, Kolmogorov's contributions to information theory and algorithmic complexity, *Ann. Probab.* **17**(3) (1989) 840–865.

[8]   C. D. Cutler, A dynamical system with integer information dimension and fractal correlation exponent, *Commun. Math. Phys.* **129**(3) (1990) 621–629.

[9]   C. D. Cutler, Measure disintegrations with respect to σ-stable monotone indices and pointwise representation of packing dimension, in *Proc. 1990 Measure Theory Conf. Oberwolfach. Supplemento Ai Rendiconti del Circolo Mathematico di Palermo, Ser. II*, Vol. 28, (1992), pp. 319–340.

[10]   C. D. Cutler, A review of the theory and estimation of fractal dimension, in *Dimension Estimation and Models* (World Scientific River Edge, NJ, 1993), pp. 1–107.

[11]   A. Day, Algorithmic randomness for amenable groups, Draft (September 2017) arXiv: 1802.03831.

[12]   A. Day and J. Miller, Randomness for non-computable measures, *Trans. Amer. Math. Soc.* **365** (2013) 3575–3591.

[13]   R. G. Downey and D. R. Hirschfeldt, *Algorithmic Randomness and Complexity* (Springer, 2010).

[14]   H. G. Eggleston, The fractional dimension of a set defined by decimal properties, *Quart. J. Math., Oxford Ser.* **20** (1949) 31–36.

[15] K. Falconer, *Fractal Geometry*, 2nd edn. (Wiley, 2003).

[16] O. Frostman, Potentiel d'équilibre et capacité des ensembles avec quelques applications à la théorie des fonctions, *Meddel. Lunds. Univ. Mat. Sem.* **3** (1935) 1–118.

[17] H. Furstenberg, Disjointness in ergodic theory, minimal sets, and a problem in Diophantine approximation, *Theory Comput. Syst.* **1**(1) (1967) 1–49.

[18] P. Gács, Exact expressions for some randomness tests, *Zeit. Math. Logik Grundlagen Math.* **26**(25–27) (1980) 385–394.

[19] P. Gács, On the relation between descriptional complexity and algorithmic probability, *Theoret. Comput. Sci.* **22**(1–2) (1983) 71–93.

[20] M. Hoyrup, The dimension of ergodic random sequences, in *STACS'12 (29th Symposium on Theoretical Aspects of Computer Science)*, **44** (2011) 567–576.

[21] Y. Kagan, Fractal dimension of brittle fracture, *J. Nonlinear Sci.* **1** (1991) 1–16.

[22] Y. Kagan and L. Knopoff, Spatial distribution of earthquakes: The two-point correlation function, *Geophys. J. Int.* **62**(2) (1980) 303–320.

[23] A. I. Khinchin, *Mathematical Foundations of Information Theory* (Dover Publications, Inc., New York, 1957).

[24] L. A. Levin, On the notion of a random sequence, *Soviet Math. Dokl* **14**(5) (1973) 1413–1416.

[25] L. A. Levin, Some theorems on the algorithmic approach to probability theory and information theory: (1971 Dissertation directed by A.N. Kolmogorov), *Ann. Pure Appl. Logic* **162**(3) (2010) 224–235.

[26] M. Li and P. Vitányi, *An Introduction to Kolmogorov Complexity and Its Applications*, 3rd edn. (Springer-Verlag, New York, 2008).

[27] E. Lindenstrauss, Pointwise theorems for amenable groups, *Invent. Math.* **146**(2) (2001) 259–295.

[28] J. Lutz and E. Mayordomo, Dimensions of points in self-similar fractals, *SIAM J. Comput.* **38**(3) (2008) 1080–1112.

[29] J. H. Lutz, Dimension in complexity classes, In *Proc. Fifteenth Annual IEEE Conf. Computational Complexity* (IEEE Computer Society, 2000), pp. 158–169.

[30] J. H. Lutz, The dimensions of individual strings and sequences, *Inform. Comput.* **187**(1) (2003) 49–79.

[31] J. H. Lutz and N. Lutz, Algorithmic information, plane Kakeya sets, and conditional dimension, in *34th Symp. Theoretical Aspects of Computer Science, STACS 2017*, March 8–11, 2017, Hannover, Germany, (2017) 1–13.

[32] B. B. Mandelbrot, Multifractal measures, especially for the geophysicist, *Pure Appl. Geophys.* **131** (1989) 5–42.

[33] P. Martin-Löf, The definition of random sequences, *Inform Control* **9**(6) (1966) 602–619.

[34] E. Mayordomo, A Kolmogorov complexity characterization of constructive Hausdorff dimension, *Inform. Process. Lett.* **84**(1) (2002) 1–3.

[35] D. Ornstein and B. Weiss, The Shannon–McMillan–Breiman theorem for a class of amenable groups, *Israel J. Math.* **44**(1) (1983) 53–60.

[36] J. Reimann, Effective mutlifractal spectra, in preparation.

[37] J. Reimann, Computability and fractal dimension, Ph.D. thesis, University of Heidelberg (2004).

[38] J. Reimann, Effectively closed sets of measures and randomness, *Ann. Pure Appl. Logic* **156** (2008) 170–182.

[39] J. Reimann and T. Slaman, Measures and their random reals, *Trans. Amer. Math. Soc.* **367**(7) (2015) 5081–5097.

[40] C. A. Rogers, *Hausdorff Measures* (Cambridge University Press, 1970).

[41] B. Y. Ryabko, Coding of combinatorial sources and Hausdorff dimension, *Dokl. Akad. Nauk SSSR* **277**(5) (1984) 1066–1070.

[42] B. Y. Ryabko, Noise-free coding of combinatorial sources, Hausdorff dimension and Kolmogorov complexity, *Problemy Peredachi Informatsii* **22**(3) (1986) 16–26.

[43] C.-P. Schnorr, *Zufälligkeit und Wahrscheinlichkeit. Eine algorithmische Begründung der Wahrscheinlichkeitstheorie* (Springer-Verlag, Berlin, 1971).

[44] C.-P. Schnorr, Process complexity and effective random tests, *J. Comput. System Sci.* **7** (1973) 376–388.

[45] P. C. Shields, *The Ergodic Theory of Discrete Sample Paths*, Graduate Studies in Mathematics, Vol. 13 (American Mathematical Society, Providence, RI, 1996).

[46] S. G. Simpson, Symbolic dynamics: entropy = dimension = complexity, *Theory Comput. Syst.* **56**(3) (2015) 527–543.

[47] L. Staiger, Complexity and entropy, in J. Gruska and M. Chytil (eds.), *Mathematical Foundations of Computer Science 1981: Proc. 10th Symp. Štrbské Pleso*, Czechoslovakia August 31 – September 4, 1981, (Springer, Berlin, 1981), pp. 508–514.

[48] L. Staiger, Combinatorial properties of the Hausdorff dimension, *J. Statist. Planning Infer.* **23**(1) (1989) 95–100.

[49] L. Staiger, Kolmogorov complexity and Hausdorff dimension, *Inform. Comput.* **103**(2) (1993) 159–194.

[50] L. Staiger, A tight upper bound on Kolmogorov complexity and uniformly optimal prediction, *Theory Comput. Syst.* **31**(3) (1998) 215–229.

[51]   L. Staiger, Constructive dimension equals Kolmogorov complexity, *Inform. Process. Lett.* **93**(3) (2005) 149–153.

[52]   L. Staiger, The Kolmogorov complexity of infinite words, *Theor. Comput. Sci.* **383**(2–3) (2007) 187–199.

[53]   S. A. Terwijn, Complexity and randomness, *Rend. Semin. Mat., Torino* **62**(1) (2004) 1–37.

[54]   V. A. Uspensky and A. Shen, Relations between varieties of Kolmogorov complexities, *Math. Syst. Theory* **29** (1996) 271–292.

[55]   V. V. V'yugin, Ergodic theorems for individual random sequences, *Theoret. Comput. Sci.* **207**(2) (1998) 343–361.

[56]   P. Walters, *An Introduction to Ergodic Theory*, Graduate Texts in Mathematics, Vol. 79, (Springer, New York, 1982).

[57]   L.-S. Young, Dimension, entropy and Lyapunov exponents, *Ergodic Theory Dynam. Syst.* **2**(1) (1982) 109–124.

[58]   A. K. Zvonkin and L. A. Levin, The complexity of finite objects and the development of the concepts of information and randomness by means of the theory of algorithms, *Russian Math. Surv.* **25**(6) (1970) 83–124.

Part III
Computable Structure Theory

Chapter 5

Computable Reducibility for Cantor Space

Russell Miller

Department of Mathematics, Queens College —
City University of New York, 65-30 Kissena Blvd.
Flushing, NY 11367, USA
Russell.Miller@qc.cuny.edu

We examine various versions of Borel reducibility on equivalence relations on the Cantor space 2^ω, using reductions given by Turing functionals on the inputs $A \in 2^\omega$. In some versions, we vary the number of jumps of A which the functional is allowed to use. In others, we do not require the reduction to succeed for all elements of the Cantor space at once, but only when applied to arbitrary finite or countable subsets of 2^ω. In others we allow an arbitrary oracle set in addition to the inputs. All of these versions, inspired largely by work on computable reducibility on equivalence relations on ω, combine to yield a rich set of options for evaluating the precise level of difficulty of a Borel reduction, or the reasons why a Borel reduction may fail to exist.

1. Introduction to Reducibility

The subject of reducibility of equivalence relations has bifurcated in recent years. Much early work was devoted to the topic of *Borel reducibility*, concerning equivalence relations on the Cantor space 2^ω of all subsets of the set ω of natural numbers. More recently, computability theorists have adapted the notion in order to address equivalence relations on ω itself. The principal notion here has borne several names, after being arrived at independently by several researchers; we find *computable reducibility* to be the most natural of these. The purpose of this chapter is to hybridize the two: we will present natural reducibilities on equivalence relations on 2^ω, some stronger than Borel reducibility and some weaker, through which the ideas cultivated by computability theorists can be applied.

Suppose that E and F are equivalence relations on the domains S and T, respectively. A *reduction* of E to F is a function $g : S \to T$ satisfying the property:

$$(\forall x_0, x_1 \in S) \; [x_0 \; E \; x_1 \iff g(x_0) \; F \; g(x_1)].$$

The point is that, if one has the ability to compute g and to decide the relation F, then one can decide E as well. Thus E may be considered to be "no harder to decide" than F, at least modulo the difficulty of computing g. In practice, the domains S and T are usually equal, with the cases $S = T = 2^\omega$ and $S = T = \omega$ being by far the most widely studied. (It is clear that, in order for a reduction to exist, F must have at least as many equivalence classes as E, and so a situation where T has lower cardinality than S would be of interest only if E partitions S into relatively few classes.)

The crucial question in this definition is how much computational power one allows the function g to have. Of course, whenever F has at least as many equivalence classes as E, some reduction g of E to F must exist, unless one refuses to allow use of the Axiom of Choice. In set theory, the standard preference has been to require g to be a Borel function, in which case S and T should be Polish spaces. We say that E is *Borel-reducible to F*, and write $E \leq_B F$, if a Borel reduction from E to F exists. Research here has focused on the situation $S = T = 2^\omega$. Computability theorists seized on the same notion under the requirement that $g : \omega \to \omega$ be Turing-computable, in which case one needs $S = T = \omega$. In this situation, by analogy, we say that E is *computably reducible to* F, and write $E \leq_0 F$, if a Turing-computable reduction from E to F exists.

The notation $E \leq_T F$ seemed inadvisable here, as it already denotes the existence of a Turing reduction from the set E to the set F. One does have $E \leq_T F$ whenever $E \leq_0 F$, but the converse fails. In fact, $E \leq_0 F$ implies that there is a many-to-one reduction from E to F as sets, and for this reason $\leq_0$ has sometimes been called m-reducibility, as well as Fokina–Friedman reducibility. We prefer the term which actually describes the complexity of the reduction involved: E is computably reducible to F if there is a reduction which is a computable function, just as Borel-reducibility requires a reduction which is a Borel function. Certain other work, such as [13], has used generalizations such as $\boldsymbol{d}$-computable reductions, i.e., reductions that are $\boldsymbol{d}$-computable functions on ω, for some Turing degree $\boldsymbol{d}$. We will use $\leq_{\boldsymbol{d}}$ to denote this $\boldsymbol{d}$-computable reducibility, and hence $\leq_0$ to denote Turing-computable reducibility (on equivalence relations on ω). Likewise, in [7], Fokina, Friedman and Törnquist studied "effectively Borel" reductions on equivalence relations on 2^ω, by which they meant reductions which are Δ^1_1. These will not arise here.

The goal of this chapter is to enable the transfer of many of the computability-theoretic results from articles such as [3, 13] to the context of Borel reducibility, i.e., of equivalence relations on Cantor space. We employ the very natural notion of a computable function on Cantor space, which yields the following definition.

Definition 1.1. Let E and F be equivalence relations on 2^ω. A *computable reduction* of E to F is a reduction $g : 2^\omega \to 2^\omega$ given by a computable function Φ (that is, an oracle Turing functional) on the reals involved:

$$(\forall A \in 2^\omega)(\forall x \in \omega)\ \chi_{g(A)}(x) = \Phi^A(x).$$

If such a reduction exists, then E is *computably reducible to* F, denoted $E \leq_0 F$.

So we require, for all reals A_0 and A_1, that $A_0 \; E \; A_1$ if and only if $\Phi^{A_0} \; F \; \Phi^{A_1}$. It is implicit here that, for every A, Φ^A should be a total function from ω into $\{0, 1\}$, equal to the characteristic function of $g(A)$, hence regarded as the set $g(A)$ itself. A computable reduction $E \leq_0 F$ yields a Medvedev reduction from the subset F of $(2^\omega)^2$ to the subset E, but not conversely.

The notion of a computable reduction on equivalence relations on Cantor space may be generalized using the jump operator, which maps each set $A \subseteq \omega$ to its *jump* $A' = \{\langle e, x \rangle : \Phi_e^A(x) \text{ halts}\}$. (Here $\langle x, y \rangle = \frac{(x+y+1)(x+y)+2y}{2}$ is the standard pairing function on ω.) This is best seen as representing the Halting Problem relative to the set A.

Definition 1.2. Let E and F be equivalence relations on 2^ω. A *jump-reduction* of E to F is a reduction $g : 2^\omega \to 2^\omega$ given by a computable function Φ (that is, an oracle Turing functional) on the jumps of the reals involved:

$$(\forall A \in 2^\omega) \; g(A) = \Phi^{(A')}.$$

Likewise, if for some computable ordinal α and some Φ, we have $g(A) = \Phi^{(A^{(\alpha)})}$, then the reduction g is said to be an *α-jump reduction*. We write $E \leq_\alpha F$ if such a reduction exists, thus generalizing the notation $E \leq_0 F$ above.

Another refinement of reducibilities on equivalence relations was introduced by Ng and the author in [13]. Studying equivalence relations on ω, they defined *finitary reducibilities*. In the context of Cantor space, it is natural to extend their notion to all cardinals $\mu < 2^\omega$ (as indeed was suggested in their article), yielding the following definitions.

Definition 1.3. For equivalence relations E and F on domains S and T, and for any cardinal $\mu < |S|$, we say that a function $g : S^\mu \to T^\mu$ is a *μ-ary reduction* of E to F if, for every

$\vec{x} = (x_\alpha)_{\alpha \in \mu} \in S^\mu$, we have

$$(\forall \alpha < \beta < \mu) \: [x_\alpha \: E \: x_\beta \iff g_\alpha(\vec{x}) \: F \: g_\beta(\vec{x})],$$

where $g_\alpha : S^\mu \to T$ are the component functions of $g = (g_\alpha)_{\alpha < \mu}$. For limit cardinals μ, a related notion applies with $<\mu$ in place of μ: a function $g : S^{<\mu} \to T^{<\mu}$ which restricts to a ν-ary reduction of E to F for every cardinal $\nu < \mu$ is called a $(<\mu)$-*ary reduction*. (For $\mu = \omega$, an ω-ary reduction is a *countable reduction*, and a $(<\omega)$-ary reduction is a *finitary reduction*.)

When $S = T = 2^\omega$ and the μ-ary reduction g is computable, we write $E \leq_0^\mu F$, with the natural adaptation $E \leq_\alpha^\mu F$ for α-jump μ-ary reductions. Likewise, when a $(<\mu)$-ary reduction g is α-jump computable, we write $E \leq_\alpha^{<\mu} F$. It is crucial in this definition that we require uniformity: it is not sufficient for the individual ν-ary reductions to be α-jump computable for all $\nu < \mu$. Rather, the function $g : S^{<\mu} \to T^{<\mu}$ itself must be α-jump computable.

When $\alpha > 0$, it is important to note that $\Phi^{((\vec{x})^{(\alpha)})}$ is required to equal $g(\vec{x})$; this allows more information in the oracle than it would if we had required $\Phi^{(x_0^{(\alpha)} \oplus x_1^{(\alpha)} \oplus \cdots)} = g(\vec{x})$, with the jumps of the individual inputs taken separately.

Notice that computable μ-ary reductions only make sense when $\mu \leq \omega$, as we have no method for running an oracle Turing machine with uncountably many reals in the oracle. For μ-ary reductions with $\mu \leq \omega$, the oracle is a single real whose columns are the μ-many inputs to the reduction. In a $(<\omega)$-ary reduction, with an input $\vec{A} \in (2^\omega)^n$, the oracle is officially equal to $\{n\} \oplus A_0 \oplus \cdots \oplus A_{n-1} \oplus \emptyset \oplus \emptyset \oplus \cdots$, meaning that the oracle does specify the size of its tuple. (However, we usually gloss over this issue and just write $A = A_0 \oplus \cdots \oplus A_{n-1}$ as the oracle.)

Beyond the basic intention of introducing computable reductions on equivalence relations on Cantor space, the principal goal of this chapter is to use these ideas to analyze the Borel reductions and nonreductions at the lowest levels in the $\leq_B$-hierarchy of Borel equivalence relations. Where Borel reductions exist, we wish to determine whether they are computable, and, if not, how many jumps away from computable they may be. Where no Borel reduction exists, we ask how close we can come to a full Borel reduction, using the notions of finitary and countable reducibility, and possibly allowing jump-reductions as well. Often our results here are based on existing results from the context of equivalence relations on ω, from sources including [3, 7, 13], as well as the relevant articles [1, 2, 5, 6, 8, 9, 12].

Intuitively, results in this direction suggest where the obstacles to Borel reductions lie. If $E \not\leq_0^\omega F$, then there is an obstruction at the level of computability, i.e., at a syntactic level. If, for all α, $E \not\leq_\alpha^\omega F$, then this obstruction extends upwards through all possible quantifier complexities, meaning that the reduction is stymied by extremely strong syntactic difficulties. On the other hand, if $E \not\leq_B F$ but $E \leq_0^\omega F$, then the obstruction is not syntactic, but rather has to do with the size of the continuum: it is simply not possible to perform the entire reduction uniformly on so many elements of the field of E. If $E \not\leq_B F$ and the best possible countable reduction is $E \leq_4^\omega F$, say, then 4-quantifier complexity is necessary, but beyond that the obstructions are cardinality-related rather than syntactic. Later in this chapter, we will also encounter situations where $E \leq_0^{<\omega} F$ but $E \not\leq_0^\omega F$. We will even meet natural cases where $E \leq_0^3 F$ but $E \not\leq_0^4 F$. These suggest more subtle syntactic issues.

A future goal is to begin to transfer the large body of knowledge about computable structures into the more general context of arbitrary countable structures on the domain ω. In particular, equivalence relations such as isomorphism, bi-embeddability, and elementary equivalence have been analyzed extensively for many classes of computable structures, and may be more broadly considered as equivalence relations on the corresponding classes of countable structures, often using jumps and/or finitary or countable reducibility.

2. Analyzing the Basic Borel Theory

The following diagram appeared in [3]. It shows all Borel reducibilities among the main Borel equivalence relations at the base of the hierarchy of Borel reducibility. (Their definitions will be given as we examine them in this section.)

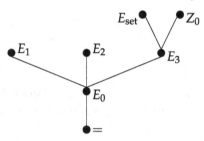

The surprising fact about this diagram is that the relation E_0 of finite difference, defined by

$$A \ E_0 \ B \iff |(A - B) \cup (B - A)| < \omega,$$

is the second-least Borel equivalence relation under $\leq_B$, even when one includes relations not shown here. Indeed, the broader principle, known as the *Glimm–Effros dichotomy*, is that for every Borel relation E on 2^ω to which equality is Borel-reducible, either E is Borel-equivalent to equality or else $E_0 \leq_B E$. That is, no Borel equivalence relation sits strictly between $=$ and E_0 under $\leq_B$, nor can it be above $=$ and Borel-incomparable with E_0. This has been extended in various ways, for instance to other Polish spaces, but here we will content ourselves with this version. An excellent description of the broader topic appears in Gao's book [8, Chapter 6]. Glimm was the first to make significant progress in this area, in [10]; Effros followed in [4] with work showing that E_0 embeds continuously into every non-smooth equivalence relation (on a Polish space) defined by the action of a Polish group; and Harrington, Kechris and Louveau gave the effective version that has come to be the usual meaning of "Glimm–Effros dichotomy" in the present day (see [11]).

As a first example of our style of investigation, therefore, we start with the fact that $= <_B E_0$. Of course, this is really two statements: there is a Borel reduction in one direction, but not in the other. The Borel reduction is in fact a computable reduction, given immediately by the program

$$\Phi^A(\langle n, k \rangle) = \begin{cases} 1 & \text{if } n \in A, \\ 0 & \text{if not.} \end{cases}$$

That is, each real A is transformed into the real whose columns are all just copies of A, so that any difference between two sets A and B becomes an infinite difference between Φ^A and Φ^B.

Thus $= <_0 E_0$: the strictness is clear, since there is not even any Borel reduction in the reverse direction. Nevertheless, we can analyze the reverse direction using finitary and countable reductions, and we find that there is a 2-jump countable reduction Γ from E_0 to $=$, given as follows. Γ has as its oracle the second jump of the join of countably many sets $A_0, A_1, \ldots$, and needs to output corresponding sets $B_0, B_1, \ldots$ for which $B_i = B_j$ if and only if $A_i \ E_0 \ A_j$. First it defines $B_0 = \emptyset$. Next, with the second jump of the join, Γ can ask whether $A_0 \triangle A_1$ is finite; if so, it sets $B_1 = \emptyset$ as well, while if not, it makes $B_1 = \{1\}$. In general, for each j, Γ asks its oracle whether A_{j+1} has finite difference with any of $A_0, \ldots, A_j$. If so, then it makes B_{j+1} equal to that B_i (noting that if there is more than one such i, then all such B_i are already equal); while if not, it makes $B_{j+1} = \{j + 1\}$. Clearly this is a countable reduction, as desired.

Of course, we also want to know whether this is the best we can do. There is no way to address larger cardinalities than ω with a Turing functional, even if CH fails, but one might hope for an n-ary or countable 1-jump or computable reduction. We now show that a 2-jump countable reduction is the best possible. Let A be any 1-generic set, and Γ be any Turing functional. Writing $A = A_0 \oplus A_1$, we know from 1-genericity that A_0 and A_1 have infinite symmetric difference, so if Γ is to be a 1-jump binary reduction of E_0 to $=$, then $\Gamma^{A'} = \Gamma^{(A_0 \oplus A_1)'}$ must be the characteristic function of a set $B_0 \oplus B_1$ with $B_0 \neq B_1$. Fix an n in the difference, say with $n \in B_0 - B_1$, so $2n \in B$ and $(2n + 1) \notin B$. Then there is an initial segment $\sigma \subseteq A'$ such $\Gamma^\sigma(2n) = 1$ and $\Gamma^\sigma(2n + 1) = 0$. For every i with $\sigma(i) = 1$, fix some $\rho_i \subseteq A$ for which $\Phi_i^{\rho_i}(i) \downarrow$, and for every i with $\sigma(i) = 0$, fix some $\rho_i \subseteq A$ such that $(\forall \tau \supseteq \rho_i) \; \Phi_i^\tau(i) \uparrow$. (Such a ρ_i must exist, by the 1-genericity of A.) Now let $\rho = \bigcup_{i < |\sigma|} \rho_i$, which is a finite initial segment of A. By our choice of each ρ_i, we see that every $C \supseteq \rho$ will have $C' \upharpoonright |\sigma| = \sigma$, and hence $\Gamma^{C'}(2n) = \Gamma^\sigma(2n) = 1$ and $\Gamma^{C'}(2n + 1) = \Gamma^\sigma(2n + 1) = 0$. However, there are many $C \supseteq \rho$ such that, with $C = C_0 \oplus C_1$, we will have $C_0 \, E_0 \, C_1$: for instance, just let $C = \rho\,\widehat{}\,00000\cdots$. But for these C we will still have $\Gamma^{C'} = D_0 \oplus D_1$ with $n \in D_0 - D_1$, so $D_0 \neq D_1$, and thus Γ does not compute a 1-jump binary reduction from E_0 to $=$.

E_1 is the equivalence relation defined on 2^ω by sets being equal on all but finitely many of their columns:

$$A_0 \, E_1 \, A_1 \iff \forall^\infty m \; A_0^m = A_1^m,$$

where A^m represents the mth column $\{k : \langle m, k \rangle \in A\}$ of A. From the Borel theory, we know that $E_0 <_B E_1$. Indeed, the Borel reduction is a computable reduction: with $\Phi^A = \{\langle m, 0 \rangle : m \in A\}$, we have $A_0 \, E_0 \, A_1$ if and only if $\Phi^{A_0} \, E_1 \, \Phi^{A_1}$. In the reverse direction, no Borel reduction exists, but there is a computable countable reduction.

Proposition 2.1. $E_1 \leq_0^\omega E_0$. *That is, there is a computable countable reduction Γ from E_1 to E_0.*

Proof. The input to Γ is a real $A = \bigoplus_n A_n$, viewed as an ω-tuple of reals A_n, and the mth column of A_n is now written A_n^m. (In fact, A_n itself is the nth column of A, but this is not a concern.) The real computed by Γ^A will be $B = \bigoplus_n B_n$, so we need $A_n \, E_1 \, A_{n'}$ if and only if $B_n \, E_0 \, B_{n'}$.

The idea behind Γ is that, if A_m and A_n differ on a given column, that column should contribute a single element to the symmetric difference $B_m \triangle B_n$. The $\langle m, n \rangle$th column $B_i^{\langle m,n \rangle}$ of each set B_i is devoted to collecting these single elements so as to satisfy the requirement

$$\mathcal{R}_{m,n}: \quad B_m^{\langle m,n \rangle} \; E_0 \; B_n^{\langle m,n \rangle} \iff A_m \; E_1 \; A_n.$$

On the sets B_i (for $i \notin \{m, n\}$), Γ will try to keep $B_i^{\langle m,n \rangle}$ equal to one of $B_m^{\langle m,n \rangle}$ and $B_n^{\langle m,n \rangle}$, depending on which of A_m and A_n may be E_1-equivalent to A_i. (For simplicity, all columns numbered $\langle m, n \rangle$ with $m \geq n$ are empty in every set B_i.)

At each stage s of the computation, for every c and i, Γ decides whether the sth element $\langle c, s \rangle$ of the cth column should belong to B_i. Suppose $c = \langle m, n \rangle$. If $m \geq n$, or if $c > s$, the answer is automatically no for all i, as noted above. Otherwise, we use the oracle to compare A_m with A_n up to s, looking for columns on which they have just now been discovered to differ, as defined here.

- For each $i < s$, if $A_m^i {\restriction} s = A_n^i {\restriction} s$ but $s \in A_m^i \triangle A_n^i$, then we say that on column i, A_m and A_n *differ at* s, meaning that s is the first stage at which it was established that these columns are distinct.
- If $A_m^s {\restriction} (s+1) \neq A_n^s {\restriction} (s+1)$, then we say that on column s, A_m and A_n *differ at* s.

So at stage s we "catch up" on one new column, and keep an eye on all previous ones.

Let s' be the greatest stage $< s$ at which we acted on behalf of the $\langle m, n \rangle$th column. If there exists a partition $P \sqcup Q$ of the set $\{0, 1, \ldots, n\}$ with $m \in P$ and $n \in Q$ such that, for every $p \in P$ and $q \in Q$, there is some t with $s' < t \leq s$ and some column $\leq s$ on which A_p and A_q differ at stage t, then we fix such a partition and define $\langle \langle m, n \rangle, s \rangle$ to lie in every B_p with $p \in P$ (including B_m), but not in any other B_i (hence not in B_n). Thus we have created a new difference (of one new element) between $B_m^{\langle m,n \rangle}$ and $B_n^{\langle m,n \rangle}$, and we say that we *acted on behalf of* $\mathcal{R}_{m,n}$ at this stage. If no such partition exists, then $\langle\langle m, n\rangle, s\rangle$ does not lie in any set B_i.

As a matter of convention, we also deem ourselves to have acted (for the first time) on behalf of $\mathcal{R}_s$ at stage s, although we did nothing specific to help satisfy it. This completes the construction.

To see that Γ is a countable reduction, consider any A and any $c = \langle m, n\rangle$ with $m < n$. If A_m and A_n lie in distinct E_1-classes, then we can partition $\{0, \ldots, n\}$ into two classes P and Q, with $m \in P$ and $n \in Q$, such that whenever $A_i \, E_1 \, A_j$ for $i < j \leq n$, then $(i \in P \iff j \in P)$. Hence, if $i \in P$ and $j \in Q$, there will be infinitely many stages at which A_i and A_j differ on some column. Therefore, there will be infinitely many stages at which we discover either this partition $P \sqcup Q$ or some other one and act on behalf of column c, putting an element into B_m but not into B_n. Thus B_m and B_n lie in distinct E_0-classes, as desired.

On the other hand, if $A_m \, E_1 \, A_n$, then these two sets differ on only finitely many columns. Therefore, there exists some s_0 such that we never act on behalf of column c after that stage. Now by stage s_0 we have only defined B_m and B_n up to their s_0th element of the dth column (that is, up to $\langle d, s_0\rangle$), and this only for $d \leq s_0$, so there are only finitely many differences between B_m and B_n so far. After stage s_0, we may add some further differences between them, on behalf of columns $d \neq c$, but only if the previous action on behalf of column d was at a stage $s' \leq s_0$, since otherwise our m and n would have to lie either both in P or both in Q, according to the rules in the construction. However, only finitely many columns (those $\leq s_0$) ever had an action taken on their behalf before stage s_0, and so only finitely many more differences will ever be added to B_m and B_n. Hence $B_m \, E_0 \, B_n$ as desired. $\qquad\square$

Next we consider E_3, the equivalence relation which holds of reals A and B if and only if, for all k, the kth columns of the two satisfy $A^k \, E_0 \, B^k$. One quickly sees a computable reduction from E_0 to E_3: just let $\Phi^A = A \oplus A \oplus \cdots$. It is known that $E_1 \not\leq_B E_3$, and in fact the two are Borel-incomparable, but there is a countable computable reduction from E_1 to E_3, since Proposition 2.1 yields $E_1 \leq_0^\omega E_0 \leq_0 E_3$. From E_3 to E_0, no Borel reduction exists, and the best we can do is to show $E_3 \leq_2^\omega E_0$. Indeed, we have $E_3 \leq_2^\omega =$, with the two-jump countable reduction to equality given by the following functional Γ. With oracle $A'' = (\bigoplus_n A_n)''$, Γ makes every $\langle\langle m, n\rangle, k\rangle \notin B_m$, and defines:

- if $(A_n^k \triangle A_m^k)$ is finite, then for all p, $\langle\langle m, n\rangle, k\rangle \notin B_p$.
- if $(A_n^k \triangle A_m^k)$ is infinite, then $\langle\langle m, n\rangle, k\rangle \in B_n$. (Here we have found that A_n and A_m are in distinct E_3-classes, so $\langle\langle m, n\rangle, k\rangle$ establishes $B_n \neq B_m$.)

  In this case, for each $p \notin \{m, n\}$, we "do no harm". Γ checks whether $(A_n^k \triangle A_p^k)$ is finite. If it is, then Γ puts $\langle\langle m, n\rangle, k\rangle \in B_p$ (since possibly A_p E_3 A_n, while definitely A_p and A_m are in distinct E_3-classes). If $(A_n^k \triangle A_p^k)$ is infinite, then Γ defines $\langle\langle m, n\rangle, k\rangle \notin B_p$ (since here the reverse holds). Notice that if A_{p_0} E_3 A_{p_1} for some $p_0 < p_1$, then either the columns $A_{p_i}^k$ both have finite difference with A_n^k, or else both have infinite difference with it. Thus no harm has been done.

So we have $E_3 \leq_2^\omega = \leq_0 E_0 \leq_0 E_1$. It remains to show that there is no 1-jump reduction (not even a binary reduction) from E_3 to E_1. (This will imply $E_3 \not\leq_1^2 E_0$ as well, since $E_0 \leq_0 E_1$.)

Here we introduce a new technique: the complexity argument. Relative to reals A and B, the relation E_3 is Π_3, whereas E_1 is only Σ_2. This immediately suggests that there should be no computable reduction of E_3 to E_1, and since one is Π and the other is Σ, there should not even be a 1-jump reduction, although the indices (in "Σ_2" and "Π_3") differ by only one.

To formalize this, let D be the set of pairs $\{(i, j) \in \omega^2 : W_i = \overline{W_j}\}$. D itself is only a Π_2 subset of ω^2, and complete at that level under 1-reducibility, but it serves our purpose, when we think of (i, j) as not just representing a decidable set W_i, but actually providing a decision procedure for

W_i. We abuse notation by writing (i, j) E_1 (i', j') to denote W_i E_1 $W_{i'}$, viewing this as the restriction of E_1 to the class of decidable sets; similarly for E_3. We now use the Σ_3-complete set $\text{Cof} = \{e \in \omega : W_e \text{ is cofinite}\}$, defining a computable function $f : \omega \to D$ such that, for all inputs e, we have $e \notin \text{Cof}$ if and only if $f(e) = (i, j)$ gives the D-index of a set E_3-equivalent to $\emptyset$ (that is, just if (i, j) E_3 (i_0, j_0), for some fixed indices with $W_{i_0} = \emptyset$ and $W_{j_0} = \omega$). This is a simple movable-marker construction: on input e, write $\overline{W_{e,s}} = \{a_{0,s} < a_{1,s} < \cdots \}$ for each stage s, and say as usual that the nth marker "moves" whenever $a_{n,s+1} \neq a_{n,s}$. We enumerate the nth column of W_i by watching this marker. If it moves at stage s, then $\langle n, s \rangle \in W_i$, while if not, then $\langle n, s \rangle \in W_j$. So, if W_e is coinfinite, every marker moves only finitely often, and every column of W_i is finite, making W_i E_3 $\emptyset$; while if $e \in \text{Cof}$, then some marker moves infinitely often, and some column of W_i is infinite, destroying this E_3-equivalence.

Now suppose there were a 1-jump binary reduction Γ of E_3 to E_1. Then, for every e, we have $e \notin \text{Cof}$ if and only if $f(e) = (i, j)$ E_3 (i_0, j_0), which holds if and only if $\Gamma^{(W_i \oplus W_{i_0})'}$ outputs two sets which are E_1-equivalent. So, using a $\emptyset'$-oracle, we can take any e, give a decision procedure for W_i (where $(i, j) = f(e)$), use our oracle to produce a decision procedure for $(W_i \oplus W_{i_0})'$, and thus run $\Gamma^{(W_i \oplus W_{i_0})'}$ on any natural-number input. This enables us to compute (below our $\emptyset'$-oracle), uniformly in (i, j), a set $(B_0 \oplus B_1)$, the output of $\Gamma^{(W_i \oplus W_{i_0})'}$, with the property that B_0 E_1 B_1 if and only if $e \notin \text{Cof}$. However, this E_1-equivalence is a Σ_2 property of the reals B_0 and B_1, hence $\Sigma_2^{\emptyset'}$, hence Σ_3, and thus the property of being coinfinite would be Σ_3^0. Since this property is known to be Π_3^0-complete, no such reduction can exist.

The preceding argument easily adapts to prove the following theorem.

Theorem 2.2. *Suppose E is an equivalence relation on 2^ω whose restriction E_D defined on the set $D = \{(i, j) \in \omega^2 :$*

$W_i = \overline{W_j}\}$ by

$$(i,j)\ E_D\ (p,q) \iff W_i\ E\ W_p$$

is Π^0_k-*complete within* D *as a set under* 1-*reducibility.* (*That is,* E_D *is* Π^0_k-*definable, and every* Π^0_k-*definable subset* $T \subseteq \omega$ *has* $(T,\overline{T}) \leq_1 (E_D, D^2 - E_D)$.) *Let* F *be any* Σ^0_n-*definable equivalence relation on* 2^ω, *with* $n \leq k$. *Then* $E \not\leq^2_{k-n} F$.

It is important here to bear in mind that elements of D are given by pairs (i,j), (p,q), etc., not just by i or p. Hence, when one addresses the question whether $W_i\ E\ W_p$, the statement $(x \in W_i \iff x \in W_p)$ is decidable, not just $\emptyset'$-decidable, although it might appear to be the latter. We enlarge on this comment in Sec. 4.

3.   Completeness Results

As we continue with our analysis of the basic Borel equivalence relations, we come next to E_2. For this it will be useful to have a notion of completeness. We give a definition which, in the base case, is semantical rather than syntactical.

Definition 3.1. A (finitary) formula $\varphi(\vec{x},\vec{A})$ with free number variables $x_0,\ldots,x_k$ and free set variables $A_0,\ldots,A_m$ is said to be Δ^0_1 if there is a Turing functional Φ such that for all sets $A_0,\ldots,A_m \subseteq \omega$, the truth of $\varphi(\vec{x},\vec{A})$ is decided, uniformly in $\vec{x} \in \omega^k$, by the function $\Phi^{A_0 \oplus \cdots \oplus A_m}$.

This is then extended through the computable ordinals in the usual way. If $\varphi(\vec{x},\vec{A})$ is Σ^0_α, then its negation is Π^0_α. A $\Sigma^0_{\alpha+1}$ formula is a computable (possibly infinitary) disjunction of Π^0_α formulas $\varphi_i(\vec{x},\vec{A})$, all with the same finite tuples $\vec{x}$ and $\vec{A}$ of free variables. For limit ordinals $\lambda < \omega^{CK}_1$, the Σ^0_λ formulas are the computable (possibly infinitary) disjunctions $\bigvee_i \varphi_i(\vec{x},\vec{A})$, where each φ_i is $\Sigma^0_{\alpha_i}$ for some $\alpha_i < \lambda$. A formula is Δ^0_α if it is equivalent to both a Σ^0_α and a Π^0_α formula.

It is not difficult to give an equivalent syntactical formulation for Δ_1^0, in which atomic formulas of the form $x_i \in A_j$ are allowed. If we allowed quantification over set variables, then we would use the symbols Σ_n^1 and Π_n^1, but in this chapter we have no need of them.

The next definition is standard, and it too has been used in the preceding sections.

Definition 3.2. A Π_α^0 equivalence relation E on 2^ω is Π_α^0-*complete* under a reducibility $\leq$ if every Π_α^0-definable equivalence relation F satisfies $F \leq E$. Here $\leq$ is usually of the form $\leq_\beta^\kappa$, where β is a computable ordinal (most often 0) and κ a cardinal (or possibly κ can be replaced by $< \kappa$, as in finitary reducibility).

In the Borel theory, the equivalence relation known as E_2 is defined using the harmonic series:

$$A \ E_2 \ B \iff \sum_{n \in A \triangle B} \frac{1}{n+1} < \infty.$$

This relation is known to be Borel-incomparable with both E_1 and E_3, but it is Σ_2^0, and so the next result is not surprising.

Proposition 3.3. $E_2 \equiv_0^\omega E_1$. *That is, there are countable computable reductions in both directions.*

Indeed, each of E_0, E_1, and E_2 is complete under $\leq_0^\omega$ among Σ_2^0-definable equivalence relations on 2^ω.

Proof. We first prove the Σ_2-completeness statement for E_1. Let E be any Σ_2^0-definable equivalence relation on 2^ω. Standard methods allow us to collapse like quantifiers, so that we may take E to be defined by using a quantifier-free formula φ with two number variables and two set variables:

$$A \ E \ B \iff (\exists x \in \omega)(\forall y \in \omega) \ \varphi(x, y, A, B).$$

Our reduction Γ accepts $A = \bigoplus_n A_n$ as its oracle and outputs $B = \bigoplus_n B_n$. For each $m < n$, the columns numbered $\langle m, n, c \rangle$ of each B_i are used to distinguish B_m and B_n under E_1 if necessary. At stage s, consider each $m < n < s$ in turn, and suppose that Γ has already made each of $B_m^{\langle m,n,0 \rangle}, \ldots, B_m^{\langle m,n,c-1 \rangle}$ distinct from the corresponding B_n-column. This will hold just if Γ has already found witnesses $y_0, \ldots, y_{c-1}$ such that $(\forall i < c)\neg\varphi(i, y_i, A_m, A_n)$, but has not yet found such a witness y_c for c. At this stage, we use the oracle set to check whether $\varphi(c, y, A_m, A_n)$ holds, where y is the least possible witness which has not yet been checked. If indeed $\varphi(c, y, A_m, A_n)$ holds, then we define $s \notin B_i^{\langle m,n,d \rangle}$ for every i and every d. If $\neg\varphi(c, y, A_m, A_n)$ holds, then we define $s \in B_m^{\langle m,n,c \rangle}$ but $s \notin B_n^{\langle m,n,c \rangle}$. Thus B_m and B_n now differ on their $\langle m, n, c \rangle$th column (and at the next stage, the value of c for this $\langle m, n \rangle$ will be incremented by 1). We also define $s \notin B_i^{\langle m,n,d \rangle}$ for every $d \neq c$ and every i. It remains to decide whether $s \in B_i^{\langle m,n,c \rangle}$ for $i \notin \{m, n\}$. Consider each such i, in increasing order, and let $S_{imn} = \{0, \ldots, i-1\} \cup \{m, n\}$ be the set of indices of higher priority (for this column) than i. For each $j \in S_{imn}$, find the greatest $x_j \leq s$ such that

$$(\forall x < x_j)(\exists y \leq s)\ \neg\varphi(x, y, A_i, A_j).$$

Then, for the least $j \in S_{imn}$ such that x_j is minimal among $\{x_k : k \in S_{imn}\}$, Γ defines $s \in B_i^{\langle m,n,c \rangle}$ if and only if $s \in B_j^{\langle m,n,c \rangle}$ (which has already been defined). At this stage, the evidence suggests that, among the indices in S_{imn}, j is most likely to have $A_j\ E\ A_i$. If there is any $k \in S_{imn}$ for which in fact $A_k\ E\ A_i$, then from some stage on, we will always choose the same j: it will have $A_j\ E\ A_i$, and the procedure ensures that then $B_i^{\langle m,n,c \rangle}$ and $B_j^{\langle m,n,c \rangle}$ differ only finitely. Other indices $k \in S_{imn}$ might also have $A_k\ E\ A_i$, but if so, then they also satisfy $A_k\ E\ A_j$, so by induction, $B_k^{\langle m,n,c \rangle}$ differs only finitely from $B_j^{\langle m,n,c \rangle}$, hence also from $B_i^{\langle m,n,c \rangle}$, exactly as required.

For each $\langle m, n \rangle$ with $m \geq n$ or $n \geq s$, we define $s \notin B_i^{\langle m,n,d \rangle}$ for every i and d. This completes the construction at stage s.

The reduction $E_1 \leq_0^\omega E_0$, from Proposition 2.1, now shows E_0 to be Σ_2^0-complete as well. To complete the proof of Proposition 3.3, we prove that $E_0 \leq_0 E_2$. This is easy, and was already known (if never actually stated) from the existing Borel reduction. We know that, for every n, $\frac{1}{2^{n+1}} + \frac{1}{2^{n+2}} + \cdot + \frac{1}{2^{n+1}} \geq \frac{1}{2}$. So, given A, define $\{2^n + 1, 2^n + 2, \ldots, 2^{n+1}\} \subseteq \Phi^A$ if and only if $n \in A$, with $\{2^n + 1, 2^n + 2, \ldots, 2^{n+1}\} \cap \Phi^A = \emptyset$ otherwise. Then Φ is the desired computable reduction of E_0 to E_1. $\square$

To finish our analysis of E_2, we state the more general lemma established above.

Lemma 3.4. $E_0 \leq_0 E_2$.

Proposition 3.5. $E_3 \leq_2^\omega =$.

Proof. The reduction Γ has as its oracle $A = (\bigoplus_n A_n)''$. It simply defines $\langle m, c \rangle \in B_n$ if and only if the columns A_m^c and A_n^c have finite symmetric difference. This works. $\square$

For those accustomed to the Borel theory, this could be a surprise. The strict inequalities

$$= \; <_B \; E_0 \; <_B \; E_3$$

both hold, and for each of them, 2-jump countable reducibility was the best possible reducibility for the reverse. So it might seem that E_3 should be farther away from equality than E_0 is. Nevertheless, there is no contradiction here: while Borel reducibility is far stronger than any countable reducibility, its absence does not stop us from composing the reductions in the reverse direction. (In general, however, for $\alpha > 0$, the composition of two α-jump computable reductions need not itself be α-jump computable.)

The equivalence relation E_{set} compares the columns of the two reals involved:

$$A \; E_{\text{set}} \; B \iff [(\forall n \exists m \; A^n = B^m) \; \& \; (\forall n \exists m \; B^n = A^m)].$$

(Notice that a given column of A or B is not required to appear exactly the same number of times as a column of each. E_{set} simply says that every column of A is equal to some column of B and vice versa.) In the Borel hierarchy, we have $E_3 <_B E_{\text{set}}$, and both of these two relations are defined by Π_3^0 formulas in A and B. We can readily give a binary computable reduction in the reverse direction.

Lemma 3.6. $E_{\text{set}} \leq_0^2 E_3$.

Proof. The reduction Γ, with oracle $A_0 \oplus A_1$, defines the $(2m)$th column of B_0 and B_1 by searching for a column of A_1 equal to A_0^m. To begin with, it sets $x \notin B_0^{2m} \cup B_1^{2m}$ for every x such that $(\forall y \leq x)[y \in A_0^m \iff y \in A_1^0]$. If it ever finds an x_0 with $[x_0 \in A_0^m \iff x_0 \notin A_1^0]$, then it makes $x_0 \in B_0^{2m}$ but $x_0 \notin B_1^{2m}$, creating a difference between the two columns. Starting with $x_0 + 1$, it then compares A_0^m with A_1^1 (starting from 0), the same way, and keeps making B_0^{2m} and B_1^{2m} the same until it finds a difference between A_0^m and A_1^1. If it ever finds a difference here, then it begins comparing A_0^m with A_1^2, and so on. Thus $B_0^{2m} \triangle B_1^{2m}$ will be finite if and only if A_0^m is equal to some column in A_1. Each column numbered $(2m + 1)$ in B_0 and B_1 is used likewise to check whether the column A_1^m appears as a column in A_0.   □

Since E_{set} and E_3 are Π_3^0 relations, however, it is more difficult to extend the idea of Lemma 3.6 beyond a binary computable reduction. To explain our approach, we now describe a ternary computable reduction.

Lemma 3.7. $E_{\text{set}} \leq_0^3 E_3$.

Proof. Consider an oracle $A = A_0 \oplus A_1 \oplus A_2$. For each (c, m, n) with $c \in \omega$, $m \leq 2$, $n \leq 2$, and $m \neq n$, the following procedure builds the $\langle m, n, c, d \rangle$th columns of the output B_0, B_1, and B_2, for every d. (For all other triples (c, m, n), these columns are all empty.) Pick p so that $\{m, n, p\} = \{0, 1, 2\}$. The $\langle m, n, c, 0 \rangle$th column is the *base column*, and we begin with the $\langle m, n, c, 1 \rangle$th column as the first *working column*; subsequently this may switch to $\langle m, n, c, k + 1 \rangle$ for $k = 1, 2, \ldots$. As in Lemma 3.6, we watch first for a difference between A_m^c and A_n^0; if we ever find one, then we watch for a difference between A_m^c and A_n^1, then A_n^2, etc. As long as no new difference appears at a stage s, we make all these columns of B_0, B_1, and B_2 agree, of course. Suppose that at stage s we find a difference between A_m^c and the current A_n^i; in this case we call s an $\langle m, n, c \rangle$-*stage*. Check whether A_m^c and the current column A_p^k of A_p show any difference up to s.

(1) If $A_m^c \upharpoonright s = A_p^k \upharpoonright s$, then we make $s \in B_m^{\langle m,n,c,k+1 \rangle}$, $s \notin B_n^{\langle m,n,c,k+1 \rangle}$ and $s \in B_p^{\langle m,n,c,k+1 \rangle}$ where k is the number of the current working column; we keep this number k fixed, but increment i by 1. (Since A_p appears as though its kth column might be a copy of A_m^c, we aim here to keep B_p E_3-equivalent to B_m, while taking a step towards making B_m and B_n E_3-inequivalent.)

(2) If $A_m^c \upharpoonright s \neq A_p^k \upharpoonright s$, then we make $s \in B_m^{\langle m,n,c,0 \rangle}$, $s \notin B_n^{\langle m,n,c,0 \rangle}$ and $s \notin B_p^{\langle m,n,c,0 \rangle}$, and increment each of i and k by 1, thus moving to a new working column for the next stage. (Here again we take a step towards making B_m and B_n E_3-inequivalent, by adding one more distinction between them in the base column. Now we try to keep B_p E_3-equivalent to B_n, since A_p^k has turned out not to match A_m^c. The three kth working columns will never receive any more elements, so all of them will be finite and will not contradict E_3-equivalence.)

If $A_m^c = A_n^i$ for some i, then there are only finitely many $\langle m, n, c \rangle$-stages in all, and no column numbered $\langle m, n, c, k \rangle$

makes any two of B_m, B_n, or B_p E_3-inequivalent. If this fails for every i, there are two cases. If some k exists with $A_m^c = A_p^k$, then we will have reached the least such k and then stayed in Step (1) at all subsequent $\langle m, n, c\rangle$-stages, thus making B_p E_3-inequivalent to B_n but allowing it still to be E_3-equivalent to B_m. (With $A_p^k = A_m^c$, which does not appear as a column of A_n, this case has A_n and A_p E_{set}-inequivalent, so this outcome is acceptable.) If no such k exists, then we executed Step (2) at infinitely many $\langle m, n, c\rangle$-stages. Each time we did, we created a further difference between $B_p^{\langle m,n,c,0\rangle}$ and $B_m^{\langle m,n,c,0\rangle}$, and also between $B_n^{\langle m,n,c,0\rangle}$ and $B_m^{\langle m,n,c,0\rangle}$, so the base column shows B_m to be E_3-inequivalent to both the others. However, $B_p^{\langle m,n,c,0\rangle} = B_n^{\langle m,n,c,0\rangle}$, and on each working column, $B_p^{\langle m,n,c,k+1\rangle}$ and $B_m^{\langle m,n,c,k+1\rangle}$ have only a finite difference, since eventually the construction reached Step (2) again and incremented k. So in this case we have not done anything to make B_p E_3-inequivalent to B_n.

This completes the construction, and the argument above makes it clear that for any column c which appears in any A_m but not in some other A_n, the outputs B_m and B_n will have infinite difference on their $\langle m, n, c, d\rangle$th column, for some d. On the other hand, if A_m E_{set} A_n, then no infinite difference between corresponding columns of B_m and B_n will ever have been created, leaving B_m E_3 B_n. Thus we have a computable ternary reduction from E_{set} to E_3. $\square$

To make this into a finitary reduction (say a $(j+2)$-ary reduction), one does a similar process for each $\langle m, n, c\rangle$, assessing whether A_m^c appears as a column of A_n. Let $\{0, 1, \dots, j+1\} = \{m, n\} \sqcup \{p_0 < p_1 < \cdots < p_{j-1}\}$. We have in each B_i, for each $\langle m, n, c\rangle$, one column $C_\sigma = B_i^{\langle m,n,c,\sigma\rangle}$ for each $\sigma \in \omega^j$. We think of $\sigma(p)$ as keeping track of the current working column for A_p, for $p = p_0, \dots, p_{j-1}$, and reverting to 0 intermittently, just as the steps in the ternary reduction occasionally reverted to the base column. For this $\langle m, n, c\rangle$, one of the columns C_σ for which

$$(\forall i < j) \, [\sigma(i) = 0 \iff A_m^c \text{ does not appear as a}$$

$$\text{column in } A_{p_i}],$$

will be the column which distinguishes B_m from B_n under E_3 (unless $A_m \, E_{\text{set}} \, A_n$). At a stage s where A_m^c distinguishes itself from the current column in A_n, we build σ_s to guess which column will be the distinguishing column. We define $\sigma_s(0)$ using p_0 exactly as for p in the ternary reduction: $\sigma_s(0) = 0$ if we want to revert to the base column for p_0 (with $\sigma_{s+1}(0) = 1 + \sigma_{s-1}(0)$ to increment the working column), or else $\sigma_s(0) = \sigma_{s-1}(0)$ if we wish to stay on the same working column. To compute $\sigma_s(1)$, we consider only those stages $s' < s$ with $\sigma_{s'}(0) = \sigma_s(0)$. The greatest such $s' < s$ gives the current working column $k = \sigma_{s'}(1)$ for p_1 (or else $\sigma_{s'}(1) = 0$, in which case the working column from the preceding such stage s'' is incremented to begin a new current working column $\sigma_s(1) = 1 + \sigma_{s''}(1)$ at stage s), and we either continue with this working column by setting $\sigma_s(1) = \sigma_{s'}(1)$, or revert to the base column by setting $\sigma_s(1) = 0$, according to the dictates of the ternary construction for m, n, and p_1. (That is, provided $\sigma_{s'}(1) > 0$, we continue if $A_m^c \upharpoonright s = A_{p_1}^{\sigma_{s'}(1)-1} \upharpoonright s$, or revert if not.) Then we define $\sigma_s(2)$ similarly, considering only those preceding stages s' at which $\sigma_{s'} \upharpoonright 2 = \sigma_s \upharpoonright 2$, and so on, until we have defined σ_s all the way up to $j - 1$. Now we can add s to $B_m^{\langle m,n,c,\sigma_s \rangle}$ but not to $B_n^{\langle m,n,c,\sigma_s \rangle}$, and add s to exactly those $B_{p_k}^{\langle m,n,c,\sigma_s \rangle}$ with $\sigma_s(k) > 0$ and $A_m^c \upharpoonright s = A_{p_k}^{\sigma_s(k)-1} \upharpoonright s$, just as in the ternary reduction. There will be exactly one σ such that $\sigma_s = \sigma$ for infinitely many s, and the reader can check, using the method from Lemma 3.7, that this does give a $(j + 2)$-ary reduction, uniformly in j. So we have proven our next result.

Proposition 3.8. $E_{\text{set}} \leq_0^{<\omega} E_3$.

The preceding proof gives the impression that it could readily be generalized to a countable computable reduction: just use the strings σ_s to define the true path through the tree $\omega^{<\omega}$. In fact, though, this method does *not* apply to a countable collection $A = \bigoplus_{n \in \omega} A_n$ of sets. Having a column C_σ for every $\sigma \in \omega^j$ worked perfectly well for finite j, but with ω in place of j we would need continuum-many columns. So this is our first example where finitary reducibility has come into focus; until now, every finitary reduction could readily be made into a countable reduction, whereas here, imitating the proof from [13, Theorem 2.6], we show that no computable countable reduction exists.

Theorem 3.9. $E_{\text{set}} \not\leq_0^\omega E_3$.

Proof. We will show that no Turing functional Γ can be a computable countable reduction. Fix such a functional Γ, which accepts any oracle $A = \bigoplus_{n \in \omega} A_n$ and outputs $B = \bigoplus_{n \in \omega} B_n$.

Let $A_0^0 = \omega$ and $A_0^{m+1} = A_1^m = [0, m]$ for each m, so A_0 and A_1 are E_{set}-inequivalent. Also set $A_n^{2m+1} = [0, m]$ for each $n > 1$ and each m. It remains to define the elements of the even-numbered columns A_n^{2m} for $n > 1$. We intend for each of these columns to be an initial segment of ω (possibly all of ω) at the end of the construction; so far, each A_n^{2m} is an initial segment of length 0, and is allowed to be extended further. We proceed according to an ω-ordering $\prec$ of the pairs $\langle i, j \rangle$ with $\langle i, j \rangle \prec \langle i, j+1 \rangle$ for all i and j. For each such pair, in order, we ask whether it is possible to extend the (currently uncapped) columns A_n^{2m} (for $n > 1$) further so as to make Γ^A output a set B with $(B_i^i \triangle B_{i+2}^i) \cap \{j, j+1, \ldots\} \neq \emptyset$. If so, then we extend these columns (finitely far) so as to make this happen, and "cap" the columns A_{i+2}^{2k} with $k \leq j$, decreeing these columns to equal the finite initial segments of ω to which they have already been extended. If not, then we extend every currently uncapped column A_{i+2}^{2k} with $k \leq j$ by one more element.

The key feature of this construction is that, at every step $\langle i, j \rangle$, every column A_{i+2}^{2k} with $k \leq j$ either is capped or receives another element. Therefore, this process defines, for each x, whether or not x lies in each even-numbered column of A_{i+2}, for each i. Thus we have now defined the entire set A. Moreover, for a fixed i, either every column A_{i+2}^{2k} (for all k) was eventually capped — if at every step $\langle i, j \rangle$ it was always possible to extend A and create one more difference between B_1^i and B_{i+2}^i — or else some column of A_{i+2} is infinite, which occurs if we reached a step at which it was no longer possible for any extension of A to create such a difference.

Suppose that, for some i, $A_1 \; E_{\text{set}} \; A_{i+2}$. Then every column of A_{i+2} is finite, which happens just if, for every j, the construction always extended A so as to make B_1^i and B_{i+2}^i differ on some number $\geq j$. If this happened, then B_1 and B_{i+2} are E_3-inequivalent, and therefore Γ was not a reduction from E_{set} to E_3. On the other hand, suppose that, for every i, A_1 and A_{i+2} are E_{set}-inequivalent. Since every column of A_{i+2} is an initial segment of ω, and every finite initial segment occurs there, each A_{i+2} must have an infinite column A_{i+2}^{2j}, and therefore $A_{i+2} \; E_{\text{set}} \; A_0$ for every i. If Γ were a reduction, we would then have $B_{i+2} \; E_3 \; B_0$ for all i, so that $B_{i+2}^i \; E_0 \; B_0^i$ for all i. But the infinite column of A_{i+2} arose because $B_1^i \triangle B_{i+2}^i$ is finite, and so $B_1^i \; E_0 \; B_0^i$ for every i. This ensures $B_1 \; E_3 \; B_0$, which again shows Γ not to be a reduction, proving the theorem. □

Finally, we consider the standard Borel equivalence relation Z_0. For a real A, the *upper density* of A is defined to equal

$$\limsup_{k \to \infty} \frac{|A \cap \{0, \ldots, k\}|}{k+1},$$

found by asking what fraction of the first k elements of ω lie in A, as $k \to \infty$. The *lower density* is defined similarly, using the lim inf, and if these two are equal, then their common value is the *density* of A. This allows us to define

$$A \; Z_0 \; B \iff A \triangle B \text{ has density } 0.$$

Z_0 is another Π^0_3 equivalence relation on 2^ω, strictly above E_3 and incomparable with E_{set} under Borel reducibility. The usual Borel reduction from E_3 to Z_0 is computable, and all other questions about Z_0 within our various hierarchies in this chapter are then answered by the following proposition.

Proposition 3.10. Z_0 *is computably countably bireducible with* E_3: $Z_0 \equiv^\omega_0 E_3$. *Hence* $Z_0 <^\omega_0 E_{\text{set}}$ *and* $Z_0 \equiv^{<\omega}_0 E_{\text{set}}$.

Proof. As we mentioned, the usual Borel reduction from E_3 to Z_0 is computable: essentially, one converts each element $\langle m, n \rangle$ of A into a finite block of elements $E_{m,n}$ in the output B, in such a way that every output B has lower density 0 (since many elements do not belong to any block $E_{m,n}$, hence do not lie in B), and has upper density $\geq \frac{1}{n}$ if and only if A^n was infinite. (The latter condition is ensured by the blocks $\langle m, n \rangle$, for all m.)

To show that $Z_0 \leq^\omega_0 E_3$, we give a functional Γ, which as usual accepts an oracle $A = \bigoplus_n A_n$ and outputs $B = \bigoplus_n B_n$. For each triple $\langle m, n, c \rangle \in \omega^3$ with $m > n$ and $c > 0$, we use column number $\langle m, n, c \rangle$ in B_m and B_n. For those s such that $\frac{|(A_m \triangle A_n) \cap \{0,1,\ldots,s\}|}{s+1} \geq \frac{1}{c}$, we place $s \in B_m^{\langle m,n,c \rangle}$ but not in $B_n^{\langle m,n,c \rangle}$, while the remaining s belongs to neither set. Thus all these columns in B_n (for all c) are finite, while B_m has an infinite column $B_m^{\langle m,n,c \rangle}$ if and only if there exists c with $\limsup_s \frac{|(A_m \triangle A_n) \cap \{0,1,\ldots,s\}|}{s+1} \geq \frac{1}{c}$. It follows that, whenever A_m and A_n are Z_0-inequivalent, B_m and B_n will be E_3-inequivalent.

Next, setting $p_0 = m$, $p_1 = n$, and $\{p_2 < p_3 < \cdots\} = \omega - \{m, n\}$, we decide, for each $i \geq 2$ in turn, whether to add the number s to the corresponding column $B_{p_i}^{\langle m,n,c \rangle}$ of B_{p_i}. Of course, if $\frac{|(A_m \triangle A_n) \cap \{0,1,\ldots,s\}|}{s+1} < \frac{1}{c}$, we leave s out of every such column, so that if $\limsup_t \frac{|(A_m \triangle A_n) \cap \{0,1,\ldots,t\}|}{t+1} < \frac{1}{c}$, every column $B_p^{\langle m,n,c \rangle}$ will be finite. If $\frac{|(A_m \triangle A_n) \cap \{0,1,\ldots,s\}|}{s+1} \geq \frac{1}{c}$, then we go through $i = 2, 3, \ldots$ in turn. For each i and each $j < i$, find the least number

$d_{j,s} \leq s$ such that

$$(\forall d \in \{d_{j,s}, d_{j,s} + 1, \ldots, s\}) \; \frac{|(A_{p_i} \triangle A_{p_j}) \cap \{0, 1, \ldots, d\}|}{d + 1} \leq \frac{1}{c \cdot 2^i}.$$

Not all of these values $d_{0,s}, d_{1,s}, \ldots, d_{i-1,s}$ need exist. If none of them exist, then $s \notin B_{p_i}^{\langle m,n,c \rangle}$. If there is at least one such value, then find the least value among them: say $d_{j_0,s}$ (for the least j_0, in case some $d_{j,s} = d_{j',s}$), and set

$$s \in B_{p_i}^{\langle m,n,c \rangle} \iff s \in B_{p_{j_0}}^{\langle m,n,c \rangle},$$

noticing that since $j_0 < i$, the right-hand side here has already been defined. (We choose to make B_{p_i} "look like" $B_{p_{j_0}}$ because the symmetric difference $A_{p_i} \triangle A_{p_{j_0}}$ has stayed small for more stages than for any other j.) This completes the construction.

The goal of the second part of the construction is to ensure that, if $A_{p_i} \; Z_0 \; A_{p_j}$ for one or more $j < i$, then $B_{p_i}^{\langle m,n,c \rangle} =^* B_{p_j}^{\langle m,n,c \rangle}$ for each such j. To see this, we induct on i. The claim we prove inductively is slightly stronger: for every $j < i$ such that $A_{p_j} \triangle A_{p_i}$ has upper density $< \frac{1}{c \cdot 2^i}$, we show that $B_{p_j}^{\langle m,n,c \rangle} =^* B_{p_i}^{\langle m,n,c \rangle}$. This is clearly enough to ensure the goal stated above. When $j = 0$ and $i = 1$, the claim is clear, so we consider $i \geq 2$, assuming inductively that the claim holds for all smaller values of i. For each j (if any) such that $A_{p_j} \triangle A_{p_i}$ has upper density $< \frac{1}{c \cdot 2^i}$, let $d_j = \lim_s d_{j,s}$, which must exist. (For all other $j < i$, no such number d_j exists, and there is nothing to prove about such a j.) Fix the j for which d_j is least; if this does not uniquely define j, fix the least such j. Now there must exist some stage s_0 such that the construction chooses this j as its j_0 at every stage $\geq s_0$. So, for $s \geq s_0$, the construction ensures that $s \in B_{p_i}^{\langle m,n,c \rangle}$ if and only if $s \in B_{p_j}^{\langle m,n,c \rangle}$, leaving these two columns to differ only finitely, as required.

Now consider any other $j' < i$ such that $A_{p_{j'}} \triangle A_{p_i}$ has upper density $< \frac{1}{c \cdot 2^i}$. It follows that $A_{p_{j'}} \triangle A_{p_j}$ has upper density $<$

$\frac{1}{c\cdot 2^{i-1}} \leq \frac{1}{c\cdot 2^{\max(j,j')}}$. By induction, the greater of j and j' will have ensured that $B_{p_{j'}}^{\langle m,n,c\rangle} =^* B_{p_j}^{\langle m,n,c\rangle}$, and hence $B_{p_{j'}}^{\langle m,n,c\rangle} =^* B_{p_i}^{\langle m,n,c\rangle}$. This completes the induction for i.

It now follows that, if $A_m \ Z_0 \ A_n$, then every column of B_m has only finite difference with the corresponding column of B_n, and so $B_m \ E_3 \ B_n$. The first part of the construction established the converse, and so this Γ is a countable computable reduction from Z_0 to E_3. The remaining statements in Proposition 3.10 now follow from Proposition 3.8 and Theorem 3.9.   □

Under computable countable reduction, therefore, the equivalence relations discussed here realize the following much simpler diagram:

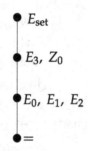

$$E_{\text{set}}$$

$$E_3, \ Z_0$$

$$E_0, \ E_1, \ E_2$$

$$=$$

Under finitary computable reduction (that is, $\leq_0^{<\omega}$), E_{set} joins the class of E_3 and Z_0.

4. Equivalence Relations Respecting Enumerations

The reader familiar with the works [3, 13] will see similarities between the constructions there and those here, but also some differences. Our proofs of Theorem 3.9 and Proposition 3.10 will strike most readers as significantly less difficult than the proofs of the corresponding results (Theorems 2.5, 2.6 and 3.3) of [13]. We now explain these differences.

Both [3] and [13] focused on Turing-computable reductions among equivalence relations on ω. They converted each standard Borel equivalence relation E into a relation E^{ce} on ω using indices of computably enumerable sets W_e (from any standard enumeration of the c.e. sets), under the following definition:

$$(\forall i, j \in \omega) \; [i \; E^{ce} \; j \iff W_i \; E \; W_j].$$

This defines $=^{ce}$, E_0^{ce}, E_1^{ce}, E_2^{ce}, E_3^{ce}, E_{set}^{ce} and Z_0^{ce}, and the combined results of [3, 13] show that the results of Sec. 2 here largely hold for the c.e. versions of these relations under computable reducibility $\leq_0$. Indeed, [13] shows that $E_{set}^{ce} \equiv_0^{<\omega} Z_0^{ce}$ but also that $E_{set}^{ce} <_0 Z_0^{ce}$, mirroring our results above for E_{set} and Z_0 on 2^ω. (Recall that $\leq_0$ denotes $\mathbf{0}$-computable reducibility on equivalence relations on ω, with the boldface "$\mathbf{0}$" naming the Turing degree of the computable sets. For computable reducibility on equivalence relations on 2^ω, we still use $\leq_0$, with the lightface "0" denoting that no jumps are allowed.)

The differences in proofs arise from the fact that for relations E^{ce}, one has only enumerations of sets to work with, rather than oracles for the sets themselves. The main reason why the authors of [3, 13] worked this way is that there is no effective enumeration of the decidable subsets of ω, while there are natural effective enumerations of the c.e. subsets. Our use of oracles overcomes this difficulty, and $\leq_0^\omega$ in particular can be restricted to the decidable sets if one likes, or to the X-decidable sets for any fixed X, giving exactly the context one might have hoped for. Here, however, we go in the other direction, adapting our definitions from Sec. 1 to the context of sets which are enumerated for us, rather than decided.

Definition 4.1. Let $\pi_1 : \omega \to \omega$ be the projection map $\pi_1(\langle x, y \rangle) = x$, using the standard pairing function $\langle x, y \rangle = \frac{(x+y+1)(x+y)+2y}{2}$. An *enumeration* of a set $S \subseteq \omega$ is a set $A \subseteq \omega$ which projects onto S, i.e., for which $\pi_1(A) = S$.

An equivalence relation E on 2^ω *respects enumerations* if

$$(\forall A, B \in 2^\omega)\ [\pi_1(A) = \pi_1(B) \implies A\ E\ B].$$

For each equivalence relation E on 2^ω, we define its *enumeration analogue* E^e to be the following equivalence relation on 2^ω, which respects enumerations:

$$A\ E^e\ B \iff \pi_1(A)\ E\ \pi_1(B).$$

It is immediately seen, for every E, that $E \leq_0 E^e$, via the reduction $A \mapsto \{\langle n, 0\rangle : n \in A\}$. In the opposite direction, we have a full 1-jump reduction from E^e to E via $A \mapsto \pi_1(A)$, of course, but a quick diagonalization argument shows that in general there need not exist even a binary computable reduction. For a more specific example, we invite the reader to show that the equality relation $=$ is Π_1^0-complete under computable countable reducibility $\leq_0^\omega$, whereas its enumeration analogue $=^e$ is Π_2^0-complete under computable finitary reducibility $\leq_0^{<\omega}$, but not under computable countable reducibility. Results in [13] are relevant here, using the relation $=^{ce}$ on ω, which is defined by setting $i =^{ce} j$ if and only if $W_i = W_j$, and which was shown there to be complete under computable finitary reducibility among Π_2^0 equivalence relations on ω.

Enumerations of sets of natural numbers are ubiquitous in mathematics and logic, and so we believe that Definition 4.1 will prove extremely useful. Moreover, it allows us to adapt the existing results in [3, 13] to prove the following theorems. (Since the ideas of these proofs are essentially identical to those of the original proofs for the E^{ce} versions, we leave them to the reader.)

Theorem 4.2. *For countable computable reducibility on the enumeration analogues of the usual Borel equivalence relations, we have*

$$=^e\ <_0^\omega\ E_0^e\ <_0^\omega\ E_3^e\ <_0^\omega\ Z_0^e,$$

with $E_1^e \equiv_0^\omega E_0^e$. Under finitary computable reducibility, the last of these two merge:

$$=^e \; <_0^{<\omega} \; E_0^e \; <_0^{<\omega} \; E_3^e \; \equiv_0^{<\omega} \; Z_0^e,$$

It was shown in [13] that 3-ary and 4-ary reducibility are distinguished by the equivalence relations E_{max}^{ce} and E_{card}^{ce}, which are the c.e. versions of the relations here.

Definition 4.3. For sets $A, B \subseteq \omega$, we define:

$$(A \; E_{\mathrm{max}} \; B \iff \max(A) = \max(B))$$
$$(A \; E_{\mathrm{card}} \; B \iff |A| = |B|).$$

Here we interpret $\max(\emptyset) = -\infty$, and $\max(A) = +\infty$ for infinite sets A. Thus all infinite sets are E_{max}-equivalent, and also E_{card}-equivalent, while $\emptyset$ forms a singleton class for each of these relations.

E_{max} and E_{card} have some properties we have not seen before in Borel equivalence relations: each has only countably many classes, including one singleton class and just one uncountable class. (Indeed E_{max} has just one infinite class.) It was shown in [13] that E_{max}^{ce} and E_{card}^{ce} are equivalent under computable reducibility (on equivalence relations on ω), and that E_{max}^{ce} is complete among Π_2^0 equivalence relations on ω under 3-ary computable reducibility, but not under 4-ary computable reducibility.

Proposition 4.4. *The relations E_{max}, E_{card}, E_{max}^e and E_{card}^e are all Π_2^0-definable, and $E_{max}^e \equiv_0 E_{card}^e$.*

Proof. The Π_2^0 definability is quick. First, $A \; E_{\mathrm{max}} \; B$ if and only if

$$(\forall n) \; [(\exists x \in A)x \geq n \iff (\exists y \in B)y \geq n],$$

and one readily converts this to a definition of $E_{\max}^e$ as well. Likewise, $A \; E_{\text{card}} \; B$ if and only if

$$(\forall n) \; [(\exists x_1 < x_2 < \cdots < x_n)[\text{all } x_i \in A]$$
$$\iff (\exists y_1 < y_2 < \cdots < y_n)[\text{all } y_i \in B]],$$

and similarly for E_{card}^e. Notice that, while this second formula certainly defines a Π_2^0 relation on A and B, it is not so easy to make the defining formula strictly finitary in the usual sense, unless one allows the use of an iterated pairing function.

To see that $E_{\max}^e \leq_0 E_{\text{card}}^e$, given A, let $\Gamma^A = \{\langle x, \langle y, z \rangle \rangle : y \geq x \; \& \; \langle y, z \rangle \in A\}$. Thus $\pi_1(\Gamma^A) = \{x : (\exists y \geq x) \; y \in \pi_1(A)\}$. For the reverse reduction, given B, use the extended pairing function $\omega^{<\omega} \to \omega$ and let

$$\Phi^B = \{\langle n, \langle x_0, y_0, \ldots, x_n, y_n \rangle \rangle : \langle x_0, y_0 \rangle, \ldots, \langle x_n, y_n \rangle$$
$$\in B \; \& \; x_0 < \cdots < x_n\}.$$

So $n \in \pi_1(\Phi^B)$ if and only if $|\pi_1(B)| > n$. For $E_{\max}^e$ and E_{card}^e, this Γ and this Φ both work. Indeed, Φ is a computable reduction from E_{card}^e not just to $E_{\max}^e$, but actually to $=^e$. $\square$

Proposition 4.5. E_{max} *is* Π_2^0-*complete under computable ternary reducibility: every* Π_2^0-*definable equivalence relation E on 2^ω has $E \leq_0^3 E_{max}$. However, E_{max}^e is not complete (in this same sense) for* $\leq_0^4$: *there do exist* Π_2^0-*definable equivalence relations E on 2^ω with $E \not\leq_0^4 E_{max}^e$.*

Since $E_{max} \leq_0 E_{max}^e$, it follows that E_{max}^e is also Π_2^0-*complete under* $\leq_0^3$ *and that E_{max} is not* Π_2^0-*complete under* $\leq_0^4$.

In light of Proposition 4.6 below, one could prove Π_2^0-completeness under $\leq_0^3$ for either $E_{\max}$ or $E_{\max}^e$, and it would follow for the other. We will imitate the work in [13], but adjust it to the context of decidable sets. That work dealt with enumerations of sets rather than with the sets themselves, but the essence of the construction is the same, and our reason for doing the proof here for $E_{\max}$ is to demonstrate the essential similarity. Roughly speaking, the similarity reflects the fact that the replacement of a positive subformula $(\exists x \in A \; R(x))$ by a subformula $(\exists \langle x, y \rangle \in A \; R(x))$ does not change the complexity. A subformula $(\exists x \notin A \; R(x)$, with R quantifier-free, would usually cause the complexity to increase when one passed from the decidable case to the case of enumerations. In Proposition 4.6 below, we will see a more subtle difference between the situations for decidability and enumerability.

Proof of Proposition 4.5. Let E be any Π_2^0 equivalence relation on 2^ω, given by a formula

$$A \; E \; B \iff \forall x \exists y \; R(A, B, x, y)$$

with $R(A, B, x, y)$ decidable by a functional $\Gamma^{A \oplus B}(x, y)$. We give a computable ternary reduction Φ from E to $E_{\max}^e$, with oracle $A = A_0 \oplus A_1 \oplus A_2$, outputting $B_0 \oplus B_1 \oplus B_2$. First, we fit the relation E into the standard Π_2^0 framework. For each $i < j$ and each s, we can use Γ to determine the greatest $x_{i,j,s} \leq s$ such that

$$(\forall x \leq x_{i,j,s})(\exists y \leq s) \; R(A_i, A_j, x, y).$$

If $x_{i,j,s+1} > x_{i,j,s}$, we say that the pair (i, j) *receives a chip* at stage $s + 1$. Thus $A_i \; E \; A_j$ if and only if the pair (i, j) receives a chip at infinitely many different stages s. For convenience we consider each of the three pairs to receive a chip at stage 0.

Φ^A uses this idea to build its outputs B_i. At stage 0, it sets $0 \in B_0$ only, $1 \in B_1$ only, and $2 \in B_2$ only. Thus all three sets currently have distinct maxima. Similarly, at each stage $s \geq 1$, if none of the three pairs $(0, 1)$, $(0, 2)$ and $(1, 2)$ receives a chip at stage s, then Φ^A specifies that none of the numbers $2s, 2s + 1$ and $2s + 2$ lies in any of B_0, B_1 and B_2, thus preserving the three maxima (all distinct, by induction) from the preceding stage. If one pair (i, j) with $i < j \leq 2$ did receive a chip, then for the least such pair, it defines $2s + 1$ to lie in B_i and $2s + 2$ to lie in B_j. If neither of the other two pairs among the three received a chip at stage s, then the third output set B_k receives no new elements. If at least one of the other two pairs did receive a chip, then it defines $2s$ to lie in B_k. None of the elements $2s, 2s+1$ and $2s+2$ will be put into any of the three output sets at any subsequent stage, so this constitutes a decision procedure for $B_0 \oplus B_1 \oplus B_2$.

Now if A_0, A_1, and A_2 lie in three distinct E-classes, then there is a stage after which no pair received any more chips. At the end of that stage, all three output sets had distinct maxima, and no further elements were ever added, so B_0, B_1 and B_2 lie in distinct $E_{\max}$-classes as required. If all three of the input sets are E-equivalent, then all three pairs received infinitely many chips, and so each of the three output sets has maximum $+\infty$. Finally, if $A_i \; E \; A_j$ but A_k lies in a different E-class than these two, then after some stage B_k never receives any more elements, hence has a finite maximum, whereas B_i and B_j each get new elements at infinitely many stages, hence both have maximum $+\infty$. Thus Φ^A is a ternary reduction from E to $E_{\max}$

To address 4-ary reducibility, we apply [13, Theorems 3.4 and 4.2], which together show that $=^{ce}$ has no computable reduction to $E^{ce}_{\max}$ (that is, under Turing-computable functions from ω to ω). The proof is a nice illustration of the connections between that topic and this one. Suppose $E_{\max}$ were complete under $\leq^4_0$ among Π^0_2-definable equivalence relations on 2^ω. We "extend" $=^{ce}$ to an E on Cantor space, by setting

$$A \; E \; B \iff [A = B \text{ or } [A \neq \emptyset \neq B \; \& \; (\forall i \in A)$$
$$(\forall j \in B) \; W_i = W_j].$$

Note that this E is Π^0_2-definable: $A = B$ is a Π^0_1 property, nonemptiness is Σ^0_1, and equality of c.e. sets is Π^0_2. By assumption, then, there is a computable 4-ary reduction Γ of E to $E_{\max}$. But now we could use this Γ to define a computable reduction of $=^{ce}$ to $E^{ce}_{\max}$, as follows. Given any indices $e_0, e_1, e_2, e_3 \in \omega$, we run $\Gamma^{\{e_0\} \oplus \cdots \oplus \{e_3\}}$, which computes a set $B = B_0 \oplus \cdots \oplus B_3$ such that $B_i \; E_{\max} \; B_j$ if and only if $\{e_i\} \; E \; \{e_j\}$, which holds just if $W_{e_i} = W_{e_j}$. Let p_i be the code number of the program which enumerates B_i by running $\Gamma^{\{e_0\} \oplus \cdots \oplus \{e_3\}}$, so that $B_i = W_{p_i}$. (In fact, we can also enumerate the complement $\overline{B_i}$, but this is unnecessary.) Then, for each $i < j \leq 3$, we have

$$p_i \, E^{ce}_{\max} \, p_j \iff W_{p_i} \, E_{\max} \, W_{p_j} \iff B_i \, E_{\max} \, B_j$$
$$\iff \{e_i\} \, E \, \{e_j\} \iff W_{e_i} = W_{e_j},$$

and so the map $\vec{e} \mapsto \vec{p}$ is a computable 4-ary reduction of $=^{ce}$ to $E^{ce}_{\max}$, which is impossible. $\square$

The main point of the next proposition is its contrast with Proposition 4.4. The behavior of $E_{\max}$ and E_{card} relative to each other is not the same as that of $E^e_{\max}$ and E^e_{card}.

Proposition 4.6. *The best reductions that hold are* $E_{max} \leq_0 E_{card}$ *and* $E_{card} \leq_0^{<\omega} E_{max}$ *(and* $E_{card} \leq_1 E_{max}$*).*

Proof. The goal of the functional Γ giving the reduction $E_{\max} \leq_0 E_{\text{card}}$ is to ensure that, for all n,

$$\max(A) \geq n \iff |B| \geq n+1 \quad (\text{where } B = \Gamma^A).$$

To this end, Γ checks whether $0 \in A$, $1 \in A, \ldots$. Each time it finds a new $n \in A$, it extends its current approximation σ_{n-1} of B by setting $\sigma_n = \sigma_{n-1}\widehat{\ }1^k$, where k is chosen so that $|\sigma_n^{-1}(1)| = n+1$. Each time $n \notin A$, it sets $\sigma_n = \sigma_{n-1}\widehat{\ }0$. This yields $B = \bigcup_n \sigma_n$ of cardinality exactly $n+1$ if $\max(A) = n$, or $B = \emptyset$ if $A = \emptyset$, or B infinite if A is infinite.

In the opposite direction, we show first that no Φ can be a countable computable reduction from E_{card} to $E_{\max}$. Given Φ, we choose $A_0 = \{0\}$ and $A_1 = \{0, 1\}$, with no elements yet in any of $A_2, A_3, \ldots$. As usual we run Φ^A with $A = \bigoplus_n A_n$. If Φ is to succeed, then at some stage s some finite portion of this oracle must yield a partial output $B = \bigoplus_n B_n$ with some k for which either

- $k \in B_0$ and $k \notin B_2$ and $(\forall j > k)(\forall i = 0, 2)$ $[j \notin B_i$ or $\Phi_s^A(\langle i, j \rangle) \uparrow]$; or
- the same with B_0 and B_2 reversed.

We freeze that finite portion of the oracle, and now add one new element (larger than the frozen part) to A_2. Now running Φ^A still yields $k \in B_0 \triangle B_2$, but must eventually give B_0 and B_2 the same maximum (possibly $+\infty$), and so eventually we find some $j > k$ with $j \in B_0$. We now freeze the finite portion of A which has been used so far in these computations. No further changes will be made to A_2.

Next we do the same process with A_1 and A_3: with $A_3 = \emptyset$, wait until Φ^A produces B_1 and B_3 with distinct maxima, then freeze A on the use of that computation, add two new large elements to A_3 (so that now $|A_3| = |A_1| = 2$), and wait until Φ^A evens up the maxima of B_1 and B_3, which must involve adding a larger element to B_1. After that, no further changes will be made to A_3.

We continue recursively, using each A_{2i} to force B_0 to include a new larger element, and using each A_{2i+1} to force B_1 to do so. Therefore $\max(B_0) = \max(B_1) = +\infty$, yet $|A_0| = 1 \neq 2 = |A_1|$. Therefore Φ was not a countable computable reduction.

The full 1-jump reduction from E_{card} to $E_{\max}$ is easy: let $n \in \Gamma^{A'}$ if and only if $|A| \geq n$. It remains to give a finitary computable reduction Ψ from E_{card} to $E_{\max}$, using an oracle $A = A_0 \oplus \cdots \oplus A_n$. For this, Ψ^A goes through $s = 0, 1, 2, \ldots$, one at a time, starting with $m_{i,-1} = -\infty$ for all $i \leq n$ For each s, Ψ determines which numbers among $s(n+1), \ldots, (s+1)(n+1) - 1$ are to be added to which sets B_i. To do so, it determines the size $m_{i,s} = |A_i \cap \{0, \ldots, n\}|$ for each i. For those i with $m_{i,s-1} = \min\{m_{j,s-1} : j \leq n\}$, it checks whether $m_{i,s} = m_{i,s-1}$: if so, then it adds no new elements to B_i; while if not, then it adds $s(n+1)$ to B_i (and possibly more elements later in this step).

If this process added no elements to any B_i, then it determines the second-smallest value in $\{m_{j,s-1} : j \leq n\}$. For those i such that $m_{i,s-1}$ has this value, it again adds no elements to B_i provided $m_{i,s} = m_{i,s-1}$, buts adds $s(n+1)$ to B_i otherwise. Again, if this process still has not added any elements to any B_i, then it proceeds with the third-smallest value in $\{m_{j,s-1} : j \leq n\}$, and so on until either some B_i is enlarged or until the values in $\{m_{j,s-1} : j \leq n\}$ run out. (Notice that, if every A_i is finite, then there will be a stage after which no more elements are ever added to any B_i.)

Now suppose that we reached a step at which $s(n+1)$ was added to some B_i. By this point, some sets B_j have been assured that no elements will be added to them at this stage. For the remaining B_k (including those B_i to which $n(s+1)$ was added, as well as all B_k with $m_{k,s-1} > m_{i,s-1}$ for those i), we determine anew the order among the maxima $m_{k,s}$. For those k such that $m_{k,s}$ is least, we add $s(n+1)$ to B_k (if it was not already there). For those k for which $m_{k,s}$ has the second-smallest value, we add $s(n+1)+1$ to B_k. Those with the third-smallest value have $s(n+1)+2$ added to B_k, and so on. Now there are at most $(n+1)$ different indices k involved, so the greatest number that can possibly be added to any B_k at stage s is $s(n+1)+n = (s+1)(n+1)-1$. No number $\leq s(n+1)+n$ will be added to any set B_i at any subsequent stage, so this is a decision procedure for the B_i. Moreover, we have ensured that that order of the maxima of the sets B_i (so far) corresponds to the order of the sizes $m_{i,s}$.

It is clear that whenever a set A_i is finite, the corresponding B_i will eventually stop receiving new elements (once the $m_{j,s}$ have reached their limiting values m_j for all $j \leq n$ with A_j finite, and once the $m_{k,s}$ with A_k infinite have all surpassed these limiting values). Moreover, we will have $m_i = m_j$ if and only if B_i and B_j have the same maximum. The infinite sets A_k will all have B_k infinite as well, and so they all satisfy $\max(B_k) = +\infty$ (and $|A_k| = +\infty$). Therefore, this is indeed a finitary computable reduction. $\square$

Corollary 4.7. $E_{max} \equiv_0^{<\omega} E_{max}^e$ *and* $E_{card} \equiv_0^{<\omega} E_{card}^e$. *Hence* E_{card} *is* Π_2^0-*complete under computable ternary reducibility* $\leq_0^3$, *but not under* $\leq_0^4$.

Proof. The general reduction $E \leq_0 E^e$ was described after Definition 4.1. The reduction Γ for $E_{max}^e \leq_0^{<\omega} E_{max}$ is given an oracle $A = A_0 \oplus \cdots \oplus A_n$, and proceeds much like the finitary reduction Ψ from E_{card} to E_{max} in Proposition 4.6. It measures the maximum $m_{i,s}$ of $\pi_1(A_i \cap \{0, \ldots, s\})$ at each stage s, and orders the sets $A_0, \ldots, A_n$ according to their current maxima, noting which maxima have changed since the previous stage. For those sets at the bottom of the list (with lowest maxima) whose maxima have not changed, it does not add any new elements to B_i. Starting with the least A_j in this order whose current maximum has $m_{j,s} \neq m_{j,s-1}$, it adds a new element, larger than any previously seen, to B_j and to each B_k with $m_{k,s-1} > m_{j,s-1}$ (or with $m_{j,s-1} = m_{k,s-1} \neq m_{k,s}$). Then it adds further new elements to these sets B_k to make sure that the current order of the B_k by maxima matches the current order of the A_k by their maxima $m_{k,s}$. This works, just as Ψ did in the proposition.

The reduction $E_{card}^e \leq_0^{<\omega} E_{card}$ now follows from

$$E_{card}^e \leq_0 E_{max}^e \leq_0^{<\omega} E_{max} \leq_0 E_{card},$$

and the rest of the corollary is the result of Proposition 4.5. $\square$

5. Noncomputable Reductions

We present here an idea for further discussion, without any immediate results. It is simple to define X-computable reductions on 2^ω, and to extend the notion to jump reductions and to finitary and countable reductions as well.

Definition 5.1. Let E and F be equivalence relations on 2^ω, fix $X \subseteq \omega$, and choose any X-computable ordinal α. We say that E is α-*jump* X-*reducible to* F, written $E \leq_{\alpha, X} F$, if there exists a Turing functional Φ such that the map

$$A \mapsto \Phi^{X \oplus A^{(\alpha)}}$$

is a reduction of E to F. Likewise, if the map

$$A = \bigoplus_i A_i \mapsto \Phi^{X \oplus A^{(\alpha)}} = \bigoplus_j B_j,$$

is a k-ary, finitary, or countable reduction of E to F (where the joins $\bigoplus_i A_i$ and $\bigoplus_j B_j$ are over the appropriate number of sets), then E is k-arily, finitarily, or *countably α-jump X-reducible to* F, written $E \leq^k_{\alpha,X} F$ or $E \leq^{<\omega}_{\alpha,X} F$ or $E \leq^\omega_{\alpha,X} F$.

As a natural example of an equivalence relation to which this can be applied, let X be any set which is not c.e. Then the equivalence relation

$$A =_X B \iff (\forall n \in X)[n \in A \iff n \in B]$$

is computably X-reducible to the equality relation on 2^ω: just write $X = \{x_0 < x_1 < \cdots\}$ and let $n \in \Gamma^{X \oplus A}$ precisely when $x_n \in A$. Indeed, any enumeration of X would suffice for this purpose. However, if Φ were a computable binary reduction of $=_X$ to equality, then we would be able to enumerate X, since then

$$n \in X \iff \exists y \in B_0 \triangle B_1, \text{ where } B_0 \oplus B_1 = \Phi^{\emptyset \oplus \{n\}}.$$

We consider the notion of X-computable reducibility to be natural for further study, especially since it allows the notion of jump-reductions to be extended to arbitrary countable ordinals α (by taking an X which can compute a copy of α). However, we will not elaborate on this notion any further here.

6. The Glimm–Effros Dichotomy

Now we examine the analogue of the Glimm–Effros dichotomy for computable reducibility. Unsurprisingly, the strict analogue fails to hold. To prove this, we use the equivalence relation $=_Y$ defined in Sec. 5, for a set Y whose complement X is computably enumerable.

Here we build the set X, along with a partial computable injective function $\psi : \omega^2 \to \omega$ satisfying the following well-known requirements, for all e:

$$\mathcal{N}_e : \quad |\overline{X}| \geq e;$$

$$\mathcal{P}_e : \quad \overline{X} \neq W_e;$$

$$\mathcal{R}_e : \quad (\exists n \geq 0) \, \psi(e, n){\downarrow} \notin X;$$

$$\mathcal{S}_e : \quad \text{defined below.}$$

The $\mathcal{N}$- and $\mathcal{P}$-requirements are completely standard for finite-injury constructions. In our construction, $\mathcal{R}_e$ first chooses a large value $\psi(e, 0)$ and protects it from entering X. If a higher-priority requirement puts $\psi(e, 0)$ into X at a stage s, then $\mathcal{R}_e$ defines $\psi(e, 1)$ to be a new large number, greater than s and protects that number instead, and so on. A standard finite-injury construction builds a computably enumerable set X satisfying all these requirements. Notice that the image of ψ is decidable, since y can only enter it at stages $\leq y$. Building X this way (without the $\mathcal{S}$-requirements) will give us a contradiction to Glimm–Effros for computable reducibility on 2^ω; we will explain the $\mathcal{S}$-requirements below when we wish to strengthen the contradiction.

For the complement $Y = \overline{X}$, we define $=_Y$ on 2^ω, the *equality relation on Y*, as in the previous section:

$$A =_Y B \iff (\forall n \in Y)[n \in A \iff n \in B].$$

It was noted there that $=_X \, \leq_0^2 \, =$ if and only if X is a computably enumerable set. Our construction makes Y properly Π_1^0, and therefore $=_Y \, \nleq_0^2 \, =$. On the other hand, we do have a computable reduction Γ from $=$ to $=_Y$. For each $(e, n) \in \text{range}(\psi)$, Γ defines $\psi(e, n) \in \Gamma^A$ if and only if $e \in A$, with Γ^A disjoint from the (decidable) complement of range(ψ). By construction, if any e has $e \in A \triangle B$, then every $\psi(e, n)$ defined will lie in $\Gamma^A \triangle \Gamma^B$, and at least one of these (finitely many) values $\psi(e, n)$ will lie in Y. Thus $= \leq_0 =_Y$ via Γ, and so $= <_0 =_Y$.

However, we also claim that E_0 is incomparable with $=_Y$ under even binary computable reducibility, and that therefore $=_Y$ reveals the failure of the Glimm–Effros dichotomy for computable reducibility on 2^ω. The essence of this claim is that E_0 is a Σ_2^0 relation, whereas $=_Y$ is Π_2^0:

$$A =_Y B \iff (\forall n)(\exists s)[n \in X_s \text{ or } [n \in A \iff n \in B]].$$

In particular, if Φ were a binary computable reduction from E_0 to $=_Y$, then, given any pair $\langle i, j \rangle$ with $W_i = \overline{W_j}$, we would have

$$i \in \text{Fin} \iff W_i \, E_0 \, \emptyset \iff (\forall n)(\exists s)[n \in X_s \text{ or}$$
$$[n \in B_0 \iff n \in B_1]],$$

where $\Phi^{W_i \oplus \emptyset} = B_0 \oplus B_1$. But given the pair (i, j), one could compute the function $\Phi^{W_i \oplus \emptyset}$ (which is total, by assumption), making the right-hand side a Π_2^0 predicate of i and j, which is impossible. (The pair (Fin, Inf) has a natural 1-reduction to (Fin $\cap$ D, Inf $\cap$ D), with $D = \{(i, j) : W_i = \overline{W_j}\}$ as before: just map each e to an index of the c.e. set $\{s : W_{e,s+1} \neq W_{e,s}\}$. The argument above then gives a 1-reduction from (Fin $\cap$ D) to a Π_2^0 set, for a contradiction.)

The $\mathcal{S}$ requirements are used to ensure that $=_Y \not\leq_0^2 E_0$, completing the proof of incomparability. $\mathcal{S}_e$ requires that, if the eth oracle Turing functional Φ_e is total on all oracles, then there should be a decidable set $D_e = W_i = \overline{W_j}$ (for some i and j) such that

$$D_e \cap Y \neq \emptyset \iff \Phi_e^{D_e \oplus \emptyset} \text{ consists of two } E_0\text{-inequivalent columns.}$$

Of course, the left-hand side just means that $D_e \neq_Y \emptyset$, so the requirements together will show that $=_Y \not\leq_0^2 E_0$.

We break up $\mathcal{S}_e$ into countably many subrequirements, so as to preserve the finite-injury nature of the construction. Each $\mathcal{S}_{e,n+1}$ inherits from $\mathcal{S}_{e,n}$ an element $x_{e,n}$ of D_e, at a stage s at which $\mathcal{S}_{e,n}$ has completed its action (if it ever does). We write $D_{e,s} = D_e \restriction s$ for the portion of D_e so far decided. This $x_{e,n}$ does not lie in X_s, so it currently appears to establish the left-hand side of $\mathcal{S}_e$. $\mathcal{S}_{e,n+1}$ waits until the symmetric difference between the columns of $\Phi_e^{(D_{e,s} {}^\frown 000000\cdots)\oplus\emptyset}$ contains $\geq n+1$ elements, extending $D_{e,s}$ by a zero at each step to encourage convergence. If the symmetric difference ever reaches this size, then $\mathcal{S}_{e,n}$ adjoins $x_{e,n}$ to X (so that it no longer establishes the left-hand side), and defines a new $x_{e,n+1}$, larger than the current stage, to hand off to $\mathcal{S}_{e,n+2}$. If the symmetric difference contains $\leq n$ elements in total at the end of the construction, then $\mathcal{S}_{e,n+1}$ has satisfied $\mathcal{S}_e$ all by itself; whereas if the symmetric difference between the columns of $\Phi_e^{D_e\oplus\emptyset}$ is infinite, then every element $x_{e,n}$ of D_e is eventually enumerated into X (hence out of Y) by its $\mathcal{S}_{e,n+1}$. Either way, $\mathcal{S}_e$ holds.

A given $\mathcal{S}_{e,n+1}$ may be injured if a higher-priority $\mathcal{P}$-requirement adds $x_{e,n}$ to X. If so, $\mathcal{S}_{e,n+1}$ simply chooses a new large $x'_{e,n}$, places it in D_e, and continues with it. This will only happen finitely often (for a given e and n), so $\mathcal{S}_{e,n}$ does accomplish its goal, possibly with finitely many false starts. All the other requirements fit together according to a standard finite-injury priority construction, and so for the set $Y = \overline{X}$, the equivalence relation $=_Y$ on 2^ω lies strictly above equality under computable reducibility, is incomparable with E_0 even under binary computable reducibility.

It is hardly surprising that, in the far more exacting context of computable reducibility, the Glimm–Effros dichotomy should fail. The surprise, after all, was that it held for Borel reducibility in the first place. Nevertheless, the arguments here, simple though they be, demonstrate that there is much more to investigate in this topic. For example, while Glimm–Effros fails for equality and E_0 under computable reducibility, it remains open whether there might be some two other equivalence relations which would be (respectively) the least and the second-least among all smooth equivalence relations on 2^ω under computable reducibility. Alternatively, it seems a bit more plausible that Glimm–Effros might hold for ω-jump reducibility, or λ-jump reducibility for some limit ordinals λ. The notions from Sec. 5 would even allow one to address λ-jump reducibility for admissible ordinals λ, which appears to be the most promising ground of all.

Acknowledgments

The author was partially supported by Grant # DMS – 1362206 from the National Science Foundation, and by a series of grants from The City University of New York PSC-CUNY Research Award Program. He wishes to acknowledge several useful suggestions and corrections by the anonymous referee.

References

[1] U. Andrews, S. Lempp, J. S. Miller, K. M. Ng, L. San Mauro and A. Sorbi, Universal computably enumerable equivalence relations, Technical Report 504, Department of Mathematics and Computer Science "Roberto Magari", University of Siena (2012).

[2] C. Bernardi and A. Sorbi, Classifying positive equivalence relations, *J. Symbolic Logic* **48**(3) (1983) 529–538.

[3] S. Coskey, J. D. Hamkins and R. Miller, The hierarchy of equivalence relations on the natural numbers under computable reducibility, *Computability* **1**(1) (2012) 15–38.

[4] E. G. Effros, Transformation groups and c^*-algebras, *Ann. Math.* (2) **81**(1) (1965) 38–55.

[5] Yu. L. Eršov, *Teoriya Numeratsii*, Matematicheskaya Logika i Osnovaniya Matematiki. [Monographs in Mathematical Logic and Foundations of Mathematics] (Nauka, Moscow, 1977).

[6] E. B. Fokina and S.-D. Friedman, On Σ_1^1 equivalence relations over the natural numbers, *Math. Logic Quart.* **58**(1–2) (2012) 113–124.

[7]   E. B. Fokina, S.-D. Friedman and A. Törnquist, The effective theory of Borel equivalence relations. *Ann. Pure Appl. Logic*, **161**(7) (2010) 837–850.

[8]   S. Gao, *Invariant Descriptive Set Theory*, Pure and Applied Mathematics (Boca Raton) (CRC Press, Boca Raton, FL, 2009).

[9]   S. Gao and P. Gerdes, Computably enumerable equivalence relations, *Studia Logica* **67**(1) (2001) 27–59.

[10] J. Glimm, Locally compact transformation groups, *Trans. Amer. Math. Soc.* **101** (1961) 124–138.

[11] L. A. Harrington, A. S. Kechris and A. Louveau, A Glimm–Effros dichotomy for Borel equivalence relations, *J. Amer. Math. Soc.* **3**(4) (1990) 903–928.

[12] I. Ianovski, R. Miller, K. M. Ng and A. Nies, Complexity of equivalence relations and preorders from computability theory, *J. Symbolic Logic* **79**(3) (2014) 859–881.

[13] R. Miller and K. M. Ng, Finitary reducibility on equivalence relations, *J. Symbolic Logic* **81**(4) (2016) 1225–1254.

Chapter 6

Logic Programming and Effectively Closed Sets

Douglas Cenzer[*,§], Victor W. Marek[†,¶] and Jeffrey B. Remmel[‡,‖]

*Department of Mathematics, University of Florida,
Gainesville, FL 32611, USA
†Department of Computer Science,
University of Kentucky, Lexington, KY 40506, USA
‡Department of Mathematics, University of California at
San Diego, La Jolla, CA 92903, USA
§cenzer@math.ufl.edu
¶marek@cs.uky.edu
‖jremmel@ucsd.edu

This paper surveys results on index problems for effectively closed sets and their applications to models of logic programs. A new notion of boundedness for trees is introduced and the complexity of index sets for the corresponding closed sets is examined. This paper focuses on the *recognition problem* in the metaprogramming of finite normal predicate logic programs. That is, let $\mathcal{L}$ be a computable first-order predicate language with infinitely many constant symbols and infinitely many n-ary predicate symbols and n-ary functions symbols for all $n \geq 1$. Then we can effectively list all the finite normal predicate logic programs $Q_0, Q_1, \ldots$ over $\mathcal{L}$. Given some property $\mathcal{P}$ of finite normal predicate logic programs over $\mathcal{L}$, we define the index set $I_{\mathcal{P}}$ to be the set of indices e such that Q_e has property $\mathcal{P}$. We classify the complexity of the index set $I_{\mathcal{P}}$ within the arithmetic hierarchy for various natural properties of finite predicate logic programs. For example, we determine the complexity of the index sets relative to all finite predicate logic programs and relative to certain special classes of finite predicate logic programs of properties such as (i) having no stable models, (ii) having no recursive stable models, (iii) having exactly c stable models for any given positive integer c, (iv) having exactly c recursive stable models for any given positive integer c, (v) having only finitely many stable models, (vii) having only finitely many recursive stable models.

1. Introduction

Effectively closed sets, or Π_1^0 classes, play a central role in computability theory and its applications. In the Cantor space $\{0, 1\}^{\mathbb{N}}$ of infinite binary sequences and in the Baire space $\mathbb{N}^{\mathbb{N}}$ of infinite sequences of natural numbers, a closed set is the set of paths $[T]$ through a tree T, and an effectively closed set is the set of infinite paths through a computable tree. An open subset of the real line $\Re$ is a countable union of open intervals, and an effectively open set is the union of a computable sequence of intervals. Then, a closed subset of $\Re$ is simply the complement of an open subset, and similarly an effectively closed set is the complement of an effectively open set. The importance of Π_1^0 classes lies in the fact that, in many areas of mathematics, the set of solutions to a given problem may be represented as a closed set, and if the problem is computable, then the set of solutions may be given as an effectively closed set. Here are a few examples. In combinatorics, we have the problem of finding a matching in a bipartite graph and the problem of finding a Hamiltonian circuit in a planar graph. There is the problem of finding a k-coloring of an infinite graph, the set of solutions is the set of k-colorings In analysis, there is the problem of finding a zero for a continuous function on the real interval. In algebra, there is the problem of finding a prime ideal in a commutative ring with unity, and in particular in a Boolean algebra. This is closely related to the classic problem in mathematical logic is of finding a complete consistent extension of an axiomatizable theory. Shoenfield [35] showed that the family of complete consistent extensions of an axiomatizable theory may be presented as an effectively closed set in the Cantor space ($\{0,1\}^{\mathbb{N}}$), and Ehrenfeucht [18] showed that any effectively closed set in $\{0,1\}^{\mathbb{N}}$ can represent such a family. Some details will be given below, since the study of stable models of logic programs may be viewed as a natural extension of the study of axiomatizable theories. For the other problems mentioned above, the reader is referred to [9–11].

This paper is focused on the connection between Π_1^0 classes and logic programming. The study of logic programming and nonmonotonic logic in general has been an important area of theoretical computer science for many years. Past research has demonstrated that logic programming with the stable model semantics and, more generally, with the answer-set semantics, is an expressive knowledge representation formalism. It can be safely stated that there is a consensus in the Knowledge Representation community that stable models are the correct generalization of the least model of a Horn program for the class of normal programs. Although the stable model semantics is considered the correct one, past research has also shown that the use of arbitrary normal logic programs admitting function symbols is not a reasonable choice for real-life programming. For example, Apt and Blair [2] proved that all arithmetic sets can be defined by using stratified programs. The import of that result is that in general, it is impossible to effectively query the unique stable model of such programs. Marek, Nerode, and Remmel [28] constructed finite predicate logic programs whose stable models could code up the paths through any infinitely branching recursive tree so that the problem of deciding whether a finite predicate logic program has a stable model is Σ_1^1-complete. Cenzer and Remmel showed in [12] that there is a closed connection between the construction of the perfect kernel of a Π_1^0 class via the iteration of the Cantor–Bendixson derivative through the ordinals and the construction of the well-founded semantics for finite predicate logic programs via Van Gelder's alternating fixed-point construction [33].

For such reasons, researchers have focused on finite predicate logic programs without function symbols. There are a number of highly effective implementations search engines to find stable models of logic programs (usually known as *Answer Set Programming* (*ASP*)) [19, 29]. Indeed, Answer set programming competitions are held [17] and the site [4] provides numerous benchmarks that can be used by designers of ASP systems. However, these search engines are, basically, limited to finite propositional programs or finite predicate programs not admitting function symbols. In addition, search engines for the well-founded semantics of fragments of first-order logic extended by inductive definitions have been implemented [14, 16].

Nevertheless, researchers have searched for *some* natural classes $\mathcal{K}$ of finite normal predicate logic programs with function symbols where programming is both useful and possible. That is, there are two key properties that one requires of such classes. First, $\mathcal{K}$ needs to be *processable*. That is, given a program $P \in \mathcal{K}$, we need to have an algorithm that identifies one or more stable models of P which can be effectively queried. In particular, we would like to be able to effectively answer questions such as whether a given atom is in a given stable model of P or whether a given atom is in all stable models of P. Second, $\mathcal{K}$ needs to be *recognizable*. That is, we need to be able to answer the query whether a given program P belongs to $\mathcal{K}$. For instance, the class of stratified programs is recognizable (one of fundamental results of Apt, Blair and Walker [3]), but it is not processable. A number of classes $\mathcal{K}$ of such programs which are both processable and recognizable have been found, see [5–7, 25, 32]. In particular, the paper [6] provides an extensive discussion of the reasons why researchers try to find classes of normal predicate logic programs admitting function symbols which are both recognizable and processable.

The goal of this paper is develop a systematic approach to study the recognition problem for the class of finite normal predicate logic programs over a computable first-order predicate language $\mathcal{L}$ with infinitely many constant symbols and infinitely many n-ary predicate symbols and n-ary functions symbols for all $n \geq 1$. Let $Q_0, Q_1, \ldots$ be an effective list all the finite normal predicate logic programs over $\mathcal{L}$. Given some property $\mathcal{P}$ of finite normal predicate logic programs over $\mathcal{L}$, we define the *index set* $I_{\mathcal{P}}$ to be the set of indices e such that Q_e has property $\mathcal{P}$. For example, suppose that $\mathcal{P}$ is the property that a finite normal predicate logic program has a recursive stable model. Then the tools of this paper will allow one to classify the complexity of $I_{\mathcal{P}}$ within the arithmetic hierarchy. In particular, we shall show that $I_{\mathcal{P}}$ is Σ_3^0-complete so that one can not effectively recognize the set of finite predicate logic programs which have recursive stable models.

Marek, Nerode, and Remmel [28] studied the stable models of infinite propositional programs. They showed that problem of finding a stable model of a recursive normal propositional logic program is essentially equivalent to finding an infinite path through an infinite recursive tree. That is, they showed that given any recursive normal propositional logic program P, one could construct a recursive tree T_P such that there is an effective one-to-one degree preserving correspondence between the set of stable models of P and the set of infinite paths through T_P. Vice versa, given any recursive tree T, they constructed a recursive normal propositional logic program P_T such that there is an effective one-to-one degree preserving correspondence between the set of stable models of P_T and the set of infinite paths through T. Such correspondences also helped to motivate the definition of various natural properties of normal propositional logic programs such as having the finite support property or the recursive finite support property (described below) since these properties correspond to natural properties of recursive trees such as being finitely branching or being highly recursive.

The main goal of this paper is to provide similar constructions when we replace recursive normal propositional logic programs by finite normal predicate logic programs. This requires us to significantly modify the original constructions in [28]. This will also respond to some suggestions from computer scientists that finite programs are a more natural object of study than infinite programs. Some initial work on index sets for finite logic programs appeared in [8].

Index sets are used here to classify the complexity of various problems. In computability theory, there is a natural enumeration T_e of the primitive recursive trees, which provides an enumeration $[T_e]$ of the Π^0_1 classes. Then the fact that a recursive binary tree may have no computable path is improved to say that the index set $\{e : [T_e]$ has a computable member$\}$ is Σ^0_3 complete, whereas $\{e : [T_e]$ is nonempty$\}$ is Π^0_2. There is also a natural enumeration Q_e of the normal finite predicate logic programs. If a logic program has the rather natural *finite support property* or *FSP*, then we will show that the index set $\text{Stab}(Q_e)$ of stable models of Q_e may be coded by a recursive tree. This leads to the result that $\{e : [Q_e]$ has a computable stable model$\}$ is Σ^0_3 complete. This paper will have a large number of similar results.

Our main result is to construct recursive functions f and g such that, for all e, (i) there is a one-to-one degree preserving correspondence between the set of stable models of Q_e and the set of infinite paths through $T_{f(e)}$ and (ii) there is a one-to-one degree preserving correspondence between the set of infinite paths through T_e and the set of stable models $Q_{g(e)}$. One can often use these two recursive functions to reduce the problem of finding the complexity of index set for a certain property $\mathcal{P}$ of finite normal predicate logic programs to the problem of computing the complexity of the index set for an appropriate property of $\mathcal{P}'$ of primitive recursive trees. We will often take the reverse point of view. That is, we shall start with a property $\mathcal{P}'$ of trees and try to find an appropriate property $\mathcal{P}$ of finite normal predicate logic programs such that the corresponding index sets are one-to-one equivalent.

The outline of this paper is as follows. Section 2 provides necessary background on normal predicate logic programs. Many of the properties such as the finite support property and the recursive finite support property which correspond to bounded trees and recursively bounded trees have appeared in the literature. However, other properties such as a program being decidable, which correspond to decidable trees, are new. This will allows us to state the main results of the paper concerning the construction of the recursive functions f and g described above. Section 3 provides background on effectively closed sets, including various notions of boundedness, such as bounded, almost always bounded, nearly bounded, and recursive versions of these notions. Here we also review the complexity results for index set for trees that we shall use, as well as prove a number of new results on the complexity of index sets for trees that will be useful for our purposes. For example, the complexity of the family of bounded recursive trees which have finitely many infinite paths is Σ_4^0-complete. In Section 4, we shall prove our main results. In Sections 5 and 6, we shall use the results of previous sections to find corresponding complexity results for index sets for finite predicate logic programs. For example, since there is a correspondence between recursively bounded trees and finite logic programs which have the finite support property, then the complexity of the family of finite logic programs which have the FSP and have finitely many stable models is Σ_4^0-complete.

2. Background on Normal Logic Programs

We shall fix a recursive language $\mathcal{L}$ which has infinitely many constant symbols $c_0, c_1, \ldots$, infinitely many variables $x_0, x_1, \ldots$, infinitely many propositional letters $A_0, A_1, \ldots$, and for each $n \geq 1$, infinitely many n-ary relation symbols $R_0^n, R_1^n, \ldots$ and n-ary function symbols $f_0^n, f_1^n, \ldots$. We note here that we shall generally use the terminology *recursive* rather than the equivalent term *computable* and likewise use *recursively enumerable* rather than *computably enumerable*. These terms have the same meaning, but the former are standard in the logic programming community which is an important audience for our paper.

A literal is an atomic formula or its negation. A ground literal is a literal which has no free variables. The Herbrand base of $\mathcal{L}$ is the set $H_{\mathcal{L}}$ of all ground atoms (atomic statements) of the language.

A (normal) logic programming clause C is of the form

$$c \leftarrow a_1, \ldots, a_n, \neg b_1, \ldots, \neg b_m, \tag{1}$$

where $c, a_1, \ldots, a_n, b_1, \ldots, b_m$ are atoms of $\mathcal{L}$. Here we allow either n or m to be zero. In such a situation, we call c the *conclusion* of C, $a_1, \ldots, a_n$ the *premises* of C, $b_1, \ldots, b_n$ the *constraints* of C and $a_1, \ldots, a_n, \neg b_1, \ldots, \neg b_m$ the *body* of C and write $\mathrm{concl}(C) = c$, $\mathrm{prem}(C) = \{a_1, \ldots, a_n\}$, $\mathrm{constr}(C) = \{b_1, \ldots, b_m\}$. A ground clause is a clause with no free variables. C is called a Horn clause if $\mathrm{constr}(C) = \emptyset$, i.e., if C has no negated atoms in its body.

A finite normal predicate logic program P is a finite set of clauses of the form (1). P is said to be a Horn program if all its clauses are Horn clauses. A ground instance of a clause C is a clause obtained by substituting ground terms (terms without free variables) for all the free variables in C. The set of all ground instances of the clauses in program P is called $\mathrm{ground}(P)$. The Herbrand base of P, $H(P)$, is the set of all ground atoms that are instances of atoms that appear in P. For any set S, we let 2^S denote the set of all subsets of S.

Given a Horn program P, we let $T_P : 2^{H(P)} \to 2^{H(P)}$ denote the usual one-step provability operator [26] associated with $\mathrm{ground}(P)$. That is, for $S \subseteq H(P)$,

$$T_P(S) = \{c : \exists_{C \in \mathrm{ground}(P)}((C = c \leftarrow a_1, \ldots, a_n)$$
$$\wedge (a_1, \ldots, a_n \in S))\}.$$

Then P has a least model Herbrand $M = T_P \uparrow_\omega (\emptyset) = \bigcup_{n \geq 0} T_P^n(\emptyset)$ where, for any $S \subseteq H(P)$, $T_P^0(S) = S$ and $T_P^{n+1}(S) = T_P(T_P^n(S))$. We denote the least model of a Horn program P by $\text{lm}(P)$.

Given a normal predicate logic program P and $M \subseteq H(P)$, we define the *Gelfond–Lifschitz reduct* of P, P_M, via the following two step process. In Step 1, we eliminate all clauses $C = p \leftarrow q_1, \ldots, q_n, \neg r_1, \ldots, \neg r_m$ of ground(P) such that there exists an atom $r_i \in M$. In Step 2, for each remaining clause $C = p \leftarrow q_1, \ldots, q_n, \neg r_1, \ldots, \neg r_m$ of ground(P), we replace C by the Horn clause $C = p \leftarrow q_1, \ldots, q_n$. The resulting program P_M is a Horn propositional program and, hence, has a least model. If the least model of P_M coincides with M, then M is called a *stable model* for P. The stable model semantics for logic programs was first developed by Gelfond and Lifschitz [21]; see also [20].

Next, we define the notion of a P-proof scheme of a normal *propositional* logic program P. Given a normal propositional logic program P, a P-proof scheme is defined by induction on its length n. Specifically, the set of P-proof schemes is defined inductively by declaring that

(I) $\langle\langle C_1, p_1 \rangle, U \rangle$ is a P-proof scheme of length 1 if $C_1 \in P$, $p_1 = \text{concl}(C_1)$, prem$(C_1) = \emptyset$, and $U = \text{constr}(C_1)$ and

(II) for $n > 1$, $\langle\langle C_1, p_1 \rangle, \ldots, \langle C_n, p_n \rangle, U \rangle$ is a P-proof scheme of length n if $\langle\langle C_1, p_1 \rangle, \ldots, \langle C_{n-1}, p_{n-1} \rangle, \bar{U} \rangle$ is a P-proof scheme of length $n - 1$ and C_n is a clause in P such that concl$(C_n) = p_n$, prem$(C_n) \subseteq \{p_1, \ldots, p_{n-1}\}$ and $U = \bar{U} \cup \text{constr}(C_n)$.

If $\mathbb{S} = \langle\langle C_1, p_1 \rangle, \ldots, \langle C_n, p_n \rangle, U \rangle$ is a P-proof scheme of length n, then we let supp$(\mathbb{S}) = U$ and concl$(\mathbb{S}) = p_n$.

Example 2.1. Let P be the normal propositional logic program consisting of the following four clauses:

$C_1 = p \leftarrow$, $C_2 = q \leftarrow p, \neg r$, $C_3 = r \leftarrow \neg q$, and $C_4 = s \leftarrow \neg t$.
Then we have the following useful examples of P-proof schemes:

(a) $\langle \langle C_1, p \rangle, \emptyset \rangle$ is a P-proof scheme of length 1 with conclusion p and empty support.

(b) $\langle \langle C_1, p \rangle, \langle C_2, q \rangle, \{r\} \rangle$ is a P-proof scheme of length 2 with conclusion q and support $\{r\}$.

(c) $\langle \langle C_1, p \rangle, \langle C_3, r \rangle, \{q\} \rangle$ is a P-proof scheme of length 2 with conclusion r and support $\{q\}$.

(d) $\langle \langle C_1, p \rangle, \langle C_2, q \rangle, \langle C_3, r \rangle, \{q, r\} \rangle$ is a P-proof scheme of length 3 with conclusion r and support $\{q, r\}$.

In this example, we see that the proof scheme in (c) had an unnecessary item, the first term, while in (d) the proof scheme was supported by a set containing q, one of atoms that were proved on the way to r. □

A P-proof scheme differs from the usual Hilbert-style proofs in that it carries within itself its own applicability condition. In effect, a P-proof scheme is a *conditional* proof of its conclusion. It becomes applicable when all the constraints collected in the support are satisfied. Formally, for a set M of atoms, we say that a P-proof scheme $\mathbb{S}$ is *M-applicable* or that *M admits $\mathbb{S}$* if $M \cap \mathrm{supp}(\mathbb{S}) = \emptyset$. The fundamental connection between proof schemes and stable models is given by the following proposition which is proved in [27].

Proposition 2.1. *For every normal propositional logic program P and every set M of atoms, M is a stable model of P if and only if*

(i) *for every $p \in M$, there is a P-proof scheme $\mathbb{S}$ with conclusion p such that M admits $\mathbb{S}$ and*

(ii) *for every $p \notin M$, there is no P-proof scheme $\mathbb{S}$ with conclusion p such that M admits $\mathbb{S}$.*

A P-proof scheme may not need all its clauses to prove its conclusion. It may be possible to omit some clauses and still have a proof scheme with the same conclusion. Thus, we define a pre-order on P-proof schemes $\mathbb{S}$, $\mathbb{T}$ by declaring that $\mathbb{S} \prec \mathbb{T}$ if

(1) $\mathbb{S}, \mathbb{T}$ have the same conclusion,
(2) every clause in $\mathbb{S}$ is also a clause of $\mathbb{T}$.

The relation $\prec$ is reflexive, transitive, and well-founded. Minimal elements of $\prec$ are *minimal proof schemes*. A given atom may be the conclusion of no, one, finitely many, or infinitely many different minimal P-proof schemes. These differences are clearly computationally significant if one is searching for a justification of a conclusion.

If P is a finite normal predicate logic program, then we define a P-proof scheme to be a ground(P)-proof scheme. Since we are considering finite normal programs over our fixed recursive language $\mathcal{L}$, we can use standard Gödel numbering techniques to assign code numbers to atomic formulas, clauses, and proof schemes. That is, we can effectively assign a natural number to each symbol in $\mathcal{L}$. Strings may be coded by natural numbers in the usual fashion. Let $\omega = \{0, 1, 2, \ldots\}$ denote the set of natural numbers and let $[x, y]$ denote the standard pairing function $\frac{1}{2}(x^2 + 2xy + y^2 + 3x + y)$ and, for $n \geq 2$, we let $[x_0, \ldots, x_n] = [[x_0, \ldots, x_{n-1}], x_n]$. Then a string $\sigma = (\sigma(0), \ldots, \sigma(n-1))$ of length n may be coded by $c(\sigma) = [n, [\sigma(0), \sigma(1), \ldots, \sigma(n-1)]]$ and also $c(\emptyset) = 0$. We define the canonical index of any finite set $X = \{x_1 < \cdots < x_n\} \subseteq \omega$ by $\mathrm{can}(X) = 2^{x_1} + 2^{x_2} + \cdots + 2^{x_n}$. We define $\mathrm{can}(\emptyset) = 0$. Then we can think of formulas of $\mathcal{L}$ as sequences of natural numbers so that the code of a formula is just the code of the sequence of numbers associated with the symbols in the formula. Then a clause C as in (1) can be assigned the code of the triple (x, y, z) where x is the code of the conclusion of C, y is the canonical index of the set of codes of prem(C), and z is the canonical index of the set of codes of constr(C). Finally, the code of a proof scheme $\mathbb{S} = \langle \langle C_1, p_1 \rangle, \ldots, \langle C_n, p_n \rangle, U \rangle$ consists of the code of a pair (s, t) where s is the code of the sequence $(a_1, \ldots, a_n)$ where a_i is the code of the pair of codes for C_i and

p_i and t is the canonical index of the set of codes for elements of U. It is then not difficult to verify that for any given finite normal predicate logic program P, the questions of whether a given n is the code of a ground atom, a ground instance of a clause in P, or a P-proof are primitive recursive predicates. The key observation to make is that since P is finite and the usual unification algorithm is effective, we can explicitly test whether a given number m is the code of a ground atom or a ground instance of a clause in P without doing any unbounded searches. It is then easy to see that, once we can determine if a number m is a code of ground instance of a clause of P in a primitive recursive fashion, then there is a primitive recursive procedure which determines whether a given number n is the code of a minimal P-proof scheme.

If P is a finite normal predicate logic program over $\mathcal{L}$, we let $N_k(P)$ be the set of all codes of minimal P-proof schemes $\mathbb{S}$ such that all the atoms appearing in all the rules used in $\mathbb{S}$ are smaller than k. Obviously $N_k(P)$ is finite. Since the predicate "minimal P-proof scheme", which holds only on codes of minimal P-proof schemes, is a primitive recursive predicate, it easily follows that we can uniformly construct a primitive recursive function h_P such that $h_P(k)$ equals the canonical index for $N_k(P)$.

A finite normal predicate logic program Q over $\mathcal{L}$ may be written out as a finite string over a finite alphabet and thus may be assigned a Gödel number $e(Q)$ in the usual fashion. The set of Gödel numbers of well-formed programs is well known to be primitive recursive (see [26]). Thus we may let Q_e be the program with Gödel number e when this exists and let Q_e be the empty program otherwise. For any property $\mathcal{P}$ of finite normal predicate logic programs, let $I(\mathcal{P})$ be the set of indices e such that Q_e has property $\mathcal{P}$.

Next, we define the notions of decidable normal logic programs and of normal logic programs which have the finite support property. Proposition 2.1 says that the presence and absence of the atom p in a stable model of a finite normal predicate logic program P depends *only* on the supports of its ground(P)-proof schemes. This fact naturally leads to a characterization of stable models in terms of propositional satisfiability. Given $p \in H(P)$, the *defining equation* for p with respect to P is the following propositional formula:

$$p \Leftrightarrow (\neg U_1 \vee \neg U_2 \vee \cdots), \qquad (2)$$

where $\langle U_1, U_2, \ldots \rangle$ is the list of all supports of minimal ground(P)-proof schemes. Here, for any finite set $S = \{s_1, \ldots, s_n\}$ of atoms, $\neg S = \neg s_1 \wedge \cdots \wedge \neg s_n$. If $U = \emptyset$, then $\neg U = \top$. Up to a total ordering of the finite sets of atoms such a formula is unique. For example, suppose we fix a total order on $H(P)$, $p_1 < p_2 < \cdots$. Then given two sets of atoms, $U = \{u_1 < \cdots < u_m\}$ and $V = \{v_1 < \cdots < v_n\}$, we say that $U \prec V$, if either (i) $u_m < v_n$, (ii) $u_m = v_n$ and $m < n$, or (iii) $u_m = v_n$, $n = m$, and $(u_1, \ldots, u_n)$ is lexicographically less than $(v_1, \cdots, v_n)$. We also define $\emptyset \prec U$ for any finite nonempty set U. We say that (2) is the *defining equation* for p relative to P if $U_1 \prec U_2 \prec \ldots$. We will denote the defining equation for p with respect to P by Eq_p^P. When P is a Horn program, an atom p may have an empty support or no support at all. The first of these alternatives occurs when p belongs to the least model of P, $\mathrm{lm}(P)$. The second alternative occurs when $p \notin \mathrm{lm}(P)$. The defining equations in such case are $p \Leftrightarrow \top$ when $p \in \mathrm{lm}(P)$ and $p \Leftrightarrow \bot$ when $p \notin \mathrm{lm}(P)$.

Let Φ_P be the set $\{Eq_p^P : p \in H(P)\}$. We then have the following consequence of Proposition 2.1.

Proposition 2.2. *Let P be a normal propositional logic program. Then the stable models of P are precisely the propositional models of the theory Φ_P.*

When P is *purely negative*, i.e., all clauses C of P have $\mathrm{prem}(C) = \emptyset$, the stable and supported models of P coincide [15] and the defining equations reduce to Clark's completion [13] of P.

Let us observe that, in general, the propositional formulas on the right-hand side of the defining equations may be infinitary.

Example 2.2. Let P be an infinite normal propositional logic program consisting of clauses $p \leftarrow \neg p_i$, for all $i \in n$. Then the defining equation for p in P is the infinitary propositional formula

$$p \Leftrightarrow (\neg p_1 \vee \neg p_2 \vee \neg p_3 \vee \cdots).$$ □

The following observation is quite useful. If U_1, U_2 are two finite sets of propositional atoms, then

$$U_1 \subseteq U_2 \text{ if and only if } \neg U_2 \models \neg U_1.$$

Here $\models$ is the propositional consequence relation. The effect of this observation is that only the inclusion-minimal supports are important.

Example 2.3. Let P be an infinite normal propositional logic program consisting of clauses $p \leftarrow \neg p_1, \ldots, \neg p_i$, for all $i \in N$. The defining equation for p in P is

$$p \Leftrightarrow [\neg p_1 \vee (\neg p_1 \wedge \neg p_2) \vee (\neg p_1 \wedge \neg p_2 \wedge \neg p_3) \vee \ldots],$$

which is infinitary. But our observation above implies that this formula is *equivalent* to the formula

$$p \Leftrightarrow \neg p_1.$$ □

Motivated by the Example 2.3, we define the *reduced defining equation* for p relative to P to be the formula

$$p \Leftrightarrow (\neg U_1 \vee \neg U_2 \vee \cdots), \qquad (3)$$

where U_i range over *inclusion-minimal* supports of minimal P-proof schemes for the atom p and $U_1 \prec U_2 \prec \cdots$. We denote this formula as rEq_p^P, and define $r\Phi_P$ to be the theory consisting of rEq_p^P for all $p \in H(P)$. We then have the following strengthening of Proposition 2.2.

Proposition 2.3. *Let P be a normal propositional program. Then stable models of P are precisely the propositional models of the theory $r\Phi_P$.*

In Example 2.3, the theory Φ_P was infinitary, but the theory $r\Phi_P$ was finitary.

Suppose that P is a normal propositional logic program P which consists of ground clauses from $\mathcal{L}$ and a is an atom in $H(P)$. Then we say that a has the *finite support property relative of P* if the reduced defining equation for a is finite. We say that P has the *finite support (FS) property* if for all $a \in H(P)$, the reduced defining equation for a is a finite propositional formula. Equivalently, a program P has the finite support property if for every atom $a \in H(P)$, there are only finitely many inclusion-minimal supports of minimal P-proof schemes for a. We say that P has the *almost always finite support (a.a.FS) property* if for all but finitely many atoms $a \in H(P)$, there are only finitely many inclusion-minimal supports of minimal P-proof schemes for a. We say that P is *recursive* if the set of codes of clauses of P is recursive and the set of codes of atoms in $H(P)$ is recursive. Note that for any finite normal predicate logic program Q, ground(Q) will automatically be a recursive normal propositional logic program. We say that P has the *recursive finite support (rec.FS) property* if P is recursive, has the finite support property, and there is a uniform effective procedure which given any atom $a \in H(P)$ produces the code of the set of the inclusion-minimal supports of P-proof schemes for a. We say that P has the *almost always recursive finite support (a.a.FS) property* if P is recursive, has the a.a.FS property, and there is a uniform effective procedure which for all but a finite set of atoms $a \in H(P)$ produces the code of the set of the inclusion-minimal supports of P-proof schemes for a. We say that a finite normal predicate logic program has the *FS* property (*rec.FS* property, *a.a.FS* property, *a.a.rec.FS* property) if ground(P) has the *FS* property (*rec.FS* property, *a.a.FS* property, *a.a.rec.FS* property).

Next, we define two additional properties of recursive normal propositional logic programs that have not been previously defined in the literature. Suppose that P is a recursive normal propositional logic program consisting of ground clauses in $\mathcal{L}$ and M is a stable model of P. Then for any atom $p \in M$, we say that a minimal P-proof scheme $\mathbb{S}$ is the *smallest minimal P-proof for p relative to M* if $\text{concl}(\mathbb{S}) = p$ and $\text{supp}(\mathbb{S}) \cap M = \emptyset$ and there is no minimal P-proof scheme $\mathbb{S}'$ such that $\text{concl}(\mathbb{S}') = p$ and $\text{supp}(\mathbb{S}') \cap M = \emptyset$ and the Gödel number of $\mathbb{S}'$ is less than the Gödel number of $\mathbb{S}$.

We say that P is *decidable* if for all $N > 0$ and any finite (possibly empty) set of ground atoms $\{a_1, \ldots, a_n\} \subseteq H(P)$ such that the code of each a_i is less than or equal to N, and any finite set of minimal P-proof schemes $\{\mathbb{S}_1, \ldots, \mathbb{S}_n\}$ such that $\text{concl}(\mathbb{S}_i) = a_i$, we can effectively decide whether there is a stable model of M of P such that

(a) $a_i \in M$ and $\mathbb{S}_i$ is the smallest minimal P-proof scheme for a_i such that $\text{supp}(\mathbb{S}_i) \cap M = \emptyset$ and
(b) for any ground atom $b \notin \{a_1, \ldots, a_n\}$ such that the code of b is less than or equal to N, $b \notin M$.

We say that a finite normal predicate logic program is decidable if $\text{ground}(P)$ is decidable.

It will turn out that under our coding of trees into finite predicate logic programs, decidable trees induce decidable programs and under our coding of finite predicate logic programs into trees, decidable programs induce decidable trees. Moreover, decidability combined with the property of having the recursive finite support property ensures that there exists processable stable models when there are stable models. That is, we have the following theorem.

Theorem 2.1. *Suppose that P is a recursive normal logic program which has the recursive finite support property and is* decidable. Then if P has a stable model, we can effectively find a recursive stable model of P.

Proof. Let $a_0, a_1, \ldots$ be a list of all elements of $H(P)$ by increasing code numbers. That is, if c_i is the code of a_i, then $c_0 < c_1 < \cdots$. We will effectively construct a list of pairs of sets (A_i, R_i) for $i \geq 0$ such that for all i, $A_i \cap R_i = \emptyset$, $\{a_0, \ldots, a_i\} \subseteq A_i \cup R_i$, $A_i \subseteq \{a_0, \ldots, a_i\}$, $A_i \subseteq A_{i+1}$, and $R_i \subseteq R_{i+1}$. Then $A = \bigcup_{i \geq 0} A_i$ will be our desired recursive stable model. Thus, we shall think of A_i as being the set of atoms that we have accepted to be in the stable model at stage i and R_i as being the set of atoms that have been rejected from being in A at stage i. Our construction will proceed in stages.

Stage 0. Consider a_0. Since P has the recursive finite support property, we can effectively find, the supports of all the minimal P-proof schemes with conclusion a_0. If U is the support of a minimal proof scheme with conclusion a_0, then the fact that the set of minimal proofs schemes of P is r.e. means that we can search through the list of minimal proof schemes of P until we find the minimal proof scheme $\mathbb{S}_U$ with the smallest possible code such the conclusion of $\mathbb{S}_U$ is a_0 and the support of $\mathbb{S}_U$ is U. Thus if $U_1, \ldots, U_k$ is the set of all supports of minimal proof schemes with conclusion a_0, then we can effectively find proof schemes $\mathbb{S}_1, \ldots, \mathbb{S}_k$ such that for each i, $\mathbb{S}_i$ is the smallest minimal proof scheme such that the conclusion of $\mathbb{S}_i$ is a_0 and the support of $\mathbb{S}_i$ is U_i. Then since P is decidable, we use our effective procedure with $N = c_0$ to determine whether there is a stable model M for which $\mathbb{S}_i$ is the smallest minimal proof scheme such that $\text{supp}(\mathbb{S}_i) \cap M$. If there is no such i, then a_0 is not in any stable model so we set $A_0 = \emptyset$ and $R_0 = \{a_0\}$. If there is such an i, then we let t_0 be the least such i and we set $A_0 = \{a_0\}$ and $R_0 = \text{supp}(\mathbb{S}_{t_0})$.

Stage $s + 1$. Assume that at stage s, we have constructed A_s and R_s such that $A_s \cap R_s = \emptyset$, $\{a_0, \ldots, a_s\} \subseteq A_s \cup R_s$, $A_s \subseteq \{a_0, \ldots, a_s\}$, and for each $a \in A_s$, we have constructed a proof scheme $\mathbb{S}_a$ such that if $A_s = \{d_1, \ldots, d_k\}$ and $N_s = c_{s+1}$, then our decision procedure associated with the decidability of P will answer "yes" when we give it N_S, the set $\{d_1, \ldots, d_k\}$ and the corresponding proof schemes $\mathbb{S}_{d_1}, \ldots, \mathbb{S}_{d_k}$. Moreover, we assume that

$$R_s = \{a_i : i \leq s \ \& \ a_i \notin A_s\} \cup \bigcup_{i=1}^{k} \mathrm{supp}(\mathbb{S}_{d_i}).$$

This means that there is at least one stable model M such that for each i, $\mathbb{S}_{d_i}$ is the least proof scheme which witnesses that d_i is in M and $(\{a_0, \ldots, a_s\} - A_s) \cap M = \emptyset$.

Now consider a_{s+1}. By the fact that P has the recursive support property, we can effectively find the finite set of supports $V_1, \ldots, V_r$ of the minimal P-proof schemes of a_{s+1} and we can find P-proof schemes $\mathbb{T}_1, \ldots, \mathbb{T}_r$ such that for each $1 \leq i \leq r$, $\mathbb{T}_i$ is the smallest possible proof scheme with conclusion a_{s+1} and support V_i. Then for each $i < r$, we can query the decision procedure associated with the decidability of P on the set $\{d_1, \ldots, d_k, a_{s+1}\}$ and the corresponding proof schemes $\mathbb{S}_{d_1}, \ldots, \mathbb{S}_{d_k}, \mathbb{T}_i$. If we get an answer "yes" for any i, then we let t_{s+1} be the least such i and we set $A_{s+1} = A_s \cup \{a_{s+1}\}$ and $R_{s+1} = R_s \cup \mathrm{supp}(\mathbb{T}_{t_{s+1}})$. Note that since $\mathbb{S}_{d_1}, \ldots, \mathbb{S}_{d_k}, \mathbb{T}_i$ are the smallest minimal P proof schemes which witness that $d_1, \ldots, d_k, a_{s+1}$ are in some fixed stable model M such that $(\{a_0, \ldots, a_{s+1}\} - A_{s+1}) \cap M = \emptyset$, we must have that $A_{s+1} \cap R_{s+1} = \emptyset$. If there is no such i, then there is no stable model M which contains a_{s+1} and is such that for each i, $\mathbb{S}_{d_i}$ is the least proof scheme that witnesses that d_i is in M and $(\{a_0, \ldots, a_s\} - A_s) \cap M = \emptyset$. In that case, we let $A_{s+1} = A_s$ and $R_{s+1} = R_s \cup \{a_{s+1}\}$. It easily follows that our inductive assumption will hold at stage $s + 1$.

This completes the construction. It is easy to see that if $A = \bigcup_{s \geq 0} A_s$ and $R = \bigcup_{s \geq 0} R_s$, then $A \cap R = \emptyset$ and $\{a_0, a_1, \ldots\} \subseteq A \cup R$. Thus A and R partition $H(P)$. It is also easy to see that A is recursive since our construction is effective and at stage s, we have determined whether $a_s \in A$. We claim that A is stable model. That is, if A is not a stable model, then either there exists an a_s such that $a_s \in A$ and a_s has no P-proof scheme admitted by A or there is an $a_t \notin A$ such that a_t has an P-proof scheme which is admitted by A. Our construction ensures that if a_s is in A, then a_s has an P-proof scheme admitted by A. Thus suppose that $a_t \notin A$. Then let $W_1, \ldots, W_k$ be the supports of the minimal proof schemes of a_t. Let a_r be the largest element in $W_1 \cup \ldots \cup W_k$. Then consider what happens at stage r. Suppose $A_r = \{e_1, \ldots, e_k\}$. Then our construction also specifies minimal P-proof schemes $\mathbb{S}_1, \ldots, \mathbb{S}_k$ such that there is a stable model M such that for $1 \le i \le k$, $\mathbb{S}_i$ is the smallest proof scheme which witnesses that e_i is in M, $\mathrm{supp}(\mathbb{S}_1) \cup \cdots \cup \mathrm{supp}(\mathbb{S}_k) \subseteq R_r$, and $\{a_0, \ldots, a_r\} - A_r \cap M = \emptyset$. Thus a_t is not in M. Let $V_1, \ldots, V_b$ be the supports of the minimal P-proof schemes of a_{t+1}. This means that $M \cap \mathrm{supp}(V_i) \ne \emptyset$ for each i. But for each i, $\mathrm{supp}(V_i) \subseteq \{a_0, \ldots, a_r\}$ and $M \cap \{a_0, \ldots, a_r\} = A_r$. Thus it must be the case that $\mathrm{supp}(V_i) \cap A_r \ne \emptyset$ for all i and hence a_t does not have a P-proof scheme admitted by A. Thus A is a stable model of P. □

We now introduce and illustrate a technical concept that will be useful for our later considerations. At first glance, there are some obvious differences between stable models of normal propositional logic programs and models of sets of sentences in a propositional logic. For example, if T is a set of sentences in a propositional logic and $S \subseteq T$, then it is certainly the case that every model of T is a model of S. Thus a set of propositional sentences T has the property that if T has a model, then every subset of T has a model. This is not true for normal propositional logic programs. That is, suppose that P_0 is a normal propositional logic program which has a stable model and a is atom which is not in the Herbrand base of P_0, $H(P_0)$. Let P be the normal propositional logic program consisting of P_0 plus the clause $C = a \leftarrow \neg a$. Then P automatically does not have a stable model. That is, consider a potential stable model M of P. If $a \in M$, then C does not contribute to P_M so that there will be no clause of P_M with a in the head. Hence, a is not in the least model of P_M so that M is not a stable model of P. On the other hand, if $a \notin M$, then C will contribute the clause $a \leftarrow$ to P_M so that a must be in the least model of P_M and, again, M is not equal to the least model of P_M. It follows that $P_0 \cup \{a \leftarrow \neg a, a \leftarrow\}$ has a stable model but $P_0 \cup \{a \leftarrow \neg a\}$ does not.

One can see from the example above that there may be a finite set of clauses in a normal propositional or predicate logic program P which prevent P from having a stable model. Our next definition captures the key property which ensures that the tree T_P which corresponds to a given finite normal predicate logic program is forced to be finite. We say that a finite normal predicate logic program Q_e over $\mathcal{L}$ has an *explicit initial blocking set* if there is an m such that

(1) for every $i \leq m$, either i is not the code of an atom of ground(P) or the atom a coded by i has the finite support property relative to P and there is at least one atom a in $H(P)$ whose code is less than or equal to m and

(2) for all $S \subseteq \{0, \dots, m\}$, either

    (a) there exists an $i \in S$ such that i is not the code of an atom in $H(P)$, or

    (b) there is an $i \notin S$ such that there exists a minimal P-proof scheme p such that concl(p) $= a$ where a is the atom of $H(P)$ with code i and supp(p) $\subseteq \{0, \dots, m\} - S$, or

    (c) there is an $i \in S$ such that every minimal P-proof scheme $\mathbb{S}$ of the atom a of $H(P)$ with code i has supp($\mathbb{S}$) $\cap S \neq \emptyset$.

The definition of a finite normal predicate logic program Q_e over $\mathcal{L}$ having an *initial blocking set* is the same as the definition of

Q_e having an explicit initial blocking set except that we drop the condition that for every $i \leq m$ which is the code of an atom $a \in H(P)$, a must have the finite support property relative to P.

3. Computability Theory and Effectively Closed Sets

First we review some basic concepts of recursion theory and Π_1^0 classes.

Let ϕ_e denote the partial recursive function which is computed by the eth Turing machine. Thus $\phi_0, \phi_1, \ldots$ is a list of all partial recursive functions. We let W_e be the set of all $x \in \omega$ such that $\phi_e(x)$ converges. Thus $W_0, W_1, \ldots$ is a list of all recursively enumerable (r.e.) sets. More generally, a recursive functional ϕ takes as inputs both numbers $a \in \omega$ and functions $x : \omega \to \omega$. The function inputs are treated as "oracles" to be called on when needed. Thus a particular computation $\phi(a_1, \ldots, a_n; x_1, \ldots, x_m)$ only uses a finite amount of information $x_i \lceil c$ about each function x_i. We shall write $\phi_e(a_1, \ldots, a_n; x_1, \ldots, x_m)$ for the recursive functional computed by the eth oracle machine. In the special case where $n = m = 1$ and x_1 is a sequence of 0s and 1s and $X = \{n : x_1(n) = 1\}$, then we shall write $\phi_e^X(a_1)$ or $\{e\}^X(a_1)$ instead of $\phi_e(a_1; x_1)$. The *jump* of a set $A \subseteq \omega$, denoted A', is the set of all e such that $\phi_e^A(e)$ converges. We let $0'$ denote the jump of the empty set. For $A, B \subseteq \omega$, we write $A \leq_T B$ if A is Turing reducible to B and $A \equiv_T B$ if $A \leq_T B$ and $B \leq_T A$.

We shall assume the reader is familiar with the usual arithmetic hierarchy of Σ_n^0 and Π_n^0 subsets of ω as well as Σ_1^1 and Π_1^1 sets, see Soare's book [31] for any unexplained notation. A subset A of ω is said to be D_n^m if it is the difference of two Σ_n^m sets. A set $A \subset \omega$ is said to be an *index set* if, for any a, b, $a \in A$ and $\phi_a = \phi_b$ imply that $b \in A$. For example, Fin $= \{a : W_a$ is finite$\}$ is an index set. We are particularly interested in the complexity of such index sets. Recall that a subset A of ω is said to be Σ_n^m-complete (respectively, Π_n^m-complete, D_n^m-complete) if A is Σ_n^m (respectively, Π_n^m, D_n^m) and any Σ_n^m (respectively, Π_n^m, D_n^m) set B is many-to-one reducible to A. For example, the set Fin $= \{e : W_e$ is finite$\}$ is Σ_2^0-complete.

If $\Sigma \subseteq \omega$, then $\Sigma^{<\omega}$ denotes the set of finite strings of letters from Σ and Σ^ω denotes the set of infinite sequences of letters from Σ. For a string $\sigma = (\sigma(0), \sigma(1), \ldots, \sigma(n-1))$, we let $|\sigma|$ denote the length n of σ. The empty string has length 0 and will be denoted by $\emptyset$. A constant string σ of length n consisting entirely of k's will be denoted by k^n. For $m < |\sigma|$, $\sigma \restriction m$ is the string $(\sigma(0), \ldots, \sigma(m-1))$. We say σ is an *initial segment* of τ (written $\sigma \prec \tau$) if $\sigma = \tau \restriction m$ for some $m < |\sigma|$. The concatenation $\sigma^\frown\tau$ (or sometimes just $\sigma\tau$) is defined by

$$\sigma^\frown\tau = (\sigma(0), \sigma(1), \ldots, \sigma(m-1), \tau(0), \tau(1), \ldots, \tau(n-1)),$$

where $|\sigma| = m$ and $|\tau| = n$. We write $\sigma^\frown a$ for $\sigma^\frown(a)$ and $a^\frown\sigma$ for $(a)^\frown\sigma$. For any $x \in \Sigma^\omega$ and any finite n, the *initial segment* $x \restriction n$ of x is $(x(0), \ldots, x(n-1))$. We write $\sigma \prec x$ if $\sigma = x \restriction n$ for some n. For any $\sigma \in \Sigma^n$ and any $x \in \Sigma^\omega$, we let $\sigma^\frown x = (\sigma(0), \ldots, \sigma(n-1), x(0), x(1), \ldots)$.

If $\Sigma \subseteq \omega$, a *tree* T over Σ^* is a set of finite strings from $\Sigma^{<\omega}$ which contains the empty string $\emptyset$ and which is closed under initial segments. We say that $\tau \in T$ is an *immediate successor* of a string $\sigma \in T$ if $\tau = \sigma^\frown a$ for some $a \in \Sigma$. We will identify T with the set of codes $c(\sigma)$ for $\sigma \in T$. Thus we say that T is recursive, r.e., etc. if $\{c(\sigma) : \sigma \in T\}$ is recursive, r.e., etc. If each node of T has finitely many immediate successors, then T is said to be *finitely branching*. Finally, a tree T is *highly recursive* if it is recursive and there is a partial recursive function f such that for any $\sigma \in T$, there are $f(\sigma)$ immediate successors of σ.

Definition 3.1. Suppose that $g : \omega^{<\omega} \to \omega$. Then we say that

(1) T is *g-bounded* if for all σ and all integers i, $\sigma^\frown i \in T$ implies $i \le g(\sigma)$.

(2) T is *almost always g-bounded* if there is a finite set $F \subseteq T$ of strings such that for all strings $\sigma \in T \setminus F$ and all integers i, $\sigma^\frown i \in T$ implies $i < g(\sigma)$.

(3) T is *nearly g-bounded* if there is an $n \geq 0$ such that for all strings $\sigma \in T$ with $|\sigma| \geq n$ and all integers i, $\sigma^\frown i \in T$ implies $i < g(\sigma)$.

(4) T is *bounded* if it is g-bounded for some $g : \omega^{<\omega} \to \omega$.

(5) T is *almost always bounded (a.a.b.)* if it is almost always g-bounded for some $g : \omega^{<\omega} \to \omega$.

(6) T is *nearly bounded* if it is nearly g-bounded for some $g : \omega^{<\omega} \to \omega$.

(7) T is *recursively bounded (r.b.)* if T is g-bounded for some recursive $g : \omega^{<\omega} \to \omega$.

(8) T *almost always recursively bounded (a.a.r.b.)* if it is almost always g-bounded for some recursive $g : \omega^{<\omega} \to \omega$.

(9) T *nearly recursively bounded (nearly r.b.)* if it is nearly g-bounded for some recursive $g : \omega^{<\omega} \to \omega$.

For any tree T, an *infinite path* through T is a sequence $(x(0), x(1), \ldots)$ such that $x \restriction n \in T$ for all n. Let $[T]$ be the set of infinite paths through T. We let $\text{Ext}(T)$ denote the set of all $\sigma \in T$ such that $\sigma \prec x$ for some $x \in [T]$. Thus, $\text{Ext}(T)$ is the set of all σ in T that lie on some infinite path through T. We say that T is *decidable* if T is recursive and $\text{Ext}(T)$ is recursive.

Here are some variations on the definition. See [10] for details.

Lemma 3.1. *For any class $P \subseteq \omega^\omega$, the following conditions are equivalent.*

(1) $P = [T]$ *for some recursive tree $T \subset \omega^{<\omega}$.*

(2) $P = [T]$ *for some primitive recursive tree T.*

(3) $P = \{x : \omega \to \omega : (\forall n)R(n, [x \restriction n])\}$, *for some recursive relation R.*

(4) $P = [T]$ *for some tree $T \subset \omega^{<\omega}$ which is Π_1^0.*

The notions of boundedness for a Π_1^0 class follow from the notions of boundedness for trees. We say that a Π_1^0-class $\mathcal{C}$ is

(1) *bounded* if $C = [T]$ for some recursive tree which is bounded,
(2) *almost always bounded* (*a.a.b.*) if $C = [T]$ for some recursive tree which is almost always bounded,
(3) *nearly bounded* if $C = [T]$ for some recursive tree which is nearly bounded,
(4) *recursively bounded* (*r.b.*) if $C = [T]$ for some highly recursive tree,
(5) *almost always recursively bounded* (*a.a.r.b.*) if $C = [T]$ for some recursive tree which is almost always recursively bounded,
(6) *nearly recursively bounded* (nearly *r.b.*) if $C = [T]$ for some recursive tree which is nearly recursively bounded, and
(7) *decidable* if $C = [T]$ for some decidable tree.

Theorem 3.1. *For any recursive tree* $T \subseteq \omega^{<\omega}$, *the following conditions hold.*

(a) $\mathrm{Ext}(T)$ *is a* Σ_1^1 *set.*
(b) *If* T *is finitely branching, then* $\mathrm{Ext}(T)$ *is a* Π_2^0 *set.*
(c) *If* T *is highly recursive, then* $\mathrm{Ext}(T)$ *is a* Π_1^0 *set.*

For any nonempty Π_1^0 class $P = [T]$, one can compute a member of P from the tree $\mathrm{Ext}(T)$, by always taking the leftmost branch in $\mathrm{Ext}(T)$.

Hence the following theorem is an immediate consequence of Theorem 3.1.

Theorem 3.2. *For any nonempty* Π_1^0 *class* $P \subseteq \omega^{<\omega}$,

(a) P *has a member which is recursive in some* Σ_1^1 *set,*
(b) *if* P *is bounded, nearly bounded, or almost always bounded, then* P *has a member which is recursive in* $\mathbf{0}''$,
(c) *if* P *is recursively bounded, nearly recursively bounded, or almost always recursively bounded, then* P *has a member which is recursive in* $\mathbf{0}'$, *and*
(d) *if* $P = [T]$, *where* T *is decidable, then* P *has a recursive member.*

If $T \subseteq \omega^{<\omega}$ is tree and $f \in [T]$, then we say that f is isolated, if there is $k > 0$ such that f is the only element of $[T]$ which extends $(f(0), \ldots, f(k))$. The complexity of isolated paths in recursive trees was determined by Kreisel.

Theorem 3.3 (Kreisel [59]). *Let P be a Π_1^0 class.*

(a) *Any isolated member of P is hyperarithmetic.*

(b) *Suppose that P is bounded, nearly bounded, or almost always bounded. Then, any isolated member of P is recursive in $\mathbf{0}'$.*

(c) *Suppose P is recursively bounded, nearly recursively bounded, or almost always recursively bounded. Then, any isolated member of P is recursive.*

A set $A \subseteq \omega$ is *low* if $A' = \mathbf{0}'$. Jockusch and Soare [22–24] proved the following important results about recursively bounded Π_1^0-classes.

Theorem 3.4 (Low Basis Theorem).

(a) *Every nonempty r.b. Π_1^0 class P contains a member of low degree.*

(b) *There is a low degree $\mathbf{a}$ such that every nonempty r.b. Π_1^0 class contains a member of degree $\leq \mathbf{a}$.*

(c) *If P is r.b., then P contains a member of r.e. degree.*

(d) *Every r.b. Π_1^0 class P contains members a and b such that any function recursive in both a and b is recursive.*

(e) *If P is s bounded Π_1^0 class, then P contains a member of Σ_2^0 degree.*

(f) *Every bounded Π_1^0 class contains a member a such that $a' \leq_T \mathbf{0}''$.*

(g) *Every bounded Π_1^0 class P contains members a and b such that any function recursive in both a and b is recursive in $\emptyset'$.*

As a warmup for the results on stable models of logic programs coming up in Section 4, here is the argument that the complete consistent extensions of an axiomatizable theory, in first-order logic, may be viewed as an effectively closed set.

Theorem 3.5 (Shoenfield [35]). *For any c.e. theory Γ of an effective first-order language $\mathcal{L}$, the class of complete consistent extensions of Γ can be represented as a Π_1^0 class in $\{0,1\}^{\mathbb{N}}$.*

Proof. Let $a_0, a_1, \ldots$ be an effective enumeration of the sentences of $\mathcal{L}$. Then, the sentence γ_i may be identified with the number i, so that a theory Γ is represented by the set $\{i : a_i \in \Gamma\}$, and a class of theories is represented by a class in $\{0,1\}^{\omega}$. Let $\Gamma \vdash_s a_i$ be the computable relation which means that there is a proof of a_i from Γ of length s. Then the class $P(\Gamma)$ of complete consistent extensions of Γ may be represented by the set of infinite paths through the computable tree T defined so that for any $\sigma = (\sigma(0), \ldots, \sigma(n-1))$, σ is in T if and only if the following conditions hold.

(1) For any $i < n$, if $\Gamma \vdash_n a_i$, then $\sigma(i) = 1$.
(2) For any $i, j < n$, if $\Gamma \vdash_n a_i \to a_j$ and $\sigma(i) = 1$, then $\sigma(j) = 1$.
(3) For any $i, j, k < n$, if $a_k = (a_i \wedge a_j)$, $\sigma(i) = 1$ and $\sigma(j) = 1$, then $\sigma(k) = 1$.
(4) For any $i, j < n$, if $\sigma(i) = 1$ and $a_j = \neg a_i$, then $\sigma(j) = 0$.
(5) For any $i, j < n$, if $a_j = \neg a_i$, then either $\sigma(i) = 1$ or $\sigma(j) = 1$.

Let x be an infinite path through T and let $\Delta_x = \{a_i : x(i) = 1\}$. Condition (1) ensures that $\Gamma \subseteq \Delta$, while conditions (1), (2), and (3) ensure that Δ is a theory. Condition (4) ensures that Δ is c $\square$

We can now apply Theorem 3.4 to logical theories.

Proposition 3.1. *For any consistent, axiomatizable first-order theory Γ in an effective langue $\mathcal{L}$, Γ has a complete consistent extension which is computable in $\mathbf{0}'$.*

Conversely, any Π_1^0 class in $\{0,1\}^{\mathbb{N}}$ represents the set of complete consistent extensions of an axiomatizable theory Γ.

Theorem 3.6. *Any Π_1^0 class $P \subseteq \{0,1\}^{\mathbb{N}}$ may be represented as the set of complete, consistent extensions of an axiomatizable theory Γ in propositional logic.*

Proof. We give the proof due to Ehrenfeucht [18]. Let the language $\mathcal{L}$ consist of a countable sequence $A_0, A_1, \ldots$ of propositional variables. For any $x \in \{0,1\}^{\mathbb{N}}$, we can define a complete consistent theory $\Delta(x)$ for $\mathcal{L}$ to be $\mathrm{Con}(\{C_i : i \in \omega\})$, where $C_i = A_i$ if $x(i) = 1$ and $C_i = \neg A_i$ if $x(i) = 0$. It is clear that every complete consistent theory of $\mathcal{L}$ is one of these. Thus, for any Π_1^0 class $P \subseteq \{0,1\}^{\mathbb{N}}$, we want a theory Γ such that $\Delta(P) = \{\Delta(x) : x \in P\}$ is the set of complete, consistent extensions of Γ.

For each finite sequence $\sigma = (\sigma(0), \ldots, \sigma(n-1))$, let $P_\sigma = C_0 \wedge C_1 \wedge \cdots \wedge C_{n-1}$, where $C_i = A_i$ if $\sigma(i) = 1$ and $C_i = \neg A_i$ if $\sigma(i) = 0$. Let the binary tree T be given such that $P = [T]$ and define the theory $\Gamma(T)$ to consist of all $P_\sigma \to A_n$ such that $\sigma \in T$ and $\sigma^\frown 0 \notin T$ and all $P_\sigma \to \neg A_n$ such that $\sigma \in T$ and $\sigma^\frown 1 \notin T$, where $|\sigma| = n$. We claim that $\Delta(P)$ is in fact equal to the set of complete consistent extensions of $\Gamma(T)$. Suppose first that $x \in P$ and let $\mathrm{Con}(\{C_i : i \in \omega\}) = \Delta(x)$. Now any $\gamma \in \Gamma(T)$ is of the form $P_\sigma \to \pm A_i$ for some $\sigma \in T$; say that $|\sigma| = n$. There are several cases. If $\sigma \neq x \lceil n$, then $\Delta(x) \vdash \neg P_\sigma$, so that we always have $\Delta(x) \vdash P_\sigma \to \pm A_n$. Thus we may suppose that $\sigma = x \lceil n$. If $\sigma^\frown 0 \notin T$, then of course $x(n) = 1$, so that $C_n = A_n \in \Delta(x)$ and therefore $\Delta(x) \vdash P_\sigma \to A_n$. Similarly, if $\sigma^\frown 1 \notin T$, then $\Delta(x) \vdash P_\sigma \to \neg A_n$. Thus $\Delta(x)$ is a complete consistent extension of $\Gamma(T)$. On the other hand, let Δ be a complete consistent extension of $\Gamma(T)$. Then, for each i, we have either $\Delta \vdash A_i$ or $\Delta \vdash \neg A_i$; let $C_i = A_i$ if $A_i \in \Delta$ and $C_i = \neg A_i$ otherwise. Define $x \in \{0,1\}^\omega$ so that $x(i) = 1$ if and only if $\Delta \vdash A_i$. Then clearly $\Delta = \Delta(x)$. It remains to be shown that $x \in P$. Now if $x \notin P$, then there is some n such that $\sigma = x \lceil n+1 \notin T$ and $x \lceil n \in T$. Then $P_\sigma = C_0 \wedge \cdots \wedge C_{n-1}$, so that $\Delta \vdash P_\sigma$, and $P_\sigma \to \neg C_i \in \Gamma(T)$, so that Δ is not consistent with $\Gamma(T)$. This contradiction proves that $\Delta = \Delta(x)$. $\qquad\square$

We now spell out the indexing for Π_1^0 classes and primitive recursive trees that we will use in this paper. Let $\pi_0, \pi_1, \ldots$ be an effective enumeration of the primitive recursive functions from ω to $\{0,1\}$ and let

$$T_e = \{\emptyset\} \cup \{\sigma : (\forall \tau \preceq \sigma)\pi_e(c(\tau)) = 1\},$$

where $c(\tau)$ is the code of τ. It is clear that each T_e is a primitive recursive tree. Observe also that if $\{\sigma : \pi_e(c(\sigma)) = 1\}$ is a primitive recursive tree, then T_e will be that tree. Thus every primitive recursive tree occurs in our enumeration $T_0, T_1, \ldots$. (Note that, henceforth, we will generally identify a finite sequence $\tau \in \omega^{<\omega}$ with its code.) Then we let $P_e = [T_e]$ be the *e*th Π_1^0 class. It follows from Lemma 3.1 every Π_1^0 class occurs in the enumeration P_e.

There is a large literature on the complexity of elements in Π_1^0-classes and index sets for primitive recursive trees. In the remainder of this section, we shall list the key results which will be needed for our applications to index sets associated with normal finite predicate logic programs.

Our first results establish the complexity of determining whether a primitive recursive tree is recursively bounded, almost always recursive bounded, bounded, almost always bounded, or decidable.

Theorem 3.7.

(a) $\{e : T_e \text{ is r.b.}\}$ *is* Σ_3^0-*complete.*
(b) $\{e : T_e \text{ is a.a.r.b.}\}$ *is* Σ_3^0-*complete.*
(c) $\{e : T_e \text{ is nearly r.b.}\}$ *is* Σ_3^0-*complete.*
(d) $\{e : T_e \text{ is bounded}\}$ *is* Π_3^0-*complete.*
(e) $\{e : T_e \text{ is a.a. bounded}\}$ *is* Σ_4^0-*complete.*
(f) $\{e : T_e \text{ is nearly bounded}\}$ *is* Σ_4^0 *complete.*
(g) $\{e : T_e \text{ is r.b. and decidable}\}$ *is* Σ_3^0-*complete.*

Proof. The only parts which are not proved by Cenzer and Remmel in [9] are parts (b) and (e). (In [9], Cenzer and Remmel used the term *almost bounded* for what we termed *nearly bounded*.)

We shall show how to modify the proofs of (c) and (f) in [9] to prove (b) and (e), respectively. Similar modifications of the proofs in [9] for index sets relative to nearly bounded and nearly recursively bounded trees can be used to establish the remaining index set results which we list in this section.

The facts that $\{e : T_e$ is a.a.r.b.$\}$ is Σ_3^0 and $\{e : T_e$ is a.a.b.$\}$ is Σ_4^0 are easily established by simply writing out the definitions.

To prove the Σ_3^0-completeness of $\{e : T_e$ is a.a.r.b.$\}$, we can use the same proof that was used by Cenzer and Remmel [9] to establish that $\{e : T_e$ is r.b.$\}$ is Σ_3^0-complete. It is easy to see that a tree T is r.b. if and only if there is a recursive function $g : \omega \to \omega$ such that if $(a_0, \ldots, a_n) \in T$, then $a_i < g(i)$ for all $i \in T$. Similarly, a tree T is a.a.r.b. if and only if there is a recursive function $g : \omega \to \omega$ such that for all but finitely many $(a_0, \ldots, a_n) \in T$, $a_i < g(i)$ for all $i \in T$. In each case, we shall call such a g a bounding function.

Now Rec $= \{e : W_e$ is recursive$\}$ is Σ_3^0-complete, see Soare's book [31]. We define a reduction f of Rec to $\{e : T_e$ is r.b.$\}$. This will be done so that $[T_{f(e)}]$ is empty if W_e is finite and $[T_{f(e)}]$ has a single element if W_e is infinite. The primitive recursive tree $T_{f(e)}$ is defined so that we put $\sigma = (s_0, s_1, \ldots, s_{k-1}) \in T_{f(e)}$ if and only if $s_0 < s_1 < \cdots < s_{k-1}$ and there exists a sequence $m_0 < m_1 < \cdots < m_{k-1}$ such that, for each $i < k$, $m_i \in W_{e,s_i} \setminus W_{e,s_i-1}$ and m_i is the least element of $W_{e,s_{k-1}} \setminus \{m_0, \ldots, m_{i-1}\}$. We observe that if W_e is finite, then $T_{f(e)}$ is also finite and therefore recursively bounded. Now fix e and suppose that W_e is infinite. Then we define a canonical sequence $n_0 < n_1 < \cdots$ of elements of W_e and corresponding sequence of stages $t_0 < t_1 < \cdots$ such that, for each i, $n_i \in W_{e,t_i} \setminus W_{e,t_i-1}$ and $(t_0, t_1, \ldots, t_i) \in T_{f(e)}$ as follows. Let n_0 be the least element of W_e and t_0 is the least stage t such that $n_0 \in W_{e,t}$. Then for each k, let n_{k+1} be the least element of $W_e \setminus W_{e,t_k}$ and t_{k+1} be the least stage t such that $n_{k+1} \in W_{e,t}$. Then, for each k, $(t_0, \ldots, t_k) \in T_{f(e)}$ and $n_k \in W_{e,t_k}$. Furthermore, we can prove by induction on k that

$$k \in W_e \to k \in W_{e,t_k}.$$

For $k = 0$, this is because $n_0 = 0$ if $0 \in W_e$. Assuming the statement to be true for all $i < k$, we see that if $k \in W_e$, then either $k \in W_{e,t_{k-1}}$, or else $n_k = k$. In either case, we have $k \in W_{e,t_k}$.

The key fact to observe is that for any $(s_0, \ldots, s_k) \in T_{f(e)}$, $s_k \le t_k$. To see this, let $(s_0, \ldots, s_k) \in T_{f(e)}$, let $(m_0, \ldots, m_k)$ be the associated sequence of elements of W_e. Suppose by way of contradiction that $s_k \ge t_k$. It follows from the definitions of $T_{f(e)}$ and of $t_0, \ldots, t_k$ that in fact $s_i = t_i$ and $m_i = n_i$ for all $i \le k$. Thus if we let $g(n) = t_n + 1$, then g will be a bounding function for $T_{f(e)}$. Now if W_e is recursive. then the sequence $t_0 < t_1 < \ldots$ is also recursive and thus $T_{f(e)}$ is recursively bounded.

Now suppose that $T_{f(e)}$ has a recursive bounding function h. Then we must have $t_k < h(k)$ for each σ of length k. It then follows from the equation above that $k \in W_e \iff k \in W_{e,h(k)}$, so that W_e is recursive. Thus $T_{f(e)}$ is *r.b* if and only if W_e recursive and, hence, $\{e : T_e \text{ is r.b.}\}$ is Σ_3^0-complete. However, note that if $h : \omega^{<\omega} \to \omega$ is a function which witnesses that $T_{f(e)}$ is almost always recursively bounded, then there will be an n such that $t_k < h(k)$ for all $k \ge n$. In that case, for all $k \ge n$, $k \in W_e \iff k \in W_{e,h(k)}$ which still implies that W_e is recursive. Thus $T_{f(e)}$ is *a.a.r.b.* if and only if W_e is recursive so that $\{e : T_e \text{ is a.a.r.b.}\}$ is also Σ_3^0-complete.

This argument is typical of the completeness arguments for the properties about cardinalities of $[T]$ or the number of recursive elements of $[T]$ that appear in the rest of the theorems in this section. That is, the completeness argument for *r.b.* trees also works for *a.a.r.b.* trees.

For the completeness argument for (d), we shall use the fact that $Cof = \{e : \omega - W_e \text{ is finite}\}$ is Σ_3^0-complete set, see [31]. We let $W_{e,s}$ denote the set of elements that are enumerated into W_e in s or fewer steps as in [31]. By definition, all $x \in W_{e,s}$ are less than or equal to s and the question of whether $x \in W_{e,s}$ is a primitive recursive predicate. Then we can define a primitive recursive function $\phi(e, m, s) = (least\ n > m)(n \notin W_{e,s} - \{0\})$.

For any given e, let U_e be the tree such that $(m) \in U_e$ for all $m \geq 0$ and $(m, s + 1) \in U_e$ if and only if m is the least element such that $\phi(e, m, s+1) > \phi(e, m, s)$. Note that when $m \geq s+1$, the least n such that $n > m$ and $n \notin W_{e,s}$ is just $m + 1$ since all elements of $W_{e,s+1}$ are less than $s + 1$. Thus the only candidates for $(m, s + 1)$ to be in U_e are $m \leq s + 1$. Thus the tree U_e will be primitive recursive. Now if $W_e - \{0\}$ is not cofinite, then for each m, there is a minimal $n > m$ such that $n \notin W_e$. It follows that $\lim_s \phi(e, m, s) = n$, so that $\phi(e, m, s + 1) > \phi(e, m, s)$ for only finitely many s, which will make U_e finitely branching. On the other hand, if $W_e - \{0\}$ is cofinite and we choose m so that $n \in W_e - \{0\}$ for all $n > m$, then it is clear that there will be infinitely many s such that $\phi(e, m, s + 1) > \phi(e, m, s)$. It follows that if m is the largest element not in $W_e - \{0\}$, then for infinitely many s, $(m, s+1)$ will be in U_e and for all $p > m$, there can be only finitely many s such that $(p, s + 1)$ is in U_e. Thus if $W_e - \{0\}$ is cofinite, then there will be exactly one node which has infinitely many successors. Clearly, there is a recursive function f such that $T_{f(e)} = U_e$. But then

$$e \in \omega - Cof \iff T_{f(e)} \text{ is bounded.}$$

Since $\omega - Cof$ is Π_3^0-complete, it follows that $\{e : T_e \text{ is bounded}\}$ is Π_3^0-complete.

Now let S be an arbitrary Σ_4^0 set and suppose that $a \in S \iff (\exists k) R(a, k)$ where R is Π_3^0. By the usual quantifier methods, we may assume that $R(a, k)$ implies that $R(a, j)$ for all $j > k$. By our argument for the Π_3^0-completeness of $\{e : T_e \text{ is bounded}\}$, there is a recursive function h such that $R(a, k)$ holds if and only if $U_{h(a,k)}$ is bounded and such that $U_{h(a,k)}$ is $a.a.$ bounded for every a and k. Now we can define a recursive function ϕ so that

$$T_{\phi(a)} = \{(0)\} \cup \{(k + 1)^{\frown}\sigma : \sigma \in U_{h(a,k)}\}.$$

If $a \in S$, then $U_{h(a,k)}$ is bounded for all but finitely many k and is $a.a.$ bounded for the remainder. Thus $U_{\phi(a)}$ is $a.a.$ bounded.

If $a \notin S$, then, for every k, $U_{h(a,k)}$ is not bounded, so that $U_{\phi(a)}$ is not *a.a.* bounded. Thus $a \in S$ if and only if $T_{\phi(a)}$ is *a.a.b* and $\{e : T_e \text{ is } a.a. \text{ bounded}\}$ is Σ_4^0-complete.

Remark. As it stands, it is clear that there are no infinite paths through $T_{\phi(a)}$ since every node $T_{\phi(a)}$ has length at most 3. The reason that we constructed the tree $T_{\phi(a)}$ to contain the node (0) is for the remaining completeness arguments which follow in this section. That is, we are now free to modify the construction to add a tree above (0) which has the required number of infinite paths or infinite recursive paths as is need to prove various completeness results for *a.a.* bounded or *a.a.* recursively bounded trees. The key observation to make is that our completeness arguments to establish the complexity for various properties concerning the number of infinite paths or infinite recursive paths through *r.b.* trees in [9] always produced bounded trees. Since the complexity results for *r.b.* trees had complexity contained in Σ_4^0, it follows that we can modify the construction by placing trees above (0) in $T_{\phi(a)}$ to show that complexity for various properties concerning the number of infinite paths or infinite recursive paths through *a.a.b* trees is Σ_4^0-complete. Thus we shall not give the details of such arguments. $\qquad\square$

Next we give several index set results concerning the size of $[T]$ for primitive recursive trees T which have various properties. These results are either proved in [9] or follow by modifying the results in [9] as described in Theorem 3.7 to prove results about *a.a.* bounded or *a.a.* recursively bounded trees. In fact, in all the results that follow, the index set results for properties relative to *a.a.* bounded trees is exactly the same as the index set results for nearly bounded trees and the index set results for properties of *a.a.r.b.* trees is exactly the same as the index set results for nearly *r.b.* trees. Thus, we shall only state the results for *a.a.* and *a.a.r.b.* trees.

Theorem 3.8.

(a) $\{e : T_e$ *is r.b. and* $[T_e]$ *is empty*$\}$ *is* Σ_2^0*-complete.*

(b) $\{e : T_e$ *is r.b. and* $[T_e]$ *is nonempty*$\}$ *is* Σ_3^0*-complete.*

(c) $\{e : T_e$ *is bounded and* $[T_e]$ *is empty*$\}$ *is* Σ_2^0*-complete.*

(d) $\{e : T_e$ *is bounded and* $[T_e]$ *is nonempty*$\}$ *is* Π_3^0*-complete.*

(e) $\{e : T_e$ *is a.a.r.b. and* $[T_e]$ *is nonempty*$\}$ *and* $\{e : T_e$ *is a.a.r.b. and* $[T_e]$ *is empty*$\}$ *are* Σ_3^0*-complete.*

(f) $\{e : T_e$ *is a.a.b. and* $[T_e]$ *is nonempty*$\}$ *and* $\{e : T_e$ *is a.a.b. and* $[T_e]$ *is empty*$\}$ *are* Σ_4^0*-complete.*

(g) $\{e : [T_e]$ *is nonempty*$\}$ *is* Σ_1^1*-complete and* $\{e : [T_e]$ *is empty*$\}$ *is* Π_1^1*-complete.*

Theorem 3.9. *For any positive integer* c,

(a) $\{e : T_e$ *is r.b. and* $\mathrm{Card}([T_e]) > c\}$, $\{e : T_e$ *is r.b. and* $\mathrm{Card}([T_e]) \leq c\}$, *and* $\{e : T_e$ *is r.b. and* $\mathrm{Card}([T_e]) = c\}$ *are all* Σ_3^0*-complete.*

(b) $\{e : T_e$ *is a.a.r.b. and* $\mathrm{Card}([T_e]) > c\}$, $\{e : T_e$ *is a.a.r.b. and* $\mathrm{Card}([T_e]) \leq c\}$, *and* $\{e : T_e$ *is a.a.r.b. and* $\mathrm{Card}([T_e]) = c\}$ *are all* Σ_3^0*-complete.*

(c) $\{e : T_e$ *is bounded and* $\mathrm{Card}([T_e]) \leq c\}$ *and* $\{e : T_e$ *is bounded and* $\mathrm{Card}([T_e]) = 1\}$ *are both* Π_3^0*-complete.*

(d) $\{e : T_e$ *is bounded and* $\mathrm{Card}([T_e]) > c\}$ *and* $\{e : T_e$ *is bounded and* $\mathrm{Card}([T_e]) = c + 1\}$ *are both* D_3^0*-complete.*

(e) $\{e : T_e$ *is a.a.b. and* $\mathrm{Card}([T_e]) > c\}$, $\{e : T_e$ *is a.a. bounded and* $\mathrm{Card}([T_e]) \leq c\}$, *and* $\{e : T_e$ *is a.a. bounded and* $\mathrm{Card}([T_e]) = c\}$ *are all* Σ_4^0*-complete.*

(f) $\{e : T_e$ *is r.b., dec. and* $\mathrm{Card}([T_e]) > c\}$, $\{e : T_e$ *is r.b., dec. and* $\mathrm{Card}([T_e]) \leq c\}$, *and* $\{e : T_e$ *is r.b., dec. and* $\mathrm{Card}([T_e]) = c\}$ *are all* Σ_3^0*-complete.*

(g) $(\{e : \mathrm{Card}([T_e]) > c\})$ *is* Σ_1^1*-complete,* $\{e : \mathrm{Card}([T_e]) \leq c\}$ *is* Π_1^1*-complete, and* $\{e : \mathrm{Card}([T_e]) = c\}$ *is* Π_1^1*-complete.*

Theorem 3.10.

(a) $\{e : T_e$ is r.b. and $[T_e]$ is infinite$\}$ is D_3^0-complete and $\{e : T_e$ is r.b. and $[T_e]$ is finite$\}$ is Σ_3^0-complete.

(b) $\{e : T_e$ is a.a.r.b. and $[T_e]$ is infinite$\}$ is D_3^0-complete and $\{e : T_e$ is a.a.r.b. and $[T_e]$ is finite$\}$ is Σ_3^0-complete.

(c) $\{e : T_e$ is bounded and $[T_e]$ is infinite$\}$ is Π_4^0-complete and $\{e : [T_e]$ is bounded and finite$\})$ is Σ_4^0-complete.

(d) $\{e : T_e$ is a.a. bounded and $[T_e]$ is infinite$\}$ is D_4^0-complete and $\{e : [T_e]$ is a.a. bounded and finite$\})$ is Σ_4^0-complete.

(e) $(\{e : [T_e]$ is infinite$\}$ is Σ_1^1-complete and $\{e : [T_e]$ is finite$\})$ is Π_1^1-complete.

(f) $\{e : T_e$ is r.b. and dec. and $[T_e]$ is infinite$\}$ is D_3^0-complete and $\{e : T_e$ is r.b. and dec. and $[T_e]$ is finite$\}$ is Σ_3^0-complete.

Theorem 3.11. $\{e : [T_e]$ is uncountable$\}$ is Σ_1^1-complete, $\{e : [T_e]$ is countable$\}$ is Π_1^1-complete, and $\{e : [T_e]$ is countably infinite$\}$ is Π_1^1-complete.

The same result holds for r.b., a.a.r.b., bounded, a.a.b. primitive recursive trees.

Next we give some index set results concerning the number of recursive elements in $[T]$ where T is a primitive recursive tree. Here we say that $[T]$ is *recursively empty* if $[T]$ has no recursive elements and is *recursively nonempty* if $[T]$ has at least one recursive element. Similarly, we say that $[T]$ has *recursive cardinality equal to* c if $[T]$ has exactly c recursive members.

Theorem 3.12.

(a) $\{e : T_e$ is r.b. and $[T_e]$ is recursively nonempty$\}$ is Σ_3^0-complete, $\{e : T_e$ is r.b. and $[T_e]$ is recursively empty$\}$ is D_3^0-complete, and $\{e : T_e$ is r.b. and $[T_e]$ is nonempty and recursively empty$\}$ is D_3^0-complete.

(b) $\{e : T_e$ is a.a.r.b. and $[T_e]$ is recursively nonempty$\}$ is Σ_3^0-complete, $\{e : T_e$ is a.a.r.b. and $[T_e]$ is recursively empty$\}$

is D_3^0-complete, and $\{e : T_e$ is a.a.r.b. and $[T_e]$ is nonempty and recursively empty$\}$ is D_3^0-complete.

(c) $\{e : T_e$ is bounded and $[T_e]$ is recursively nonempty$\}$ is D_3^0-complete, $\{e : T_e$ is bounded and $[T_e]$ is recursively empty$\}$ is Π_3^0-complete, and $\{e : T_e$ is bounded and $[T_e]$ is nonempty and recursively empty$\}$ is Π_3^0-complete.

(d) $\{e : T_e$ is a.a. bounded and $[T_e]$ is recursively nonempty$\}$, $\{e : T_e$ is a.a. bounded and $[T_e]$ is recursively empty$\}$, and $\{e : T_e$ is a.a. bounded and $[T_e]$ is nonempty and recursively empty$\}$ are all Σ_4^0-complete.

(e) $\{e : [T_e]$ recursively nonempty$\}$ is Σ_3^0-complete, $\{e : [T_e]$ recursively empty$\}$ is Π_3^0-complete, and $\{e : [T_e]$ is nonempty and recursively empty$\}$ is Σ_1^1-complete.

Theorem 3.13. *Let c be a positive integer.*

(a) $\{e : T_e$ is r.b. and $[T_e]$ has recursive cardinality $> c\}$ is Σ_3^0-complete, $\{e : T_e$ is r.b. and $[T_e]$ has recursive cardinality $\leq c\}$ is D_3^0-complete, and $\{e : T_e$ is r.b. and $[T_e]$ has recursive cardinality $= c\}$ is D_3^0-complete.

(b) $\{e : T_e$ is a.a.r.b. and $[T_e]$ has recursive cardinality $> c\}$ is Σ_3^0-complete, $\{e : T_e$ is a.a.r.b. and $[T_e]$ has recursive cardinality $\leq c\}$ is D_3^0-complete, and $\{e : T_e$ is a.a.r.b. and $[T_e]$ has recursive cardinality $= c\}$ is D_3^0-complete.

(c) $\{e : T_e$ is bounded and $[T_e]$ has recursive cardinality $> c\}$ is Π_3^0-complete, $\{e : T_e$ is bounded and $[T_e]$ has recursive cardinality $\leq c\}$ is D_3^0-complete, and $\{e : T_e$ is bounded and $[T_e]$ has recursive cardinality $= c\}$ is D_3^0-complete.

(d) $\{e : T_e$ is a.a. bounded and $[T_e]$ has recursive cardinality $> c\}$, $\{e : T_e$ is a.a.b. and $[T_e]$ has recursive cardinality $\leq c\}$, and $\{e : T_e$ is a.a.b. and $[T_e]$ has recursive cardinality $= c\}$ are all Σ_4^0-complete.

(e) $\{e : [T_e]$ has recursive cardinality $> c\}$ is Σ_3^0-complete, $\{e : [T_e]$ has recursive cardinality $\leq c\}$ is Π_3^0-complete, and $\{e : [T_e]$ has recursive cardinality $= c\}$ is D_3^0-complete.

Theorem 3.14. $\{e : [T_e]$ *has finite recursive cardinality*$\}$ *is* Σ^0_4- *complete and* $\{e : [T_e]$ *has infinite recursive cardinality*$\}$ *is* Π^0_4- *complete. The same result is true for r.b., a.a.r.b., bounded, and a.a.b. primitive recursive trees.*

Given a primitive recursive tree $[T]$, we say that $[T]$ is *perfect* if it has no isolated elements. Cenzer and Remmel also proved a number of index set results for primitive recursive trees T such that $[T]$ is perfect. Here is one example.

Theorem 3.15.

(a) $\{e : T_e$ *is r.b. and* $[T_e]$ *is perfect*$\}$ *and* $\{e : T_e$ *is r.b. and* $[T_e]$ *is nonempty and perfect*$\}$ *are* D^0_3-*complete.*

(b) $\{e : T_e$ *is a.a.r.b. and* $[T_e]$ *is perfect*$\}$ *and* $\{e : T_e$ *is a.a.r.b. and* $[T_e]$ *is nonempty and perfect*$\}$ *are* D^0_3-*complete.*

(c) $\{e : T_e$ *is bounded and* $[T_e]$ *is perfect*$\}$ *and* $\{e : T_e$ *is bounded and* $[T_e]$ *is nonempty and perfect*$\}$ *are* Π^0_4-*complete.*

(d) $\{e : T_e$ *is a.a.b and* $[T_e]$ *is perfect*$\}$ *and* $\{e : T_e$ *is a.a.b. and* $[T_e]$ *is nonempty and perfect*$\}$ *are* D^0_4-*complete.*

(e) $\{e : [T_e]$ *is perfect*$\}$ *and* $\{e : [T_e]$ *is nonempty and perfect*$\}$ *are* Σ^1_1-*complete.*

These results can now be applied to logical theories by the following uniform versions of Theorems 3.5 and 3.6. The eth axiomatizable propositional theory may be defined as follows. Let $p_0, p_1, \ldots$ be an effective enumeration of the sentences of propositional logic, let $\Delta_e = \{p_i : i \in W_e\}$, and let $\Gamma_e = \{p : \Delta_e \vdash p\}$.

Theorem 3.16. *There are primitive recursive functions f and g such that, for any primitive recursive tree T_e, $[T_e]$ represents the propositional theory $\Gamma_{f(e)}$, and for any propositional theory Γ_e, $[T_{g(e)}]$ represents the propositional theory Γ_e.*

Then we can apply the index set results for Π^0_1 classes to obtain the following index set results for logical theories.

Theorem 3.17.

(a) $\{e : \Gamma_e$ *has a complete consistent extension*$\}$ *is* Σ_3^0 *complete.*
(b) $\{e : \Gamma_e$ *has infinitely many computable complete consistent extensions*$\}$ *is* Π_4^0 *complete.*
(c) $\{e : \Gamma_e$ *has uncountably many complete consistent extensions*$\}$ *is* Σ_1^1 *complete.*

Proof. Here is a proof of part (a). First we see that the set

$$A = \{e : \Gamma_e \text{ has a complete consistent extension}\}$$

is indeed a Σ_3^0 set, since $e \in A \iff g(e) \in \{i : [T_i] \neq \emptyset\}$, which is Σ_3^0 by Theorem 3.8. For the completeness, observe that $[T_e]$ is nonempty if and only if $\Gamma_{f(e)}$ has a complete consistent extension. Since $\{e : [T_e] \neq \emptyset\}$ is Σ_3^0 complete by Theorem 3.8, it follows that $\{e : \Gamma_e$ has a complete consistent extension$\}$ is also Σ_3^0 complete.   □

4. The Main Theorems

The goal of this section is to prove the main results of the paper on the correspondence between the set of the stable models of a finite normal predicate logic program and the set of infinite paths through a tree T.

Theorem 4.1. *There is a uniform effective procedure which given any recursive tree $T \subseteq \omega^{<\omega}$ produces a finite normal predicate logic program P_T such that the following conditions hold.*

(1) *There is an effective one-to-one degree-preserving correspondence between the set of stable models of P_T and the set of infinite paths through T.*
(2) *T is bounded if and only if P_T has the FS property.*
(3) *T is recursively bounded if and only if P_T has the rec.FS property.*

(4) T *is decidable and recursively bounded if and only if P_T is decidable and has the rec.FS property.*

Theorem 4.2. *There is a uniform recursive procedure which given any finite normal predicate logic program P produces a primitive recursive tree T_P such that the following hold.*

(1) *There is an effective one-to-one degree-preserving correspondence between the set of stable models of P and the set of infinite paths through T_P.*

(2) *P has the FS property or P has an explicit initial blocking set if and only if T_P is bounded.*

(3) *If P has a stable model, then P has the FS property if and only if T_P is bounded.*

(4) *P has the rec.FS property or an explicit initial blocking set if and only if T_P is recursively bounded.*

(5) *If P has a stable model, then P has the rec.FS property if and only if T_P is recursively bounded.*

(6) *P has the a.a.FS property or P has an explicit initial blocking set if and only if T_P is nearly bounded.*

(7) *If P has a stable model, then P has the a.a.FS property if and only if T_P is nearly bounded.*

(8) *P has the a.a.rec.FS property or an explicit initial blocking set if and only if T_P is nearly recursively bounded.*

(9) *If P has a stable model, then P has the a.a.rec.FS property if and only if T_P is nearly recursively bounded.*

(10) *If P has a stable model, then P is decidable if and only if T_P is decidable.*

Let $\{e\}^B$ denotes the function computed by the eth oracle machine with oracle B. If $A \subseteq \omega$, we write $\{e\}^B = A$ if $\{e\}^B$ is a characteristic function of A. If f is a function $f \colon \omega \to \omega$, then we let $\mathrm{gr}(f) = \{\langle x, f(x) \rangle \colon x \in \omega\}$. Given an normal finite predicate logic program P and a recursive tree $T \subseteq \omega^{<\omega}$, we say that there is an effective one-to-one degree preserving correspondence between the set of stable models of P and the set of infinite paths through T_P if there are indices e_1 and e_2 of oracle Turing machines such that

(i) $(\forall M \in \mathrm{Stab}(P))(\{e_1\}^M = f_M \in [T])$, and
(ii) $(\forall f \in [T])(\{e_2\}^{\mathrm{gr}(f)} = M_f \in \mathrm{Stab}(P))$, and
(iii) $(\forall M \in \mathrm{Stab}(P))(\forall f \in [T])(\{e_1\}^M = f$ if and only if $\{e_2\}^{\mathrm{gr}(f)} = M)$.

Condition (i) says that the stable models of P uniformly produce infinite paths through the tree T via an algorithm with index e_1 and condition (ii) says that the infinite paths through the tree T, uniformly produce stable models of P via an algorithm with index e_2. Finally, condition (iii) asserts that our correspondence is one-to-one and if $\{e_S\}^M = f$, then f is Turing equivalent to M. In what follows, we will not explicitly construct the indices e_1 and e_2, but our constructions will make it clear that such indices exist.

4.1. *Proof of Theorem 4.1*

Suppose that T is a recursive tree contained in $\omega^{<\omega}$. Note that by definition, the empty sequence, whose code is 0, is in T.

A classical result, first explicit in [1, 36], but known a long time earlier in equational form, is that every r.e. relation can be computed by a suitably chosen predicate over the least model of a finite predicate logic Horn program. An elegant method of proof due to Shepherdson (see [34] for references) uses the representation of recursive functions by means of finite register machines. When such machines are represented by Horn programs in the natural way, we get programs in which every atom can be proved in only finitely many ways; see also [30]. Thus we have the following proposition.

Proposition 4.1. *Let $r(\cdot, \cdot)$ be a recursive relation. Then there is a finite predicate logic program P_r computing $r(\cdot, \cdot)$ such that every atom in the least model M_r of P_r has only finitely many minimal proof schemes and there is a recursive procedure such* *that given an atom a in Herbrand base of P_r produces the code of the set of P_r-proof schemes for a. Moreover, the least model of P_r is recursive.*

It follows that there exist the following three normal finite predicate logic programs such that the set of ground terms in their underlying language are all of the form 0 or $s^n(0)$ for $n \geq 1$ where 0 is a constant symbol and s is a unary function symbol. We shall use n has an abbreviation for the term $s^n(0)$ for $n \geq 1$.

(I) There exists a finite predicate logic Horn program P_0 such that for a predicate tree($\cdot$) of the language of P_0, the atom tree(n) belongs to the least Herbrand model of P_0 if and only if n is a code for a finite sequence σ and $\sigma \in T$.

(II) There is a finite predicate logic Horn program P_1 such that for the predicate $seq(\cdot)$ of the language of P_1, the atom $seq(n)$ belongs to the least Herbrand model of P_1 if and only if n is the code of a finite sequence $\alpha \in \omega^{<\omega}$.

(III) There is a finite predicate logic Horn program P_2 which correctly computes the following recursive predicates on codes of sequences.

   (a) samelength($\cdot, \cdot$). This succeeds if and only if both arguments are the codes of sequences of the same length.

   (b) diff($\cdot, \cdot$). This succeeds if and only if the arguments are codes of sequences which are different.

   (c) shorter($\cdot, \cdot$). This succeeds if and only if both arguments are codes of sequences and the first sequence is shorter than the second sequence.

   (d) length($\cdot, \cdot$). This succeeds when the first argument is a code of a sequence and the second argument is the length of that sequence.

   (e) notincluded($\cdot, \cdot$). This succeeds if and only if both arguments are codes of sequences and the first sequence is not an initial segment of the second sequence.

(f) num($\cdot$). This succeeds if and only if the argument is either 0 or $s^n(0)$ for some $n \geq 1$.

Now let P^- be the finite predicate logic program which is the union of programs $P_0 \cup P_1 \cup P_2$. We denote its language by $\mathcal{L}^-$ and we let M^- be the least model of P^-. By Proposition 4.1, this program P^- is a Horn program, M^- is recursive, and for each ground atom a in the Herbrand base of P^-, we can explicitly construct the set of all P^--proof schemes of a. In particular, tree(n) $\in M^-$ if and only if n is the code of node in T.

Our final program P_T will consist of P^- plus clauses (1)–(7) given below. We assume no predicate that appears in the head of any of these clauses is in the language $\mathcal{L}^-$. However, we do allow predicates from P^- to appear in the body of clauses (1)–(7). It follows that for any stable model of the extended program, its intersection with the set of ground atoms of $\mathcal{L}^-$ will be M^-. In particular, the meaning of the predicates listed above will always be the same.

We are ready now to write the additional clauses which, together with the program P^-, will form the desired program P_T. First of all, we select three new unary predicates:

(i) path($\cdot$), whose intended interpretation in any given stable model M of P_T is that it holds only on the set of codes of sequences that lie on an infinite path through T. This path will correspond to the path encoded by the stable model of M,

(ii) notpath($\cdot$), whose intended interpretation in any stable model M of P_T is the set of all codes of sequences which are in T but do not satisfy path($\cdot$), and

(iii) control($\cdot$), which will be used to ensure that path($\cdot$) always encodes an infinite path through T.

This given, the final seven clauses of our program are the following.

(1) path(X) $\longleftarrow$ tree(X), $\neg$notpath(X)
(2) notpath(X) $\longleftarrow$ tree(X), $\neg$path(X)

(3) path(0) $\longleftarrow$

(4) notpath(X) $\longleftarrow$ tree(X), path(Y), tree(Y),
samelength(X,Y), diff(X,Y)

(5) notpath(X) $\longleftarrow$ tree(X), tree(Y), path(Y), shorter(Y,X),
notincluded(Y,X)

(6) control(X) $\longleftarrow$ path(Y), length(Y,X)

(7) control(X) $\longleftarrow$ ¬control(X), num(X)

Clearly, $P_T = P^- \cup \{(1), \ldots, (7)\}$ is a finite program.

We should note that technically, we must ensure that all the predicates that we use in our normal finite predicate logic program P_T come from our fixed recursive language $\mathcal{L}$. The predicates we have used in P_T were picked mainly for mnemonic purposes, but since we are assuming that $\mathcal{F}$ has infinitely many constant symbols and infinitely many n-ary relation symbols and n-ary functions symbols for each n, there is no problem to substitute our predicate names by corresponding predicate names that appear in $\mathcal{L}$.

Our goal is to prove the following.

(A) T is a finitely branching recursive tree if and only if every element of $H(P_T)$ has only finitely many minimal proof schemes. Thus, T is finitely branching if and only if P_T has the FS property.

(B) T is highly recursive if and only if for every atom a in $H(P_T)$, we can effectively find the set of all minimal P_T-proof schemes of a.

(C) There is a one-to-one degree preserving correspondence between $[T]$ and $\mathrm{Stab}(P_T)$.

First, we prove (A) and (B). When we add clauses (1)–(7), we note that no atom of $\mathcal{L}^-$ is in the head of any of these new clauses. This means that no ground instance of such a clause can be present in a minimal P_T-proof scheme whose conclusion is an atom of $\mathcal{L}^-$. This means that the minimal P_T-proof schemes whose conclusion is an atom p of $\mathcal{L}^-$ can involve only clauses from P^-. Thus, for any ground atom a of $\mathcal{L}^-$, a will have no minimal P_T-proof scheme if $a \notin M^-$ and we can effectively compute the finite set of P_T-proof schemes for a if $a \in M^-$.

Next, consider the atoms appearing in the heads of clauses (1)–(7). These are atoms of the following three forms:

(i) path(t),
(ii) notpath(t), and
(iii) control(t)

The ground terms of our language are of form n, where $n \in \omega$, that is, of the form 0 or $s^n(0)$ for $n \geq 1$. Note that all clauses that have path(X) or notpath(X) have tree(X) in the body. Thus for atoms of the form path(t) and notpath(t), the only ground terms which possess a P_T-proof scheme must be those for which t is a code of a sequence of natural numbers belonging to T. The reason for this is that predicates of the form tree(t) from $\mathcal{L}^-$ fail if t is not the code of sequence in T. The only exception is clause (3) whose head is path(0) and 0 is the code of the empty sequence which is in every tree T by definition. This eliminates from our consideration ground atoms of the form path(t) and notpath(t) with $t \notin T$. Similarly, the only ground atoms of the form control(t) which possess a proof scheme are atoms of the form control(n) where n is a natural number. Thus we are left with these cases:

(a) path($c(\sigma)$) where $\sigma \in T$,
(b) control(n) where $n \in \omega$, and
(c) notpath($c(\sigma)$) where $\sigma \in T$.

Case (a). Atoms of the form path($c(\sigma)$) where $\sigma \in T$.

There are only two type ground clauses C with path($\cdot$) in the head, namely, those that are ground instances of clauses of type (1) and (3). Clause (3) is a Horn clause. This implies that a minimal P_T-proof scheme which derives path(0) and uses (3) must be of the form $\langle\langle \text{path}(0), (3), \emptyset \rangle\rangle$. Next, consider a minimal P_T-proof scheme of path($c(\sigma)$) which contains clause (1).

In such a case, the minimal P_T-proof will consist of a minimal P^--proof scheme of tree($c(\sigma)$) which will have empty support followed by the triple $\langle \text{path}(c(\sigma)), (1)^*, \{\text{notpath}(c(\sigma))\}\rangle$ where $(1)^*$ is the result of substituting $c(\sigma)$ for X in clause (1). Since we are assuming that tree($c(\sigma)$) has only finitely many P^--proof schemes and we can effectively find them, it follows that path($c(\sigma)$) has only finitely many minimal P_T-proof schemes and we can effectively find them.

Case (b). Atoms of the form control(n) where $n \in \omega$.

There are only two types of ground instances of clauses with the atom control(n) in the head, namely, ground instances of clauses (6) and (7). The only minimal P_T-proof schemes of control(n) that use a ground instance of clause (7) must consist of a minimal P^--proof scheme of num(n) followed by the triple $\langle \text{control}(n), (7)^*, \{\text{control}(n)\}\rangle$ where $(7)^*$ is the result of substituting n for X in (7). Since we are assuming that num(n) has only finitely many minimal P^--proof schemes and we can effectively find them, we can effectively find all minimal P_T-proof schemes of control(n) that uses a ground instance of (7). If we have a minimal P_T-proof scheme $\mathbb{T}$ with conclusion control(n) that uses a ground instances of clause (6), then the last triple of $\mathbb{T}$ must be of the form

$$\langle \text{control}(n), \text{control}(n) \leftarrow \text{path}(c(\tau)), \text{length}(c(\tau), n), A\rangle$$

where $c(\tau)$ is the code of node in T of length n. Moreover, in $\mathbb{T}$, this triple must be preceded by some interweaving of minimal P_T-proof schemes for path($c(\tau)$) and length($c(\tau), n$). Now we effectively find the finite set of minimal P^--proof schemes for length($c(\tau), n$) and we can effectively find the set of all P_T-minimal proof schemes for path($c(\tau)$). Moreover, it must be the case that A is the support of the minimal P_T-scheme of path($c(\tau)$) that was interweaved with one of the minimal proof schemes for length($c(\tau)), n$) to create $\mathbb{T}$. Since the support of any proof scheme for path($c(\tau)$) where $|\tau| \geq 1$ is just

$\{\text{notpath}(c(\tau))\}$, it follows that $A = \{\text{notpath}(c(\tau))\}$ if $|\tau| \geq 1$ and $A = \emptyset$ if $|\tau| = 0$. Now if T is finitely branching, there will only be finitely many choices for τ since to derive $\text{path}(c(\tau))$, τ must be in T. Hence there will be only finitely many choices of $\mathbb{T}$. On the other hand, if T is not finitely branching, then there will be an n such that there are infinitely many nodes $\tau \in T$ of length n for some $n > 0$ so that there will be infinitely many different supports of minimal P_T-proof schemes for $\text{control}(n)$. If T is highly recursive, then we can effectively find all $\tau \in T$ of length n so that we can effectively find all such $\mathbb{T}$. Similarly, if P_T has the rec.FS property, then for $n > 0$, we can read off all the nodes in T of length n from the supports of the minimal P_T-proof schemes of $\text{control}(n)$ so that T will be highly recursive. Thus T is finitely branching if and only if there are finitely many minimal P_T-proof schemes for $\text{control}(n)$ for each $n \geq 0$. Similarly, if T is highly recursive, then we can effectively find all the minimal P_T-proof schemes for $\text{control}(n)$ for each $n \geq 0$ and if P_T has the rec. FS property, then T is highly recursive.

Case (c). Atoms of the form $\text{notpath}(c(\sigma))$.

Here we have to take into account clauses (2), (4), and (5). First consider a minimal P_T-proof scheme of $\text{notpath}(c(\sigma))$ which contains a ground instance of clause (2). In such a case, the last element of the minimal P_T-proof scheme will consist of a minimal P^--proof scheme of $\text{tree}(c(\sigma))$ which will have empty support followed by the triple

$$\langle \text{notpath}(c(\sigma)), (2)^*, \{\text{path}(c(\sigma))\} \rangle,$$

where $(2)^*$ is the result of substituting $c(\sigma)$ for X in (2). Since we are assuming that $\text{tree}(c(\sigma))$ has only finitely many minimal P^--proof schemes and we can effectively find them, it follows that $\text{notpath}(c(\sigma))$ has only finitely many minimal P_T-proof schemes that use a ground instance of clause (2) and we can effectively find them.

Next, consider a P_T-proof scheme $\mathbb{T}$ with conclusion $\text{notpath}(c(\sigma))$ which contains a ground instance of clause (4).

Then there must exists a $\tau \in T$ of length $|\sigma|$ such that the last triple in the proof scheme is of the form

$$\langle c(\sigma), (4)^*, A \rangle, \tag{4}$$

where $(4)^*$ is the result of substituting $c(\sigma)$ for X and $c(\tau)$ for Y in (4). Then $\mathbb{T}$ must consist of an interweaving of minimal P^--proof schemes for tree$(c(\sigma))$, tree$(c(\tau))$, samelength$(c(\sigma), c(\tau))$, and diff$(c(\sigma), c(\tau))$ and a minimal P_T-proof scheme path$(c(\tau))$ with support A. In each case, there are only finitely many such minimal P_T-proofs schemes of these atoms and we can effectively find them. Thus for each $\tau \in T$ of length $|\sigma|$, we can effectively find all the minimal P_T-proof schemes of notpath$(c(\sigma))$ that end in a triple of the form of (4). Now if T is finitely branching, it follows that there will be only finitely many minimal P_T-proof schemes that use a ground instance of clause (4) and, if T is highly recursive, then we can effectively find all $\tau \in T$ of length $|\sigma|$ so that we can effectively find all minimal P_T-proof schemes that use a ground instance of clause (4).

Finally, consider a P_T-proof scheme $\mathbb{T}$ with conclusion notpath$(c(\sigma))$ which contains ground instance of clause (5). Then there must exists a $\tau \in T$ whose length is less than the length of σ and which is not an initial segment of σ such that the last triple in the proof scheme is of the form

$$\langle c(\sigma), (5)^*, A \rangle, \tag{5}$$

where $(5)^*$ is the result of substituting $c(\sigma)$ for X and $c(\tau)$ for Y in (5). Then $\mathbb{T}$ must consist of an interweaving of minimal P^--proof schemes for tree$(c(\sigma))$, tree$(c(\tau))$, shorter$(c(\tau), c(\sigma))$, and notincluded$(c(\tau), c(\sigma))$ and a minimal P_T-proof scheme of path(τ) with support A. In each case, there are only finitely many minimal P_T-proofs schemes of these atoms and we can effectively find them. Thus for each τ whose length is less than the length of σ and which is not an initial segment of σ, we can effectively find all the minimal P_T-proof schemes of notpath($c(\sigma)$) that end in a triple of the form of (5). Now if T is finitely branching, it follows that there will be only finitely many minimal P_T-proof schemes that use a ground instance of clause (5) and, if T is highly recursive, then we can effectively find all $\tau \in T$ of length $|\sigma|$ so that we can effectively find all minimal P_T-proof schemes that use a ground instance of clause (5).

Thus, we have proved that if T is finitely branching, then every ground atom possesses only finitely many minimal P_T-proof schemes and if T is highly recursive, then for every ground atom $a \in H(P_T)$, we can effectively find the set of all many minimal P_T-proof schemes of a. Thus if T is finitely branching, then P_T has the FS property and if T is highly recursive, then T has the rec.FS property. On the other hand, we have shown by our analysis in (b) that if P_T has the FS property, then T must be finitely branching and if P_T has the rec.FS property, then T is highly recursive. This proves (A) and (B) and establishes parts (2) and (3) of Theorem 4.1.

To prove (C), we shall establish a "normal form" for the stable models of P_T. Each such model must contain M^-, the least model of P^-. In fact, the restriction of a stable model of P_T to $H(P^-)$ is M^-. Given any $\beta = (\beta(0), \beta(1), \ldots) \in \omega^\omega$, recall that $\beta \restriction n = (\beta(0), \beta(1), \ldots, \beta(n-1))$. Then we let

$$M_\beta = M^- \cup \{\text{control}(n) : n \in \omega\} \cup \{\text{path}(0)\}$$

$$\cup \{\text{path}(c(\beta \restriction n)) : n \in \omega\}$$

$$\cup \{\text{notpath}(c(\sigma)) : \sigma \in T \text{ and } \sigma \nprec \beta\}. \qquad (6)$$

We claim that M is a stable model of P_T if and only if $M = M_\beta$ for some $\beta \in [T]$.

First, assume that M is a stable model of P. Thus M is the least model of the Gelfond–Lifschitz transform $(\text{ground}(P_T))_M$. We know that the atoms of $\mathcal{L}^-$ in M constitute M^-. First observe that by clause (3), $\text{path}(0) \in M$. Thus we can not use clause (2) to derive that $\text{notpath}(0)$ is in M. Moreover, it is easy to see that we cannot use clauses of the form (4) or (5) to derive that notpath(0) is in M so that it must be the case that notpath(0) $\notin M$. Next, suppose that $\sigma \in T$ and the length of σ is greater than or equal to 1. It is easy to see from clauses (1) and (2) that it cannot be the case that neither path($c(\sigma)$) and notpath($c(\sigma)$) are in M. Since clauses of the form of (1) are the only clauses that we can use to derive that path($c(\sigma)$) is in the least model of (ground(P_T))$_M$ when $|\sigma| \geq 1$, it follows that it cannot be the case that both path($c(\sigma)$) and notpath($c(\sigma)$) are in M. Thus exactly one of path($c(\sigma)$) and notpath($c(\sigma)$) must be in M for all $\sigma \in T$. Next, we claim that control(n) $\in M$ for all n. That is, if control(n) $\notin M$ for some n, then the Gelfond–Lifschitz transform of the ground clause control(n) $\leftarrow \neg$control(n), num(n) from (7) would be control(n) $\leftarrow$ num(n) which would force control(n) to be in M. Since control(n) $\in M$, the only way that one could derive that control(n) is in the least model of (ground(P_T))$_M$ is via a proof scheme that uses a ground instance of clause (6). This means that for each $n \geq 0$, there must be a $\tau^{(n)} \in T$ of length n such that path($c(\tau^{(n)})$) $\in M$. But then we can use clause (4) to show that if σ is a node in T of length n which is different from $\tau^{(n)}$, then notpath($c(\sigma)$) $\in M$. But now the clauses of type (5) will force that it must be the case that if $m < n$, then $\tau^{(m)}$ must be an initial segment of $\tau^{(n)}$. Thus the path τ where $\tau^{(n)} \sqsubseteq \tau$ for all n is an infinite path through T and $M = M_\tau$. Note that this shows that if $[T]$ is empty, then P has no stable model.

To complete the argument for (C), we have to prove that $\beta \in [T]$ implies that M_β is a stable model of P. Let lm(M_β) be the least model of (ground(P_T))$_{M_\beta}$. The presence of clauses (1) and (2) in P_T implies that $\{$path($c(\beta \upharpoonright (n))$) : $n \in \omega\} \cup \{$notpath($c(\sigma)$) : $\sigma \in T \setminus \{\beta \upharpoonright (n) : n \in \omega\}\} \subseteq$ lm(M_β). Then clause (6) can be used to show that for all n, control(n) also belongs to lm(M_β). Since $M^- \subseteq$ lm(M_β), it follows that $M_\beta \subseteq$ lm(M_β).

Next, we must prove that lm(M_β) $\subseteq M_\beta$. We know that since none of the heads of rules (1)–(7) involve predicates in $H(P^-)$, it must be the case that $\mathrm{lm}(M_\beta) \cap H(P^-) = M^-$. The only ground clauses from (1) that are in $(\mathrm{ground}(P_T))_{M_\beta}$ are clauses of the form

$$\mathrm{path}(c(\beta \restriction n)) \leftarrow \mathrm{tree}(c(\beta \restriction n)).$$

These are the only clauses of $(\mathrm{ground}(P_T))_{M_\beta}$ which have $\mathrm{path}(c(\sigma))$ in the head for $\sigma \in T$ so that $\{\mathrm{path}(c(\sigma)) : \sigma \in T\} \cap lm(M_\beta) \subseteq M_\beta$. Since M_β contains all ground clauses of the form $\mathrm{control}(n)$, the only clauses that we have to worry about are clauses with $\mathrm{notpath}(c(\sigma))$ in the head for $\sigma \in T$. The only ground clause from (2) that are in $(\mathrm{ground}(P_T))_{M_\beta}$ are clauses of the form

$$\mathrm{notpath}(c(\sigma)) \leftarrow \mathrm{tree}(c(\sigma))$$

where $\sigma \in T - \{\beta^{(n)} : n \geq 0\}$. Thus, the conclusions of all such clauses are in M_β. Hence we are reduced to considering ground clauses of the form (4) and (5). Since all such clauses must have an atom $\mathrm{path}(c(\tau))$ in the body, the only way we can use these clauses is to derive $\mathrm{notpath}(c(\sigma))$ in its head is if $\tau \in \{\beta^{(n)} : n \geq 0\}$. But then it is easy to see that this forces $\sigma \notin \{\beta^{(n)} : n \geq 0\}$. Thus, the only atoms $\mathrm{notpath}(c(\sigma)) \in \mathrm{lm}(M_\beta)$ that can be derived from $(\mathrm{ground}(P_T))_{M_\beta}$ are those such that $\sigma \in T - \{\beta^{(n)} : n \geq 0\}$. Thus $\mathrm{lm}(M_\beta) \subseteq M_\beta$. This proves part (1) of Theorem 4.1.

Finally, consider part (4) of Theorem 4.1. By part (3), we know that T is highly recursive if and only if P_T has the rec.FS property. We must show that if T is decidable and recursively bounded, then P_T is decidable. So suppose we are given a set of ground atoms $\{a_1, \ldots, a_n\}$ and corresponding minimal P_T-proof schemes $\mathbb{S}_i$ of a_i. For these atoms to be contained in a stable model of P_T, it must be the case that the ground atoms in the language of P^- must all be in M^- and there corresponding proof schemes must be the least minimal proofs schemes for P^-. This we can check recursively. The remaining atoms are of the form $\mathrm{path}(c(\sigma))$, $\mathrm{notpath}(c(\tau))$, and $\mathrm{control}(n)$. It must be the case that atoms of the form $\mathrm{path}(c(\sigma))$ and $\mathrm{notpath}(c(\tau))$ among

$\{a_1, \ldots, a_n\}$ must be consistent with being the initial segment of the path through T. If that is not the case, then it is clear that $\{a_1, \ldots, a_n\}$ is not contained in a stable model of P_T. If it is the case, let α be the longest string σ such that $\mathrm{path}(c(\sigma)) \in \{a_1, \ldots, a_n\}$. Now if $\alpha \notin \mathrm{Ext}(T)$, then again $\{a_1, \ldots, a_n\}$ is not contained in a stable model of P_T. If it is, then let m be the maximum of all n such that $\mathrm{control}(n) \in \{a_1, \ldots, a_n\}$ and $|\tau|$ such that $\mathrm{notpath}(c(\tau)) \in \{a_1, \ldots, a_n\}$. Since T is recursively bounded, then we can effectively find all strings of length m which extend α. Now if there is a string β of length m such that $\alpha \prec \beta$, $\beta \in \mathrm{Ext}(T)$, and there is no initial segment γ of β such that $\mathrm{notpath}(c(\gamma)) \in \{a_1, \ldots, a_n\}$, then it will be the case that $\{a_1, \ldots, a_n\}$ is contained in a stable model. For each such β and all $\delta \in T$ of length less than or equal to m, the only minimal proof schemes of ground atoms of the form $\mathrm{path}(c(\delta))$, $\mathrm{notpath}(c(\delta))$, and $\mathrm{control}(n)$ for $n \leq m$ depend only on the ground atoms $\mathrm{path}(c(\gamma))$ for γ contained in β. Thus by our analysis of Cases (a)–(c) above, we can compute the appropriate minimal proofs schemes and then check if the corresponding minimal P_T-proof schemes equals $\{\mathbb{S}_1, \ldots, \mathbb{S}_n\}$. Thus P_T is decidable.

On the other hand, suppose that P_T has the rec.FS property and P_T is decidable. Then given a node $\beta = (\beta_1, \ldots, \beta_n) \in T$. It is easy to see that for any path $\pi \in \omega^\omega$ which extends T, the elements of M_π which mention only β, nodes of length $\leq |\beta|$, and the elements $0, s^1(0) \ldots, s^{|\beta|}(0)$ are the same. Thus let

$$M_\beta = M^- \cup \{\mathrm{control}(n) : n \leq |\beta|\} \cup \{\mathrm{path}(0)\}$$
$$\cup \{\mathrm{path}(c(\alpha)) : \alpha \sqsubseteq \beta\}$$
$$\cup \{\mathrm{notpath}(c(\sigma)) : \sigma \in T, |\sigma| \leq |\beta|, \text{ and } \sigma \not\prec \beta\}. \quad (7)$$

Then M_β is finite and our analysis shows that we can effectively find all the minimal P_T-proofs schemes $\mathbb{S}_1, \ldots, \mathbb{S}_r$ which mention only β, nodes of length $\leq |\beta|$, and the elements

$0, s^1(0), \ldots, s^{|\beta|}(0)$ which have conclusions in M_β. By the decidability of P_T, we know whether there is a stable model which contains M_β and has $\mathbb{S}_1, \ldots, \mathbb{S}_r$ as the corresponding minimal P_T-proof schemes for elements in M_β. If there is such a stable model, then β must be an initial segment of some $\pi \in [T]$ so that $\beta \in \text{Ext}(T)$. If there is no such stable model, then there is no infinite path $\pi \in [T]$ such that $\beta \sqsubseteq \pi$ so that $\beta \notin \text{Ext}(T)$. Thus, if P_T is decidable and has the rec.FS property, then T is decidable and highly recursive. This completes the proof of Theorem 4.1

4.2.   *Proof of Theorem* 4.2

Suppose that we are given a normal finite predicate logic program P. Then by our remarks in the previous section, the Herbrand base $H(P)$ will be primitive recursive, ground(P) will be a primitive recursive program and, for any atom $a \in H(P)$, the set of minimal P-proof schemes with conclusion a is primitive recursive. We should note, however, that it is not guaranteed that the Support(a) which is the set of can(X) such that X is the support of a minimal P-proof scheme of a is recursive. However, it is the case that Support(a) is an r.e. set.

Our basic strategy is to encode a stable model M of ground(P) by a path $f_M = (f_0, f_1, \ldots)$ through the complete ω-branching tree $\omega^{<\omega}$ as follows.

(1) First, for all $i \geq 0$, $f_{2i} = \chi_M(i)$. That is, at the stage $2i$, we encode the information about whether or not the atom encoded by i belongs to M. Thus, in particular, if i is not the code of ground atom in $H(P)$, then $f_{2i} = 0$.
(2) If $f_{2i} = 0$, then we set $f_{2i+1} = 0$. But if $f_{2i} = 1$ so that $i \in M$ and i is the code of a ground atom in $H(P)$, then we let f_{2i+1} equal $q_M(i)$ where $q_M(i)$ is the least code for a minimal P-proof scheme $\mathbb{S}$ for i such that the support of $\mathbb{S}$ is disjoint from M. That is, we select a minimal P-proof

scheme $\mathbb{S}$ for i, or to be precise for the atom encoded by i, such that $\mathbb{S}$ has the smallest possible code of any P-proof scheme $\mathbb{T}$ such that $\mathrm{supp}(\mathbb{T}) \cap M = \emptyset$. If M is a stable model, then, by Proposition 2.1, at least one such P-proof scheme exists for i.

Clearly, $M \leq_T f_M$ since it is enough to look at the, values of f_M at even places to read off M. Now given an M-oracle, it should be clear that for each $i \in M$, we can use an M-oracle to find $q_M(i)$ effectively. This means that $f_M \leq_T M$. Thus the correspondence $M \mapsto f_M$ is an effective degree-preserving correspondence. It is trivially one-to-one.

Next, given a program P, we construct a primitive recursive tree $T_P \subseteq \omega^\omega$ such that $[T_P] = \{f_M : M \in \mathrm{stab}(P)\}$. Let N_k be the set of all codes of minimal P-proof schemes $\mathbb{S}$ such that all the atoms appearing in all the rules used in $\mathbb{S}$ are smaller than k. Obviously N_k is finite. It follows from our remarks in the previous section that since P is a normal finite predicate logic program, the predicate "minimal P-proof scheme" which holds only on codes of minimal P-proof schemes is a primitive recursive predicate. This means that there is a primitive recursive function h such that $h(k)$ equal the canonical index for N_k. Moreover, given the code of sequence $\sigma = (\sigma(0), \ldots, \sigma(k)) \in \omega^{<\omega}$, there is a primitive recursive function which will produce canonical indexes of the sets $I_\sigma = \{i : 2i \leq k \wedge \sigma(2i) = 1\}$ and $O_\sigma = \{i : 2i \leq k \wedge \sigma(2i) = 0\}$.

For any given $k \geq 2$, we let $\overline{k} = \max(\{2j + 1 : 2j + 1 < k\}$ and if $\sigma = (\sigma(0), \ldots, \sigma(k))$ is an element of $\omega^{<\omega}$, then we let $\overline{\sigma} = (\sigma(0), \ldots, \sigma(\overline{k}))$. If $k = 1$ and $\sigma = (\sigma(0))$, then we let $\overline{k} = 0$ and $\overline{\sigma} = \emptyset$. In what follows, we shall identify each atom in $H(P)$ with its code. Then we define T_P by putting a node $\sigma = (\sigma(0), \ldots, \sigma(k))$ into T_P if and only if the following five conditions are met:

(a) $\forall i (2i + 1 \leq \overline{k} \wedge \sigma(2i) = 0 \Rightarrow \sigma(2i + 1) = 0)$,

(b) $\forall i (2i + 1 \leq \bar{k} \wedge \sigma(2i) = 1 \Rightarrow \sigma(2i + 1) = q)$ where q is a code for a minimal P-proof scheme $\mathbb{S}$ such that $\mathrm{concl}(\mathbb{S}) = i$, $\mathrm{supp}(\mathbb{S}) \cap I_{\bar{\sigma}} = \emptyset$, and there is no number $j < \sigma(k)$ such that j is a code for a minimal P-proof scheme $\mathbb{T}$ with conclusion i such that $\mathrm{supp}(\mathbb{S}) = \mathrm{supp}(\mathbb{T})$,

(c) $\forall i (2i + 1 \leq \bar{k} \wedge \sigma(2i) = 1 \Rightarrow$ there is no code $c \in N_{\lfloor k/2 \rfloor}$ of a minimal P-proof scheme $\mathbb{T}$ such that $\mathrm{conc}(\mathbb{T}) = i$, $\mathrm{supp}(\mathbb{T}) \subseteq O_{\bar{\sigma}}$ and $c < \sigma(2i + 1)$ (here $\lfloor \cdot \rfloor$ is the number-theoretic "floor" function),

(d) $\forall i (2i + 1 \leq \bar{k} \wedge \sigma(2i) = 0 \Rightarrow$ there is no code $c \in N_{\lfloor k/2 \rfloor}$ of a minimal P-proof scheme $\mathbb{U}$ such that $\mathrm{concl}(\mathbb{U}) = i$ and $\mathrm{supp}(\mathbb{U}) \subseteq O_{\bar{\sigma}})$, and

(e) if $k = 2i + 1$ and $\sigma(2i) = 1$, then $\sigma(2i + 1) = q$ where q is a code for a minimal P-proof scheme $\mathbb{S}$ such that $\mathrm{concl}(\mathbb{S}) = i$ and there is no number $j < \sigma(k)$ such that j is a code for a minimal P-proof scheme $\mathbb{T}$ with conclusion i such that $\mathrm{supp}(\mathbb{S}) = \mathrm{supp}(\mathbb{T})$.

The first thing to observe that each of the conditions (a)–(e) requires that we check only a bounded number of facts about codes that have an explicit bound in terms of the code of σ. This implies that T_P has a primitive recursive definition. It is immediate from our conditions defining T_P that if $\sigma \in T_P$ and $\tau \prec \sigma$, then $\tau \in T_P$. Thus T_P is a primitive recursive tree. Conditions (a) and (b) ensure that the set of all paths π through T_P meet the minimal conditions to be of the form f_M for some stable model. That is, condition (a) ensures that if $\pi(2i) = 0$, then $\pi(2i+1) = 0$. Condition (b) ensures that if $\pi(2i) = 1$, then $\pi(2i+1)$ is the code of a minimal P-proof scheme with conclusion i and there is no smaller code of a minimal P-proof scheme of i with the same support. Conditions (c), (d) and (e) are carefully designed to ensure that T_P has the properties that we want. First, condition (c) limits the possible infinite paths through T_P. We claim that if π is an infinite path through T_P and $\pi(2i) = 1$, then $\pi(2i+1) = r$ where r is smallest code of a minimal P-proof scheme with conclusion i whose support does not intersect $M_\pi = \{j : \pi(2j) = 1\}$. That is, if $\pi(2i+1)$ is the code of minimal P-proof scheme with conclusion i whose support is disjoint from M_π which is greater than r, then there will be some $k > 2i+1$ such that $c \in N_{\lfloor k/2 \rfloor}$ in which case condition (d) would not allow $(\pi(0), \ldots, \pi(k+2))$ to be put into T_P. Similarly, if $\pi(2i+1)$ is the code of minimal P-proof scheme $\mathbb{S}$ with conclusion i whose support is not disjoint from M_π, then there will be some $k > 2i+1$ such that $\mathrm{supp}(\mathbb{S}) \cap I_{(\pi(0),\ldots,\pi(k))} \neq \emptyset$ in which case condition (b) would not allow $(\pi(0), \ldots, \pi(k+2))$ to be put into T_P. Likewise, condition (d) ensures that if $\pi(2i) = 0$, there can be no minimal P-proof scheme $\mathbb{S}$ with conclusion i whose support is disjoint from M_π since otherwise for large enough k, condition (e) would not allow $(\pi(0), \ldots, \pi(k))$ to be put into T_P. Finally, condition (e) is designed to ensure that T_P is finitely branching if and only if P has the FSP property or has an explicit initial blocking set. Note that for a node $(\sigma(0), \ldots, \sigma(2i), \sigma(2i+1))$ where $\sigma(2i) = 1$, $\sigma(2i+1)$ can be the code of *any* minimal P-proof scheme $\mathbb{S}$ with conclusion i for which there is no smaller number which codes a proof scheme with the same conclusion and same support. For example, we do not require $\mathrm{supp}(\mathbb{S}) \cap I_\sigma = \emptyset$. However, if $\mathrm{supp}(\mathbb{S}) \cap I_\sigma \neq \emptyset$, then condition (b) will ensure that there are no extensions of σ in T.

Our next goal is to show that every $f \in [T_P]$ is of the form f_M for a suitably chosen stable model M of P. It is clear that if M is stable model of P, then for all k, $(f_M(0), \ldots, f_M(k))$ satisfies conditions (a)–(e) so that $f_M \in [T_P]$. Thus $\{f_M : M \in \mathrm{Stab}(P)\} \subseteq [T_P]$.

Next, assume that $\beta = (\beta(0), \beta(1), \ldots)$ is an infinite path through T_P and $M_\beta = \{i : \beta(2i) = 1\}$. Then we must prove that (I) M_β is a stable model of P and (II) $f_{(M_\beta)} = \beta$.

For (I), suppose that M_β is not a stable model of P. Let $\mathrm{lm}(M_\beta)$ be the least model of Gelfond–Lifschitz transform $\mathrm{ground}(P)_{M_\beta}$ of $\mathrm{ground}(P)$ relative to M_β. Then by Proposition 2.1, it must be the case that either

(i) there is $j \in M_\beta \setminus \mathrm{lm}(M_\beta)$ or
(ii) there is $j \in \mathrm{lm}(M_\beta) \setminus M_\beta$.

If (i) holds, then let i be the least $j \in M_\beta \setminus \mathrm{lm}(M_\beta)$ and consider the string $\beta \upharpoonright (2i + 3) = (\beta(0), \ldots, \beta(2i + 3))$. For $\beta \upharpoonright (2i + 3)$ to be in T, it must be the case that $\beta(2i + 1)$ is a code of a minimal proof scheme $\mathbb{S}$ such that $\mathrm{concl}(\mathbb{S}) = i$ and $\mathrm{supp}(\mathbb{S}) \cap I_{\beta \upharpoonright (2i+1)} = \emptyset$. But since $i \notin lm(M_\beta)$, there must be some n belonging to $M_\beta \cap \mathrm{supp}(\mathbb{S})$. Clearly, it must be the case that $n > i$. Choose such an n. Then $\beta \upharpoonright (2n) \notin T$ because $\mathrm{supp}(\mathbb{S}) \cap I_{\beta \upharpoonright (2n)} \neq \emptyset$ which contradicts our assumption that $\beta \in [T]$. Thus (i) cannot hold.

If (ii) holds, then let i be the least $j \in \mathrm{lm}(M_\beta) \setminus M_\beta$ and consider again $\beta \upharpoonright (2i + 3)$. Since $i \in \mathrm{lm}(M_\beta)$, there must be a proof scheme $\mathbb{T}$ such $\mathrm{concl}(\mathbb{T}) = j$ and $\mathrm{supp}(\mathbb{T}) \cap M_\beta = \emptyset$. But then there is an $n > 2i+1$ large enough so that $\mathrm{supp}(\mathbb{T}) \subseteq O_{\beta \upharpoonright (n)}$. But then $\beta \upharpoonright (n)$ does not satisfy the condition (e) of our definition to be in the tree which again contradicts our assumption that $\beta \in [T]$. Thus (ii) also cannot hold so that M_β must be a stable model of P.

Thus we need only to verify claim (II), namely, that $\beta = f_{(M_\beta)}$. Now if $\beta \neq f_{(M_\beta)}$, then it must be the case that for some $i \in M_\beta$, there is a code c of a minimal proof scheme $\mathbb{S}$ such that $\mathrm{concl}(\mathbb{S}) = i$, $\mathrm{supp}(\mathbb{S}) \cap M_\beta = \emptyset$ and $c < \beta(2i + 1)$. But then there is an $n > 2i+1$ large enough so that $\mathrm{supp}(\mathbb{S}) \subseteq O_{\beta \upharpoonright (n)}$ and hence $\beta \upharpoonright (n)$ does not satisfy condition (d) of our definition to be in T. Hence, if $\beta \neq f_{(M_\beta)}$, then $\beta \upharpoonright (n) \notin T_P$ for some n and so $\beta \notin [T_P]$. This completes the proof of (II) and hence part (1) of the theorem holds.

Next, consider parts (2)–(10). Note that the tree T_P has the property that if $\beta^{(n)} \in T$, then

(†) for every i such that $2i \leq n$, $\beta(2i) \in \{0, 1\}$ and
(‡) for every i such that $2i+1 \leq n$, $\beta(2i+1)$ is either 0 or it is a code of a minimal proof scheme $\mathbb{S}$ such that $\mathrm{concl}(\mathbb{S}) = i$ and

no $j < \beta(2i+1)$ is the code of a minimal P-proof scheme of i with the same conclusion and the same support.

Thus if P has a finite number of supports of minimal P-proof schemes for each i, then T_P will be automatically finitely branching. Next, suppose that P has the additional property that there is a recursive function h whose values at i encode all the supports of minimal P-proof schemes for i. Say the possible supports of minimal P-proof schemes for i are $S_1^i, \ldots, S_{\ell_i}^i$. Then for each $1 \le j \le \ell_i$, we can effectively find the smallest code c_j^i of a minimal P-proof scheme for i with support S_i. Thus for each i, we can use h to compute $c_1^i, \ldots, c_{\ell_i}^i$. But then we know the possible values of $\sigma(2i+1)$ for any $\sigma \in T_P$ must come from $0, c_1^i, \ldots, c_{\ell_i}^i$ so that T_P is recursively bounded. Next, observe that if P has the a.a. FS support property, then it will be the case that for all sufficient large i, there will be only a finite number of supports of minimal P-proof schemes of i so that T_P will be nearly bounded. Similarly, if P has the a.a. rec.FS support property, then it will be the case that for all sufficient large i, we can effectively find the supports of all minimal P-proof schemes of i so that as above, we can effectively find the possible values of $\sigma(2i+1)$ and, hence, T_P will be nearly recursively bounded.

Next, suppose that P does not have the FS property. Let i be the least atom such that there exist infinitely many supports of P-proof schemes with conclusion i. Now suppose that there is a node $\sigma = (\sigma(0), \ldots \sigma(2i+1))$ of length $2i+1$ in T_P. It is easy to check that it will also be the case that $\sigma^* = (\sigma(0), \ldots, \sigma(2i-1), 1, r)$ is a node in T_P where r is any code of a minimal P-proof scheme $\mathbb{S}$ of i such that there is no smaller code q of a minimal P-proof scheme $\mathbb{T}$ of i such that $\mathrm{supp}(\mathbb{S}) = \mathrm{supp}(\mathbb{T})$. Thus if T_P has a node of length $2i+1$, then T_P will not be infinitely branching. Note that if P has a stable model, then T_P has a node of length $2i+1$ so that T_P is finitely branching if and only if P has the FS property. If T_P does not have any node of length $2i+1$, then it is easy to check that our conditions ensure that $\{0, \ldots, i-1\}$ is an explicit initial blocking set for P. Thus T_P

is finitely branching if and only if P has the FS property or P has an explicit initial blocking set.

Now suppose that P does not have the *a.a.* FS property. Then there will be infinitely many i which are codes of ground atoms of P such that there exist infinitely many supports of P-proof schemes with conclusion i. Now suppose that there is a node $\sigma = (\sigma(0), \ldots \sigma(2i + 1))$ of length $2i + 1$ in T_P. Then again, $\sigma^* = (\sigma(0), \ldots, \sigma(2i - 1), 1, r)$ is a node in T_P where r is any code of a minimal P-proof scheme $\mathbb{S}$ of i such that there is no smaller code q of a minimal P-proof scheme $\mathbb{T}$ of i such that $\mathrm{supp}(\mathbb{S}) = \mathrm{supp}(\mathbb{T})$. Thus, if T_P has a node of length $2i + 1$, then T_P will have a node of length $2i$ which has infinitely many successors in T_P. Note that if P has a stable model, then T_P has a node of length $2i + 1$ for all i so that T_P is nearly bounded if and only if P has the *a.a.* FS property. If T_P does not have any node of length $2i+1$, then it is easy to check that our conditions ensure that $\{0, \ldots, i - 1\}$ is an initial blocking set for P. Thus T_P is nearly bounded if and only P has the *a.a.* FS property or P has an initial blocking set.

Next, assume that T_P is finitely branching. By König's lemma, either T_P is finite or T_P has an infinite path. If T_P has an infinite path, then there will be nodes of length $2i + 1$ in T_P for all i. Hence for each i, there will be nodes of the form $\sigma^* = (\sigma(0), \ldots, \sigma(2i-1), 1, r)$ in T_P where r is any code of a minimal P-proof scheme $\mathbb{S}$ of i such that there is no smaller code q of a minimal P-proof scheme $\mathbb{T}$ of i such that $\mathrm{supp}(\mathbb{S}) = \mathrm{supp}(\mathbb{T})$. Thus if T_P is highly recursive, then for all i, we can find all the codes r is any code of a minimal P-proof scheme $\mathbb{S}$ of i such that there is no smaller code q of a minimal P-proof scheme $\mathbb{T}$ of i such that $\mathrm{supp}(\mathbb{S}) = \mathrm{supp}(\mathbb{T})$ because we can compute the set of nodes of length $2i + 1$ as a function of i. Thus T_P is highly recursive if and only if P has the recursive FS property or P has an explicit initial blocking set. Similarly if P has a stable model, then T_P must have an infinite path so that T_P is highly recursive if and only if P has the rec. FS property.

Next, assume that T_P is nearly bounded. Thus there is an $m \geq 0$ such that each node of length greater than or equal to m has only finitely many successors in T_P. If T_P has nodes of length $2i$ for all $i \geq 0$, there will be nodes of the form $\sigma^* = (\sigma(0), \ldots, \sigma(2i-1), 1, r)$ in T_P where r is any code of a minimal P-proof scheme $\mathbb{S}$ of i such that there is no smaller code q of a minimal P-proof scheme $\mathbb{T}$ of i such that $\text{supp}(\mathbb{S}) = \text{supp}(\mathbb{T})$. Hence if $2i \geq m$, then it must be the case that there are only finitely many supports of minimal P-proof schemes of the atom a coded by i. Clearly, T_P has infinite path, then there will be a node a length $2i$ for all i so that P must have the *a.a.* FS property. Similarly, if T_P is nearly recursively bounded and T_P has nodes of length $2i$ for all i, then P will have the *a.a.* rec.FS property. Thus if T_P is nearly bounded, then either there will be some fixed n such that T_P has no nodes of length $2n$ in which case T_P has an initial blocking set or T_P has nodes of length $2n$ for all $n \geq 0$ in which P has the *a.a.* FS property. Similarly, if T_P is nearly recursively bounded, then either there will be some fixed n such that T_P has no nodes of length $2n$ in which case T_P has an initial blocking set or T_P has nodes of length $2n$ for all $n \geq 0$ in which P has the *a.a.* rec.FS property. Thus T_P is nearly bounded (nearly recursively bounded) if and only if P has an initial blocking set or P has the *a.a.* FS property (*a.a.* rec.FS property). In particular, if P has a stable model, then T_P is nearly bounded (nearly recursively bounded) if and only if P has the *a.a.* FS property (*a.a.* rec.FS property). Thus parts (2)–(9) of the theorem hold.

For (10), note that if P is decidable, then for any finite set of ground atoms $\{a_1, \ldots, a_n\} \subseteq H(P)$ and any finite set of minimal P-proof schemes $\{\mathbb{S}_1, \ldots, \mathbb{S}_n\}$ such that $\text{concl}(\mathbb{S}_i) = a_i$, we can effectively decide whether there is a stable model of M of P such that

(i) $a_i \in M$ and $\mathbb{S}_i$ is the smallest minimal P-proof scheme for a_i such that $\text{supp}(\mathbb{S}_i) \cap M = \emptyset$ and

(ii) for any ground atom $b \notin \{a_1, \ldots, a_n\}$ such that code of b is strictly less than the maximum of the codes of $a_1, \ldots, a_n$, $b \notin M$.

But this is precisely what we need to decide to determine whether a given node in T_P can be extended to an infinite path through T_P. Thus if P is decidable, then T_P is decidable. On the other hand, suppose T_P is decidable and we are given a set of atoms $\{a_1 < \cdots < a_n\} \subseteq H(P)$ and any finite set of minimal P-proof schemes $\{\mathbb{S}_1, \ldots, \mathbb{S}_n\}$ such that $\text{concl}(S_i) = a_i$. Then let $\sigma = (\sigma(0), \ldots, \sigma(2a_n + 3))$ be such that $\sigma(2a_n+2) = \sigma(2a_n+3) = 0$ and for $i \leq a_n$, $\sigma(2i) = \sigma(2i+1) = 0$ if $i \notin \{a_1 < \cdots < a_n\}$ and $\sigma(2i) = 1$ and $\sigma(2i+1) = c(S_i)$. Then there is an infinite path of T_P that passes through σ if and only if there is a a stable model of M of P such that

(i) $a_i \in M$ and $\mathbb{S}_i$ is the smallest minimal P-proof scheme for a_i such that $\text{supp}(\mathbb{S}_i) \cap M = \emptyset$ and
(ii) for any ground atom $b \notin \{a_1, \ldots, a_n\}$ such that code of b is strictly less than the maximum of the codes of $a_1, \ldots, a_n$, $b \notin M$.

Thus P is decidable if and only if T_P is decidable. This completes the proof of Theorem 4.2. □

5. Complexity of Index Sets for Normal Finite Predicate Logic Programs

In this section, we shall prove our results on the complexity of index sets associated with various properties of normal finite predicate logic programs, normal finite predicate logic programs which have the FS property, and normal finite predicate logic programs which have the recursive FS property. In what follows, we shall abbreviate the normal finite predicate logic program Q_e has the FS (rec.FS) property by saying Q_e has the FSP (rec.FSP). We shall also abbreviate the property that the set of stable models of Q_e is empty (nonempty) by saying Q_e is empty (nonempty).

Theorem 5.1.

(a) $\{e : Q_e \text{ has an initial blocking set}\}$ and $\{e : Q_e \text{ has an explicit initial blocking set}\}$ are Σ_2^0-complete.
(b) $\{e : Q_e \text{ has the rec.FSP}\}$ is Σ_3^0-complete.
(c) $\{e : Q_e \text{ has the FSP}\}$ is Π_3^0-complete.
(d) $\{e : Q_e \text{ has the rec.FSP and is dec.}\}$ is Σ_3^0-complete.

Proof. In each case, it easy to see that the index set is of the required complexity by simply writing out the definition.

Let $A = \{e : Q_e \text{ has an explicit initial blocking set}\}$ and Fin $= \{e : W_e \text{ is finite}\}$. We know that Fin is is Σ_2^0-complete, see [31]. Thus to show that A is Σ_2^0-complete, we need only show that Fin is many-to-one reducible to A. Recall that $W_{e,s}$ is the set of all elements x less than or equal to s such that $\phi_e(x)$ converges s or fewer steps. It follows that for any e, $N_e = \{s : W_{e,s} - W_{e,s-1} \neq \emptyset\}$ and the set S_e of all codes of pairs (x, y) such that $x, y \in N_e$, $x < y$, and there is no $z \in N_e$ such that $x < z < y$ are recursive sets. Then by Proposition 3.1, we can uniformly construct a normal finite predicate logic Horn program P_e^- whose set of atoms is $\{s^n(0) : n \geq 0\}$ and which contains two predicates $N(x)$ and $S(x, y)$ such that $N(s^x(0))$ holds if and only if $x \in N_e$ and $S(s^x(0), s^y(0))$ holds if and only if $[x, y] \in S_e$. Let E be a unary predicate symbol that does not appear in P_e^-. Then we let P_e, the normal finite predicate logic program that consists of P_e^- plus the following two predicate logic clauses:

(a) $E(x) \leftarrow N(x), \neg E(x)$ and
(b) $E(x) \leftarrow N(y), S(x, y)$.

The clauses in (a) and (b) will generate the following clauses in ground(P_e):

(A) $E(s^n(0)) \leftarrow N(s^n(0)), \neg E(s^n(0))$ for all $n \geq 0$ and
(B) $E(s^m(0)) \leftarrow N(s^n(0)), S(s^m(0), s^n(0))$ for all $m, n \geq 0$.

Now suppose that W_e is infinite and $N_e = \{n_0 < n_1 < \cdots\}$. Then we claim that P_e has a stable model M_e which consists of the the least model of P_e^- plus $\{E(s^{n_i}(0)) : i \geq 0\}$. That is, the presence of $N(s^n(0))$ in the body of the clauses in (A) and the presence of $N(s^n(0))$ and $S(s^m(0), s^n(0))$ in the body of the clauses in (B) ensure that the only atoms of the form $E(a)$ that can possibly be in any stable model of P_e are elements of the form $E(s^n(0))$ where $n \in N_e$. But if W_e is infinite, then the Horn clauses of type (B) ensure that $\{s^{n_i}(0) : i \geq 0\}$ will be in every stable model. This, in turn, means that none of the clauses of type (A) for $n \in N_e$ will contribute to the Gelfond–Lifschitz transform $(P_e)_{M_e}$. It follows that $(P_e)_{M_e}$ consists of P_e^- plus all the clauses in (B) plus all the clauses of the form $s^n(0) \leftarrow N(s^n(0))$ such that $n \notin N_e$. It is then easy to see that M_e is the least model of $(P_e)_{M_e}$ so that M_e is a stable model of P_e. Thus if W_e is infinite, then P_e does not have explicit initial blocking set.

Next suppose that W_e is finite. Then N_e is finite, say $N_e = \{n_0 < \cdots < n_r\}$. Then we will not be able to use a clause of type (B) to derive $E(s^{n_r}(0))$. Thus the only clause that could possibly derive $E(s^{n_r}(0))$ would be the clause

$$C = E(s^{n_r}(0)) \leftarrow N(s^{n_r}(0)), \neg E(s^{n_r}(0)).$$

But then there can be no stable model M of P_e. That is, if $s^{n_r}(0) \in M$, then clause C will not be in $(P_e)_M$ so that there will be no way to derive $E(s^{n_r}(0))$ from $(P_e)_M$. On the other hand, if $s^{n_r}(0) \notin M$, then clause C will contribute the clause $E(s^{n_r}(0)) \leftarrow N(s^{n_r}(0))$ to $(P_e)_M$ so that $E(s^{n_r}(0))$ will be in the least model of $(P_e)_M$. It follows that $\{E(0), E(s(0)), \ldots, E(s^{n_r}(0))\}$ together with all that atoms of P_e^- whose code is less than the code of $E(s^{n_r}(0))$ will be an explicit initial blocking set for P_e.

Thus we have shown that P_e has an explicit initial blocking set if and only if W_e is finite. Hence, the recursive function f such that $Q_{f(e)} = P_e$ shows that Fin is many-to-one reducible to A and, hence, A is Σ^0_2-complete. Then same proof will show that $B = \{e : Q_e \text{ has an initial blocking set}\}$ is Σ^0_2-complete.

We claim that the completeness of the remaining parts of the theorem are all a consequence of Theorem 4.1. That is, recall that $T_0, T_1, \ldots$ is an effective list of all primitive recursive trees. Then let g be the recursive function such that $Q_{g(e)} = P_{T_e}$ where P_{T_e} is the normal finite predicate logic program constructed from T_e as in the proof of Theorem 4.1. Then g shows that

(1) $\{e : T_e \text{ is r.b.}\}$ is many-to-one reducible to $\{e : Q_e \text{ has the rec.FSP}\}$,

(2) $\{e : T_e \text{ is bounded}\}$ is many-to-one reducible to $\{e : Q_e \text{ has the FSP}\}$,

(3) $\{e : T_e \text{ is r.b. and decidable}\}$ is many-to-one reducible to $\{e : Q_e \text{ has the rec.FSP and is decidable}\}$.

Hence the completeness results for parts (b), (c), and (d) immediately follow from our completeness results for $\{e : T_e \text{ is r.b.}\}$, $\{e : T_e \text{ is bounded}\}$, and $\{e : T_e \text{ is r.b. and decidable}\}$ given in Section 2. □

Next we observe that some of the complexity results for normal finite predicate logic programs do not match the corresponding complexity for trees. That is, König's lemma tells us that an infinite finitely branching tree must have an infinite path through it. It follows that $[T] = \emptyset$ for a primitive recursive finitely branching tree if and only if T is finite. This means the properties that T is bounded and empty and T is recursively bounded and empty are Σ^0_2 properties since T being finite is a Σ^0_2 predicate for primitive recursive trees. König's lemma is a form of the Compactness Theorem for Propositional Logic which we have observed fails for normal propositional logic programs. Indeed, given any normal finite predicate logic program Q_e, we can simply take an atom a which does not occur in ground(Q_e) and add the clause $C = a \leftarrow \neg a$. Then the program $Q_e \cup \{C\}$ does not have a stable model but will have the FS property if and only if Q_e has the FS property and will have the rec.FS property if and only if Q_e has the rec.FS property. Thus there is a recursive function h such that:

(1) $Q_{h(e)}$ does not have a stable model.
(2) Q_e has the FS property if and only if $Q_{h(e)}$ has the FS property.
(3) Q_e has the rec.FS property if and only if $Q_{h(e)}$ has the rec.FS property.

It follows that $\{e : Q_e$ has the FSP$\}$ is many-to-one reducible to $\{e : Q_e$ has the FSP and is empty$\}$ and $\{e : Q_e$ has the rec.FSP$\}$ is many-to-one reducible to $\{e : Q_e$ has the rec.FSP and is empty$\}$. Thus it follows from Theorem 3.7 that:

(1) $\{e : Q_e$ has the FSP and $\mathrm{Stab}(Q_e)$ is empty$\}$ is complete for Π_3^0 sets.
(2) $\{e : Q_e$ has the rec.FSP and $\mathrm{Stab}(Q_e)$ is empty$\}$ is complete for Σ_3^0 sets.

To see that $\{e : Q_e$ has the rec.FSP and is empty$\}$ is Π_3^0, we can appeal to Theorem 4.2 which constructs the a finitely branching tree T_{Q_e} such that there is a one-to-one effective degree-preserving correspondence between the stable models of Q_e and $[T_{Q_e}]$. It follows that Q_e has the FSP and is empty if and only if Q_e has the FSP and T_{Q_e} is finite which is a Π_3^0 predicate because Q_e having the FS property is Π_3^0 predicate and T_{Q_e} being finite is a Σ_2^0 predicate. Similarly, Q_e has the rec.FSP and is empty if and only if Q_e has the rec.FSP and T_{Q_e} is finite which is a Σ_3^0 predicate because Q_e having the rec.FS property is Σ_3^0 predicate and T_{Q_e} being finite is a Σ_2^0 predicate. Thus, we have proved the following theorem.

Theorem 5.2.

(a) $\{e : Q_e$ *has the rec.FSP and* $\mathrm{Stab}(Q_e)$ *is empty*$\}$ *is* Σ_3^0-*complete*.
(b) $\{e : Q_e$ *has the FSP and is empty*$\}$ *is* Π_3^0-*complete*.

The method of proof for parts (b), (c), and (d) in Theorem 5.1 can be used to prove many results about properties of stable models of normal finite predicate logic programs Q_e where $\text{Stab}(Q_e)$ is not empty. That is, one can prove that the desired index is in the proper complexity class by simply writing out the definition or by using Theorem 4.2. For example, Theorem 2.6(b) says that $\{e : T_e \text{ is } r.b. \text{ and } [T_e] \text{ is nonempty}\}$ is Σ_3^0-complete. We claim that this theorem automatically implies that $\{e : Q_e \text{ has the rec.FSP and is nonempty}\}$ is also Σ_3^0-complete. First, we claim that the fact that

$$\{e : Q_e \text{ has the rec.FSP and is nonempty}\}$$

is Σ_3^0 follows from Theorem 4.2. That is, by Theorem 4.2, Q_e has the rec.FSP and is nonempty if and only if T_{Q_e} is $r.b.$ and $[T_{Q_e}]$ is nonempty. But this latter question is Σ_3^0 question so the former question is a Σ_3^0 question. Thus Theorem 4.2 allows us to reduce complexity bounds about normal finite predicate logic programs P which have stable models to complexity bounds to their corresponding trees T_P where $[T_P]$ is nonempty. Then we can then use Theorem 4.1 plus the theorems on index sets for trees given in Section 2 to establish the necessary completeness results. For example, to show that $\{e : Q_e \text{ has the rec.FSP and is nonempty}\}$ is Σ_3^0-complete, we use Theorem 4.1 and the fact that $\{e : T_e \text{ is } r.b. \text{ and } [T_e] \text{ is nonempty}\}$ is Σ_3^0-complete. That is, it follows from the proof of Theorem 4.1 that there is a recursive function f such that $Q_{f(e)} = P_{T_e}$. Hence $e \in \{e : T_e \text{ is } r.b. \text{ and } [T_e] \text{ is nonempty}\}$ if and only if $f(e) \in \{e : Q_e \text{ has the rec.FSP and is nonempty}\}$. Thus $\{e : Q_e \text{ has the rec.FSP and is nonempty}\}$ is Σ_3^0-complete.

One can use the same techniques to prove that the following theorem follows from the corresponding index sets results on trees given in Section 2.

Theorem 5.3.

(a) $\{e : Q_e$ has the rec.FSP and $Stab(Q_e)$ is nonempty$\}$ is Σ_3^0-complete.

(b) $\{e : Q_e$ has the FSP and $Stab(Q_e)$ is nonempty$\}$ is Π_3^0-complete.

(c) $\{e : Stab(Q_e)$ is nonempty$\}$ is Σ_1^1-complete.

Note that since $\{e : Q_e$ is empty$\}$ is the complement of the Σ_1^1-complete set $\{e : Q_e$ is nonempty$\}$, we have the following corollary.

Corollary 5.1. $\{e : Stab(Q_e)$ is empty$\}$ is Π_1^1-complete.

Next we want to consider the properties of being $Stab(Q_e)$ being infinite or finite.

Theorem 5.4.

(a) $\{e : Q_e$ has the rec.FSP and $Stab(Q_e)$ is infinite$\}$ is D_3^0-complete and $\{e : Q_e$ has the rec.FSP and $Stab(Q_e)$ is finite$\}$ is Σ_3^0-complete.

(b) $\{e : Q_e$ has the FSP and $Stab(Q_e)$ is infinite$\}$ is Π_4^0-complete and $\{e : Q_e$ has the FSP and $Stab(Q_e)$ finite$\})$ is Σ_4^0-complete.

(c) $(\{e : Stab(Q_e)$ is infinite$\}$ is Σ_1^1-complete and $\{e : Stab(Q_e)$ is finite$\})$ is Π_1^1-complete.

Proof. To prove the upper bounds in each case, we do the following. Given a normal finite predicate logic program Q_e, let a and $\bar{a}$ be two atoms which do not occur in ground(P). Then let R_e be the normal finite predicate logic program which arises from Q_e by adding a to body of every clause in Q_e and adding the following two clauses:

$C_1 = a \leftarrow \neg\bar{a}$ and
$C_2 = \bar{a} \leftarrow \neg a.$

Then we claim that exactly one of a or $\bar{a}$ must be in every stable model M of R_e. That is, if neither a or $\bar{a}$ are in M, then C_1 and C_2 will contribute $a \leftarrow$ and $\bar{a} \leftarrow$ to $(R_e)_M$ so that both a and $\bar{a}$ will be in the least model of $(R_e)_M$. If both a and $\bar{a}$ are in M, then C_1 and C_2 will contribute nothing to $(R_e)_M$ so that neither a nor $\bar{a}$ will be in the least model of $(R_e)_M$ since then there will be no clauses of $(R_e)_M$ with either a or $\bar{a}$ in the head of the clause. It follows that R_e will have two types of stable models M, namely $M = \{\bar{a}\}$ or $M = M^* \cup \{a\}$ where M^* is stable model of Q_e. Thus $\mathrm{Stab}(Q_e)$ is finite if and only if $\mathrm{Stab}(R_e)$ is finite. Clearly Q_e has the FS (rec.FS) property if and only if R_e has the FS (rec.FS) property. By Theorem 4.2, there is a recursive function g such that $T_{g(e)} = T_{R_e}$ as constructed in the proof of Theorem 4.2. Then we know $\mathrm{Stab}(R_e)$ is finite if and only if $[T_{g(e)}]$ is finite and R_e has the FS (rec.FS) property if and only if $T_{g(e)}$ is bounded (r.b). Then for example, it follows that $\{e : Q_e$ has the FSP and is finite$\}$ is many-to-one reducible to $\{e : T_e$ is r.b and finite$\}$ which is Σ_3^0. In this way, the complexity bounds follows from the complexity bounds in Theorem 2.8.

To establish the completeness results in each case, we can proceed as follows. We can use the construction of Theorem 4.1 to construct a normal finite predicate logic program P_{T_e} such that $[T_e]$ is finite if and only if $\mathrm{Stab}(P_{T_e})$ is finite and T_e is bounded (r.b.) if and only if $P_{T_e^*}$ has the FS (rec.FS) property. Thus there is a recursive function f such that $Q_{f(e)} = P_{T_e}$. Then, for example, it follows that f shows that $\{e : T_e$ is r.b. and $[T_e]$ is finite$\}$ is many-to-one reducible to $\{e : Q_e$ has the FSP and is finite$\}$. Thus $\{e : Q_e$ has the FSP and is finite$\}$ is Σ_3^0-complete since $\{e : T_e$ is r.b and $[T_e]$ is finite$\}$ is Σ_3^0-complete. In this way, we can use completeness results of Theorem 2.8 to establish the completeness of each part of the theorem. $\qquad\square$

By combining the completeness results of Theorem 2.9 with Theorems 4.1 and 4.2, we can use the same method of proof to prove the following theorems.

Theorem 5.5. $\{e:\mathrm{Stab}(Q_e)$ *is uncountable*$\}$ *is* Σ_1^1-*complete and* $\{e:\mathrm{Stab}(Q_e)$ *is countable*$\})$ *and* $\{e:Q_e$ *is countably infinite*$\}$ *are* Π_1^1-*complete.*
   The same results holds for rec.FSP and FSP programs.

Theorem 5.6. *For any positive integer c,*

(a) $\{e:Q_e$ *has the rec.FSP and* $\mathrm{Card}(\mathrm{Stab}(Q_e)) > c\}$, $\{e:Q_e$ *has the rec.FSP and* $\mathrm{Card}(\mathrm{Stab}(Q_e)) \leq c\}$, *and* $\{e:Q_e$ *has the rec.FSP and* $\mathrm{Card}(\mathrm{Stab}(Q_e)) = c\}$ *are all* Σ_3^0-*complete.*

(b) $\{e:Q_e$ *has the FSP and* $\mathrm{Card}(\mathrm{Stab}(Q_e)) \leq c\}$ *and* $\{e:Q_e$ *has the FSP and* $\mathrm{Card}(\mathrm{Stab}(Q_e)) = 1\}$ *are both* Π_3^0-*complete.*

(c) $\{e:Q_e$ *has the FSP and* $\mathrm{Card}(\mathrm{Stab}(Q_e)) > c\}$ *and* $\{e:Q_e$ *has the FSP and* $\mathrm{Card}(\mathrm{Stab}(Q_e)) = c+1\}$ *are both* D_3^0-*complete.*

(d) $\{e:Q_e$ *has the rec.FSP and is dec. and* $\mathrm{Card}(\mathrm{Stab}(Q_e)) > c\}$, $\{e:Q_e$ *has the rec.FSP and is dec. and* $\mathrm{Card}(\mathrm{Stab}(Q_e)) \leq c\}$, *and* $\{e:Q_e$ *has the rec.FSP and is dec. and* $\mathrm{Card}(\mathrm{Stab}(Q_e)) = c\}$ *are all* Σ_3^0-*complete.*

(f) $\{e : \mathrm{Card}(\mathrm{Stab}(Q_e)) > c\}$ *is* Σ_1^1-*complete and* $\{e : \mathrm{Card}(\mathrm{Stab}(Q_e)) \leq c\}$ *and* $\{e : \mathrm{Card}(\mathrm{Stab}(Q_e)) = c\}$ *are* Π_1^1-*complete.*

Next we give some index set results concerning the number of recursive stable models of a normal finite predicate logic program Q_e. Here we say that $\mathrm{Stab}(Q_e)$ is *recursively empty* if $\mathrm{Stab}(Q_e)$ has no recursive elements and is *recursively non-empty* if $\mathrm{Stab}(Q_e)$ has at least one recursive element. Similarly, we say that a $\mathrm{Stab}(Q_e)$ has *recursive cardinality equal to c* if $\mathrm{Stab}(Q_e)$ has exactly c recursive members.

Theorem 5.7.

(a) $\{e : Q_e$ *has the rec.FSP and* $\mathrm{Stab}(Q_e)$ *is recursively empty*$\}$ *is* D_3^0-*complete,* $\{e:Q_e$ *has the rec.FSP and* $\mathrm{Stab}(Q_e)$ *is recursively nonempty*$\}$ *is* Σ_3^0-*complete, and* $\{e:Q_e$ *has the*

rec.FSP and $\text{Stab}(Q_e)$ *is nonempty and recursively empty*}
is D_3^0-*complete.*

(b) {$e : Q_e$ *has the FSP and* $\text{Stab}(Q_e)$ *is recursively non-empty*}
is D_3^0-*complete,* {$e : Q_e$ *has the FSP and* $\text{Stab}(Q_e)$ *is recursively empty*} *is* Π_3^0-*complete, and* {$e : Q_e$ *has the FSP and* $\text{Stab}(Q_e)$ *is nonempty and recursively empty*} *is* Π_3^0-*complete.*

(c) {$e : \text{Stab}(Q_e)$ *is recursively nonempty*} *is* Σ_3^0-*complete,* {$e : \text{Stab}(Q_e)$ *is recursively empty*} *is* Π_3^0-*complete, and* {$e : \text{Stab}(Q_e)$ *is nonempty and recursively empty*} *is* Σ_1^1-*complete.*

Proof. We say that a normal finite predicate logic program Q_e has an *isolated stable model* M, if there is a finite set of ground atoms $a_1, \ldots, a_n, b_1, \ldots, b_m$ such that $a_i \in M$ for all i and $b_j \notin M$ for all j and there is no other stable model M' such that $a_i \in M'$ for all i and $b_j \notin M'$ for all j. Thus isolated stable models are determined by a finite amount of positive and negative information. We say that a finite predicate logic program Q_e is *perfect* if $\text{Stab}(Q_e)$ is nonempty and it has no isolated elements.

To prove the upper bounds in each case, we do the following. Given a normal finite predicate logic program Q_e, let a and $\bar{a}$ be two atoms which do not occur in $\text{ground}(P)$. Jockusch and Soare [23] constructed a recursively bounded primitive recursive tree such that $[T] \neq \emptyset$ and $[T_e]$ has no recursive elements. It then follows from Theorem 3.2 that $[T]$ can have no isolated elements so that $[T]$ is perfect. By Theorem 4.1, there is a normal finite predicate logic program U such that U has the rec.FS property and there is a one-to-one degree-preserving correspondence between $[T]$ and $\text{Stab}(U)$. Thus $\text{Stab}(U)$ has no recursive or isolated elements. Now suppose that we are given a normal finite predicate logic program Q_e. Then make a copy V of the normal finite predicate logic program U such that V has no predicates which are in common with Q_e. Let a and $\bar{a}$ be two atoms which do not appear in either V or Q_e and let S_e be the normal finite predicate logic program which arises from U and Q_e by adding a to the body of every clause in Q_e, adding $\bar{a}$ to the body of every clause in V, and adding the following two clauses:

$$C_1 = a \leftarrow \neg\bar{a} \text{ and}$$
$$C_2 = \bar{a} \leftarrow \neg a.$$

Then we claim that exactly one of a or $\bar{a}$ must be in every stable model M of S_e. That is, if neither a or $\bar{a}$ are in M, then C_1 and C_2 will contribute $a \leftarrow$ and $\bar{a} \leftarrow$ to $(S_e)_M$ so that both a and $\bar{a}$ will be in the least model of $(S_e)_M$. If both a and $\bar{a}$ are in M, then C_1 and C_2 will contribute nothing to $(S_e)_M$ so that neither a nor $\bar{a}$ will be in the least model of $(S_e)_M$ since then there will be no clauses of $(S_e)_M$ with either a or $\bar{a}$ in the head of the clause. It follows that S_e will have two types of stable models M, namely $M = M_1 \cup \{\bar{a}\}$ or $M = M_2 \cup \{a\}$ where M_1 is stable model of V and M_2 is stable model of Q_e. Since V has the rec. FS property, is perfect, and has no recursive stable models, it follows that

(1) Q_e has the rec. FS (FS) property if and only if S_e has the rec. FS (FS) property.
(2) Q_e is perfect if and only if S_e is perfect.
(3) The only recursive stable models of S_e are of the form $M \cup \{a\}$ where M is a recursive stable model of Q_e.

By Theorem 4.2, there is a recursive function k such that $T_{k(e)} = T_{S_e}$ as constructed in the proof of Theorem 4.2 such that $T_{k(e)}$ is bounded (r.b.) if and only if S_e has the FS (rec.FS) property and there is an effective one-to-one degree-preserving correspondence between $\text{Stab}(S_e)$ and $[T_{k(e)}]$. It follows that $T_{k(e)}$ is bounded (r.b.) if and only if Q_e has the FS (rec.FS) property and there is an effective one-to-one degree-preserving correspondence between the recursive elements of $\text{Stab}(S_e)$ and the recursive elements of $[T_{k(e)}]$.

Then for example, it follows that

$$\{e : Q_e \text{ has the rec.FSP and is recursively empty}\}$$

is many-to-one reducible to $\{e : T_e \text{ is } r.b \text{ and } [T_e] \text{ is recursively empty}\}$ which is D_3^0. Thus

$$\{e : Q_e \text{ has the rec.FSP and is recursively empty}\}$$

is D_3^0. In this way, the upper bounds on the complexity of each index set in the theorem follow from the corresponding complexity bound of the corresponding property of trees in Theorem 2.10.

The completeness results for each part of the theorem follow from Theorem 4.1 and the corresponding completeness results in Theorem 2.10 as before.   □

The same method of proof can be used to prove the following three theorems.

Theorem 5.8. *Let c be a positive integer.*

(a) $\{e : Q_e$ *has the rec.FPS and* $\text{Stab}(Q_e)$ *has rec. cardinality* $> c\}$ *is* Σ_3^0-*complete,* $\{e : Q_e$ *has the rec.FPS and* $\text{Stab}(Q_e)$ *has rec. cardinality* $\leq c\}$ *is* D_3^0-*complete, and* $\{e : Q_e$ *has the rec.FPS and* $\text{Stab}(Q_e)$ *has rec. cardinality* $= c\}$ *is* D_3^0-*complete.*

(b) $\{e : Q_e$ *has the FPS and* $\text{Stab}(Q_e)$ *has rec. cardinality* $> c\}$ *is* Π_3^0-*complete,* $\{e : Q_e$ *has the FPS and* $\text{Stab}(Q_e)$ *has rec. cardinality* $\leq c\}$ *is* D_3^0-*complete, and* $\{e : Q_e$ *has the FPS and* $\text{Stab}(Q_e)$ *has rec. cardinality* $= c\}$ *is* D_3^0-*complete.*

(c) $\{e : \text{Stab}(Q_e)$ *has rec. cardinality* $> c\}$ *is* Σ_3^0-*complete,* $\{e : \text{Stab}(Q_e)$ *has rec. cardinality* $\leq c\}$ *is* Π_3^0-*complete, and* $\{e : \text{Stab}(Q_e)$ *has rec. cardinality* $= c\}$ *is* D_3^0-*complete.*

(d) $\{e : Q_e$ *is dec. and has the rec.FPS and* $\text{Stab}(Q_e)$ *has rec. cardinality* $> c\}$ *is* Σ_3^0-*complete,* $\{e : Q_e$ *is dec. and has the rec.FPS and* $\mathrm{Stab}(Q_e)$ *has rec. cardinality* $\leq c$} *is* D_3^0-*complete, and* $\{e \ : \ Q_e$ *is dec. and has the rec.FPS and* $\mathrm{Stab}(Q_e)$ *has rec. cardinality* $= c$} *is* D_3^0-*complete.*

Theorem 5.9. $\{e : \mathrm{Stab}(Q_e)$ *has finite rec. cardinality*} *is* Σ_4^0-*complete and* $\{e : \mathrm{Stab}(Q_e)$ *has infinite rec. cardinality*} *is* Π_4^0-*complete.*

The same results are true for programs which have the rec.FS property and the FS property.

Theorem 5.10.

(a) $\{e \ : \ Q_e$ *has the rec.FSP and* $\mathrm{Stab}(Q_e)$ *is perfect*} *is* D_3^0-*complete.*

(b) $\{e \ : \ Q_e$ *has the FSP and* $\mathrm{Stab}(Q_e)$ *is perfect*} *is* Π_4^0-*complete.*

(d) $\{e : \mathrm{Stab}(Q_e)$ *is perfect*} *is* Σ_1^1-*complete.*

The results and methods laid out above may be used, with some additional effort, to prove results about *a.a.* FSP and *a.a* rec.FSP programs.

First, we present index set results for normal finite predicate logic programs which have the *a.a.* rec.FS property.

Theorem 5.11.

(a) $\{e : Q_e$ *has the a.a. rec.FSP*} *is* Σ_3^0-*complete.*

(b) $\{e \ : \ Q_e$ *has the a.a. rec.FSP and* $\mathrm{Stab}(Q_e)$ *is nonempty*} *and* $\{e \ : \ Q_e$ *has the a.a. rec.FSP and* $\mathrm{Stab}(Q_e)$ *is empty*} *are* Σ_3^0-*complete.*

(c) $\{e : Q_e$ *has the a.a. rec.FSP and* $\mathrm{Card}(\mathrm{Stab}(Q_e)) > c$}, $\{e : \mathrm{Stab}(Q_e)$ *has the a.a. rec.FSP and* $\mathrm{Card}(\mathrm{Stab}(Q_e)) \leq c$}, *and* $\{e : Q_e$ *has the a.a. rec.FSP and* $\mathrm{Card}(\mathrm{Stab}(Q_e)) = c$} *are all* Σ_3^0-*complete.*

(d) $\{e \ : \ Q_e$ *has the a.a. rec.FSP and* $\mathrm{Stab}(Q_e)$ *is infinite*} *is* D_3^0-*complete and* $\{e : Q_e$ *has the a.a. rec.FSP and* $\mathrm{Stab}(Q_e)$ *is finite*} *is* Σ_3^0-*complete.*

(e) $\{e : Q_e$ has the a.a. rec.FSP and $\text{Stab}(Q_e)$ is uncountable$\}$ is Σ_1^1-complete, $\{e : Q_e$ has the a.a. rec.FSP and $\text{Stab}(Q_e)$ is countable$\}$, and $\{e : Q_e$ has the a.a. rec.FSP and $\text{Stab}(Q_e)$ is countably infinite$\}$ are Π_1^1-complete.

(f) $\{e : Q_e$ has the a.a. rec.FSP and $\text{Stab}(Q_e)$ is recursively nonempty$\}$ is Σ_3^0-complete, $\{e : Q_e$ has the a.a. rec.FSP and $\text{Stab}(Q_e)$ is recursively empty$\}$ is D_3^0-complete, and $\{e : Q_e$ has the a.a. rec.FSP and $\text{Stab}(Q_e)$ is nonempty and recursively empty$\}$ is D_3^0-complete.

(g) $\{e : Q_e$ has the a.a. rec.FPS and $\text{Stab}(Q_e)$ has recursive cardinality $> c\}$ is Σ_3^0-complete, $\{e : Q_e$ has the a.a. rec.FPS and $\text{Stab}(Q_e)$ has recursive cardinality $\leq c\}$ is D_3^0-complete, and $\{e : Q_e$ has the a.a. rec.FPS and $\text{Stab}(Q_e)$ has cardinality $= c\}$ is D_3^0-complete.

(h) $\{e :$ has the a.a. rec.FSP and $\text{Stab}(Q_e)$ has finite recursive cardinality$\}$ is Σ_4^0-complete and $\{e :$ has the a.a. rec.FSP and $\text{Stab}(Q_e)$ has infinite recursive cardinality$\}$ is Π_4^0-complete.

(i) $\{e : Q_e$ has the a.a. rec.FSP and $\text{Stab}(Q_e)$ is perfect$\}$ is D_3^0-complete.

Proof. It is not difficult to obtain the upper bounds on the complexity of each set using the previous methods.

The completeness for each of the index sets in our theorem can be proved as follows. Given a normal finite predicate logic program Q_e, we construct a normal finite predicate logic program Y_e as follows. Let L_e denote the underlying language of of Q_e and L_e^* be the language which contain 0, s, and a predicate $R^*(z, x_1, \ldots, x_n)$ for every predicate $R(x_1, \ldots, x_n)$ and a predicate $A^*(x)$ for every proposition atom A in L where none of 0, s, R^*, and A^* occur in L_e. To ease notation, we shall let $\bar{0} = 0$ and $\bar{n} = s^n(0)$ for each $n \geq 1$. Then by Proposition 3.1, there is a normal finite predicate logic Horn program Q^- with a recursive least model M^- whose language contains the constant symbol 0 plus all the constant symbols of L_e and the function symbol

s plus all the function symbols form L_e and whose set of predicate symbols are disjoint from the language L_e^* which includes the predicates num$(\cdot)$, noteq$(\cdot,\cdot)$, and term$(\cdot)$ such that for any ground terms t_1 and t_2,

(1) num(t_1) holds in M^- if and only if $t_1 = \bar{n}$ for some $n \geq 0$,
(2) noteq(t_1, t_2) holds in M^- if and only if there exist natural numbers n and m such that $n \neq m$ and $t_1 = \bar{n}$ and $t_2 = \bar{m}$, and
(3) term(t_1) holds in model M^- if and only if t_1 is a ground term in L_e.

Moreover, we can assume that Q^- has the rec. FS property. Then let Y_e be the program Q^- plus all clauses $C^*(x)$ that arise from a clause $C \in Q_e$ by adding the predicate num(x) to the body where x the first variable of the language that does not occur in C, term(t) to the body for each term that occurs in C, and by replacing each predicate $R(t_1, \ldots, t_n)$ that occurs in C by $R^*(x, t_1, \ldots, t_n)$ and each propositional atom A that occurs in C by $A^*(x)$. The idea is that as x varies over $\{\bar{n} : n \geq 0\}$, these clauses will produce infinitely many copies of the program Q_e. More precisely, we let $Q_e^{\bar{n}}$ denote the set of all clauses of the form $C^*(\bar{n})$. $Q_e^{\bar{n}}$ is essentially an exact copy of Q_e except that we have extended all predicates and propositional atoms to have an extra term corresponding to $\bar{n}$ and each clause contains the predicate num$(\bar{n})$ and term(t) in the body for each term in the original clause. Since none of the clauses $C^*(x)$ have any predicates from Q^- in their heads, it will be the case that in every stable model M of Y_e, M restricted to the ground atoms of Q^- will just be M^-. Thus in particular,

(1) num(t_1) holds in M if and only if $t_1 = \bar{n}$ for some $n \geq 0$,
(2) noteq(t_1, t_2) holds in M if and only if there exist natural numbers n and m such that $n \neq m$ and $t_1 = \bar{n}$ and $t_2 = \bar{m}$, and
(3) term(t_1) holds in M if and only if t_1 is a ground term in L_e.

Now if P is any ground(Q_e)-proof scheme, then we let $P^{\bar{n}}$ be the result of adding num($\bar{n}$) to each clause in P and term(t) to each clause if t occurs in P and replacing each predicate $R(t_1, \ldots, t_n)$ that occurs in P by $R^*(\bar{n}, t_1, \ldots, t_n)$ and each propositional atom A that occurs in P by $A^*(\bar{n})$. It is easy to see that all minimal ground(Y_e)-proof schemes that derive atoms outside of ground(Q^-) must consist of an interweaving of ground(Q^-)-proof schemes of num($\bar{n}$) and term(t) for each term t in L_e that occurs in the proof scheme of the form $P^{\bar{n}}$ for some ground(Q_e)-proof scheme P. It follows that if Q_e has the rec.FS (FS) property, then Y_e has the rec.FS (FS) property. However, if Q_e does not have the rec.FS, then it cannot be that Y_e has the *a.a* rec.FS property since if we could effectively find all the minimal supports of Y_e-proof schemes for all but finitely many atoms, then there would be some n in which we could find all "$Q^{\bar{n}}$-proof schemes" for any atom which contains $\bar{n}$, which would allow us to effectively find all Q_e-proof schemes for any ground atom of L. Similarly, if Q_e does not have the FS property, then the Y_e does not have the *a.a.* FS property. Thus, Q_e has the rec.FS (FS) property if and only if Y_e has the *a.a.* rec.FS (FS) property.

Next we want to add a finite number of predicate clauses Y_e to produce a normal finite predicate logic program Z_e which restricts the stable models to be essentially the same relative to the atoms of ground($Q^{\bar{n}}$) for all $n \geq 0$. To this end, we let a be an atom that does not appear in Y_e and for each predicate $R^*(z, x_1, \ldots, x_n)$ of Y_e, we add a clause

$$C_{R^*} = a \leftarrow R^*(y, x_1, \ldots, x_n), \neg R^*(z, x_1, \ldots, x_n),$$

$$\text{noteq}(y, z), \text{term}(x_1), \ldots, \text{term}(x_n), \neg a$$

and for each propositional atom A of L_e, we add a clause

$$C_{A^*} = a \leftarrow A^*(y), \neg A^*(z), \text{noteq}(x, y) \neg a.$$

First we observe that a cannot be in any stable model of M of Z_e. That is, if $a \in M$, then none of the clauses C_{R^*} and C_{A^*}

will contribute anything to $\text{ground}(Z_e)_M$. Thus no clauses with a in the head will be in $\text{ground}(Z_e)_M$ so that a will not be in the least model of M and $M \neq \text{ground}(Z_e)_M$.

Now suppose that M is a stable model of Z_e and $a \notin M$. Then it is easy to see from the form of our rules that for any predicate $R(x_1, \ldots, x_n)$ of L_e, M can only contain ground atoms of the form $R^*(t_0, t_1, \ldots, t_n)$ where $t_0 = \bar{n}$ for some $n \geq 0$ and $t_1, \ldots, t_n$ are ground terms in L_e. Similarly, for each propositional atom A of L_e and ground term t, $A(t)$ is in M implies $t = \bar{n}$ for some $n \geq 0$. We claim that for any predicate $R(x_1, \ldots, x_n)$ and any ground terms $t_1, \ldots, t_n$ in L_e, either $D_{R, t_1, \ldots, t_n} = \{R^*(\bar{n}, t_1, \ldots, t_n) : n \geq 0\}$ is contained in M or is completely disjoint from M. That is, if there is an $n \neq m$ such that $R^*(\bar{n}, t_1, \ldots, t_n) \in M$ but $R^*(\bar{m}, t_1, \ldots, t_n) \notin M$, then the clause C_{R^*} will contribute the clause

$$\bar{C}_{R^*} = a \leftarrow R^*(\bar{n}, t_1, \ldots, t_n), \text{noteq}(\bar{n}, \bar{m})$$

to $\text{ground}(Z_e)_M$ so that a would be in M since M is a model of $\text{ground}(Z_e)_M$ and, hence, M is not a stable model. Similarly, for each propositional atom A in L_e either $D_A = \{A^*(\bar{n}) : n \geq 0\}$ is contained in M or is completely disjoint from M. That is, if there is an $n \neq m$ such that $A^*(\bar{n}) \in M$ but $A^*(\bar{m}) \notin M$, then the clause C_{A^*} will contribute the clause

$$\bar{C}_{A^*} = a \leftarrow A^*(\bar{n}), \text{noteq}(\bar{n}, \bar{m})$$

to $\text{ground}(Z_e)_M$ so that a would be in M and M is not a stable model. It follows that the stable models of Z_e are in one-to-one correspondence with the stable models of Q_e. That is, if U is a stable model of Q_e, then there is a stable model $V(U)$ of Z_e such that

(1) $M^- \subseteq V(U)$;
(2) for all predicate symbols $R(x_1, \ldots, x_n)$ in L_e, and ground terms $t, t_1, \ldots, t_n$ in L_e^*, $R^*(t, t_1, \ldots, t_n) \in V(U)$ if and only if $t = \bar{m}$ for some $m \geq 0$, $t_1, \ldots, t_n \in L_e$, and $R(t_1, \ldots, t_n) \in U$; and if

(3) for all propositional letters A in L_e and ground terms t in L_e^*, $A^*(t) \in V(U)$ if and only if $t = \bar{m}$ for some $m \geq 0$ and $A \in U$.

In addition, it is easy to prove by induction on the length of proof schemes that every stable model of V of Z_e is of the form $V(U)$ where

(1) for all predicate symbols $R(x_1, \ldots, x_n)$ and ground terms $t_1, \ldots, t_n$ in L_e, $R(t_1, \ldots, t_n) \in U$ if and only if $R(\bar{0}, t_1 \ldots, t_n) \in V$, and

(2) for all atom propositional letters A in L_e, $A \in U$ if and only if $A^*(\bar{0}) \in V$.

It follows that there is an effective one-to-one degree-preserving correspondence between the $\text{Stab}(Q_e)$ and $\text{Stab}(Z_e)$. Now let ℓ be a recursive function such that $Q_{\ell(e)} = Z_e$.

Next we observe that our theorem states that the complexity of every property of normal finite predicate logic programs which have a.a. rec.FS property is the same as the corresponding complexity of the same property of normal finite predicate logic programs with just the recursive FS property. For example, in Section 3, we proved that for every positive integer c, $X = \{e : Q_e$ has the rec.FSP and $\text{Card}(\text{Stab}(Q_e)) = c\}$ is Σ_3^0-complete while we want to prove that $Y = \{e : Q_e$ has the a.a. rec.FSP and $\text{Card}(\text{Stab}(Q_e)) = c\}$ is Σ_3^0-complete. Now ℓ shows that X is many-to-one reducible to Y so that, since we have already shown that X is Σ_3^0, it must be the case that Y is Σ_3^0-complete. All the other completeness results follows from the corresponding completeness results in the same manner. □

Next we present results on the cardinality and recursive cardinality of the set of stable models of a.a FSP and a.a rec. FSP programs.

Theorem 5.12. *For any positive integer* c,

(1) $\{e : Q_e$ *has the a.a. FSP*$\}$ *is* Σ_4^0*-complete.*

(2) $\{e : Q_e$ *has the a.a. FSP and* $\mathrm{Stab}(Q_e)$ *is empty*$\}$ *and* $\{e : Q_e$ *has the a.a. FSP and* $\mathrm{Stab}(Q_e)$ *is nonempty*$\}$ *are* Σ_4^0*-complete.*

(3) $\{e : Q_e$ *has the a.a. FSP and* $\mathrm{Stab}(Q_e) > c\}$, $\{e : Q_e$ *has the a.a. FSP and* $\mathrm{Stab}(Q_e) \le c\}$, *and* $\{e : Q_e$ *has the a.a. FSP and* $\mathrm{Stab}(Q_e) = c\}$ *are* Σ_4^0*-complete.*

(4) $\{e : Q_e$ *has the a.a. FSP and* $\mathrm{Stab}(Q_e)$ *is finite*$\}$ *and* $\{e : Q_e$ *has the a.a. FSP and* $\mathrm{Stab}(Q_e)$ *is infinite*$\}$ *are* Σ_4^0*-complete.*

(5) $\{e : Q_e$ *has the a.a. FSP and* $\mathrm{Stab}(Q_e)$ *is countable*$\}$ *and* $\{e : Q_e$ *has the a.a. FSP and* $\mathrm{Stab}(Q_e)$ *is countably infinite*$\}$ *are* Π_1^1*-complete and* $\{e : Q_e$ *has the a.a. FSP and* $\mathrm{Stab}(Q_e)$ *is uncountable*$\}$ *are* Σ_1^1*-complete.*

(6) $\{e : Q_e$ *has the a.a. FSP and* $\mathrm{Stab}(Q_e)$ *is recursively empty*$\}$, $\{e : Q_e$ *has the a.a. FSP and* $\mathrm{Stab}(Q_e)$ *recursively nonempty*$\}$, *and* $\{e : Q_e$ *has the a.a. FSP and* $\mathrm{Stab}(Q_e)$ *is nonempty and recursively empty*$\}$ *are* Σ_4^0*-complete.*

(7) $\{e : Q_e$ *has the a.a. FSP and* $\mathrm{Stab}(Q_e)$ *has recursive cardinality* $c\}$, $\{e : Q_e$ *has the a.a. FSP and* $\mathrm{Stab}(Q_e)$ *has recursive cardinality* $\le c\}$, *and* $\{e : Q_e$ *has the a.a. FSP and* $\mathrm{Stab}(Q_e)$ *has recursive cardinality* $= c\}$ *are* Σ_4^0*-complete.*

Proof. Again the upper bounds on the complexity are not hard to verify.

For the completeness results in part (5) of the theorem, we can follow the same strategy as in the proof of Theorem 5.11. By Theorem 5.5, we know that $X = \{e : Q_e$ has the FSP and $\mathrm{Stab}(Q_e)$ is uncountable$\}$ is Π_1^1-complete while we want to prove that $Y = \{e : Q_e$ has the a.a. FSP and $\mathrm{Card}(\mathrm{Stab}(Q_e)$ is uncountable$\}$ is Π_1^1-complete. Now recursive ℓ such that $Z_e = Q_{\ell(e)}$ constructed in Theorem 5.11 shows that X is many-to-one reducible to Y so that Y is complete for Π_1^1 sets. All the other completeness results in part (5) follows from the corresponding completeness results in Theorem 5.5 the same manner.

However, we cannot follow that same strategy as in Theorem 5.11 in the remaining parts of theorem because the completeness results for normal finite predicate logic programs with the FS property do not match the completeness results for normal finite predicate logic programs with *a.a.* FS property. Instead we shall outline the modifications that are needed to prove an analog of Theorem 4.1 that can be used the prove the completeness result for normal finite predicate logic programs which have the *a.a.* FS property from the corresponding completeness results for *a.a* bounded trees.

First, recall the construction of the trees that were used to prove part (d) of Theorem 3.7. We defined a primitive recursive function $\phi(e, m, s) = (least\ n > m)(n \notin W_{e,s} - \{0\})$. For any given e, we let V_e be the tree such that $(m), (m, 0), (m, 1) \in U_e$ for all $m \geq 0$ and $(m, s + 2) \in V_e$ if and only if m is the least element such that $\phi(e, m, s+1) > \phi(e, m, s)$. This is only a slight modification of the tree U_e defined in that the proof of part (d) of Theorem 3.7 in that we have ensured that $(m, 0), (m, 1) \in V_e$ are always in U_e and so that we are forced to shift the remaining nodes to right by one. It will still be the case that if $W_e - \{0\}$ is cofinite, then there is exactly one node in V_e that has infinitely many successors and V_e is bounded otherwise. Clearly there is a recursive function f such that $T_{f(e)} = V_e$. But then

$$e \in \omega - Cof \iff T_{f(e)} \text{ is bounded.}$$

Next, let S be an arbitrary complete Σ_4^0 set and suppose that $a \in S \iff (\exists k)(R(a, k))$ where R is Π_3^0. By the usual quantifier methods, we may assume that $R(a, k)$ implies that $R(a, j)$ for all $j > k$. By the Π_3^0-completeness of $\{e : T_e \text{ is bounded}\}$, there is a recursive function h such that $R(a, k)$ holds if and only if $V_{h(a,k)}$ is bounded and such that $V_{h(a,k)}$ is *a.a.* bounded for every a and k. Now we can define a recursive function ψ so that

$$T_{\psi(a,e)} = \{(0)\} \cup \{(k+1)^\frown \sigma : \sigma \in V_{h(a,k)}\} \cup \{0^\frown \sigma : \sigma \in T_e\}.$$

Thus we have two parts of the tree $T_{\psi(a,e)}$. That is, above the node (0), we have a copy of T_e and we shall call this part of the tree $\text{First0}(T_{\psi(a,e)})$. We shall refer to the remaining part of $T_{\psi(a,e)}$ as $\text{NotFirst0}(T_{\psi(a,e)})$. Now, if $a \in S$, then $V_{h(a,k)}$ is bounded for all but finitely many k and is $a.a.$ bounded for the remainder. Thus $\text{NotFirst0}(T_{\psi(a,e)})$ is $a.a.$ bounded. If $a \notin S$, then, for every k, $V_{h(a,k)}$ is not bounded, so that $\text{NotFirst0}(T_{\psi(a,e)})$ is not $a.a.$ bounded. Thus $a \in S$ if and only if $\text{NotFirst0}(T_{\psi(a,e)})$ is $a.a.$ bounded. Hence if T_e is $r.b.$ or bounded, then $a \in S$ if and only if $T_{\psi(a,e)}$ is $a.a.$ bounded.

Next, we describe a normal finite predicate logic program $Q_{a,e}$ such that there is a one-to-one effective correspondence between $\text{Stab}(Q_{a,e})$ and $[T_{\psi(a,e)}]$. Our construction will just be a slight modification of the construction in Theorem 4.1. First, we shall need some additional predicates on sequences. That is, we let the predicate $\text{first0}(c(\sigma))$ be true if and only if σ is a sequence which starts with 0 and the predicate $\text{notfirst0}(c(\sigma))$ be true if and only if σ is a nonempty sequence which does not start with 0. We let the predicate $\text{third0}(c(\sigma))$ be true if and only if σ is a sequence of length ≥ 3 whose third element is 0 and we let the predicate $\text{notthird0}(c(\sigma))$ be true if and only if σ is a sequence of length ≥ 3 whose third element is not 0. We shall also require a predicate $\text{length12}(\cdot)$ which holds only on codes of sequences of length 1 or 2 and $\text{length3}(\cdot)$ which holds only on codes of sequences of length 3. Finally, we shall need a predicate $\text{agree12}(\cdot,\cdot)$ which holds only on pairs of codes $(c(\sigma),c(\tau))$ where σ and τ are of length 3 and σ and τ agree on there first two entries.

As in the proof of Theorem 4.1, there exists the following three normal finite predicate logic programs such that the set of ground terms in their underlying language are all of the form $s^n(0)$ where 0 is a constant symbol and s is a unary function symbol. We shall use n has an abbreviation for the term $s^n(0)$.

(I) There exists a finite predicate logic Horn program P_0 such that for a predicate $\text{tree}(\cdot)$ of the language of P_0, the atom tree(n) belongs to the least Herbrand model of P_0 if and only if n is a code for a finite sequence σ and $\sigma \in T_{\psi(a,e)}$.

(II) There is a finite predicate logic Horn program P_1 such that for a predicate seq($\cdot$) of the language of P_1, the atom seq(n) belongs to the least Herbrand model of P_1 if and only if n is the code of a finite sequence $\alpha \in \omega^{<\omega}$.

(III) There is a finite predicate logic Horn program P_2 which correctly computes the following recursive predicates on codes of sequences.

   (a) samelength($\cdot, \cdot$). This succeeds if and only if both arguments are the codes of sequences of the same length.

   (b) diff($\cdot, \cdot$). This succeeds if and only if the arguments are codes of sequences which are different.

   (c) shorter($\cdot, \cdot$). This succeeds if and only if both arguments are codes of sequences and the first sequence is shorter than the second sequence.

   (d) length($\cdot, \cdot$). This succeeds when the first argument is a code of a sequence and the second argument is the length of that sequence.

   (e) notincluded($\cdot, \cdot$). This succeeds if and only if both arguments are codes of sequences and the first sequence is not the initial segment of the second sequence.

   (f) first0($\cdot$). This succeeds if and only if the argument is the code of a sequence which starts with 0.

   (g) notfirst0($\cdot$). This succeeds if and only if the argument is the code of a nonempty sequence which does not start with 0.

   (h) third0($\cdot$). This succeeds if and only if the argument is the code of a sequence of length ≥ 3 whose third element is 0.

   (i) notthird0($\cdot$). This succeeds if and only if the argument is the code of a sequence of length ≥ 3 whose third element is not 0.

(j) agree12$(\cdot, \cdot)$. This succeeds if and only if the arguments are codes of a sequences of length 3 which agree on the first two elements.

(k) length12$(\cdot)$. This succeeds if and only if the argument is a code of a sequence of length 1 or 2.

(l) length3$(\cdot)$. This succeeds if and only if the argument is a code of a sequence of length 3.

(m) num$(\cdot)$. This succeeds if and only if the argument is either 0 or $s^n(0)$ for some $n \geq 1$.

(n) greater0$(\cdot)$. This succeeds if and only if the argument is $s^n(0)$ for some $n \geq 1$.

Now let P^- be the normal finite predicate logic program which is the union of programs $P_0 \cup P_1 \cup P_2$. We denote its language by $\mathcal{L}^-$ and we let M^- be the least model of P^-. By Proposition 3.1, this program P^- is a Horn program and for each ground atom b in the Herbrand base of P^-, we can explicitly construct the set of all P^--proof schemes of b. In particular, tree$(n) \in M^-$ if and only if n is the code of node in $T_{\psi(a,e)}$.

Our final program P_T will consist of P^- plus clauses (1)–(9) given below. We assume that these additional clauses do not contain any of predicates of the language $\mathcal{L}^-$ in the head. However, predicates from P^- do appear in the body of clauses (1)–(9). Therefore, whatever stable model of the extended program we consider, its trace on the set of ground atoms of $\mathcal{L}^-$ will be M^-. In particular, the meaning of the predicates listed above will always be the same.

We are ready now to write the additional clauses which, together with the program P^-, will form the desired program $Q_{a,e}$. First of all, we select three new unary predicates:

(i) path$(\cdot)$, whose intended interpretation in any given stable model M of $Q_{a,e}$ is that it holds only on the set of codes of sequences that lie an infinite path through $T_{\psi(a,e)}$ that starts with 0. This path will correspond to the path encoded by the stable model of M.

(ii) notpath($\cdot$), whose intended interpretation in any stable model M of $Q_{a,e}$ is the set of all codes of sequences which are in $T_{\psi(a,e)}$ but do not satisfy path($\cdot$).

(iii) control($\cdot$), which will be used to ensure that path($\cdot$) always encodes an infinite path through $T_{\psi(a,e)}$.

Next we include the same seven sets of clauses as we did in Theorem 4.1 so that stable models code paths through the tree T_e which sits above the node 0. This requires that we modify those clause so that we restrict ourselves to the sequences that satisfy first0(X).

This given the final seven clauses of our program are as follows:

(1) path(X) $\longleftarrow$ first0(X), tree(X), $\neg$notpath(X)

(2) notpath(X) $\longleftarrow$ first0(X), tree(X), $\neg$path(X)

(3) path($c(0)$) $\longleftarrow$

(4) notpath(X) $\longleftarrow$ first0(X), tree(X), path(Y), first0(Y), tree(Y), samelength(X, Y), diff(X, Y)

(5) notpath(X) $\longleftarrow$ first0(X), tree(X), first0(Y), tree(Y), path(Y) shorter(Y, X), notincluded(Y, X)

(6) control(X) $\longleftarrow$ first0(Y), path(Y), length(Y, X)

(7) control(X) $\longleftarrow$ greater0(X), num(X), $\neg$control(X)

Next we add the clauses we use an additional predicate In(X) which is used to ensure that the final program $Q_{a,e}$ has the a.a.FSP if and only if the tree $T_{\psi(a,e)}$ is *a.a.* bounded.

(8) path(0) $\longleftarrow$ /* Recall 0 is the code of the empty sequence */

(9) notpath(X) $\longleftarrow$ notfirst0(X), tree(X)

(10) In(X) $\longleftarrow$ notfirst0(X), tree(X), length12(X)

(11) In(X) $\longleftarrow$ notfirst0(X), tree(X), length3(X), third0(X), notfirst0(Y), tree(Y), length3(Y), notthird0(Y), $\neg$In(Y)

(12) control(0) $\longleftarrow$

Clearly, $Q_{a,e} = P^- \cup \{(1), \ldots, (12)\}$ is a finite predicate logic program.

As in the proof of Theorem 4.1, we can establish a "normal form" for the stable models of $Q_{a,e}$. Each such model must contain M^-, the least model of P^-. In fact, the restriction of a stable model of P_T to $H(P^-)$ is M^-. Given any $\beta = (0, \beta(1), \beta(2), \ldots) \in \omega^\omega$, we let

$$M_\beta = M^- \cup \{\text{control}(n) : n \in \omega\} \cup \{\text{path}(0)\}$$
$$\cup \{\text{path}(c((0, \beta(1), \ldots, \beta(n)))) : n \geq 1\}$$
$$\cup \{\text{notpath}(c(\sigma)) : \sigma \in T_{\psi(a,e)} \text{ and } \sigma \not\prec \beta\}$$
$$\cup \{\text{In}(c((m,n))) : m > 0 \text{ and } n \geq 0\}$$
$$\cup \{\text{In}(c(m,n,0)) : m > 0 \text{ and } n \geq 0\}.$$

We claim that M is a stable model of $Q_{a,e}$ if and only if $M = M_\beta$ for some $\beta \in [T_{\psi(a,e)}]$.

First let us consider the effect of the clauses (8)–(12). Clearly, clause (8) forces that path(0) must be in every stable model of $Q_{a,e}$ and the clauses in (9) force that notpath($c(\sigma)$) is in every stable model of $Q_{a,e}$ for all $\sigma \in T_{\psi(a,e)}$ which do not start with 0. Since all the clauses (1)–(6) require first0($c(\sigma)$) to be true, the only minimal $Q_{a,e}$-proof schemes for notpath($c(\sigma)$) for $\sigma \in T_{\psi(a,e)}$ which do not start with 0 must use the Horn clause of type (9). Thus the minimal $Q_{a,e}$-proof schemes with conclusion notpath($c(\sigma)$) (where σ does not start with 0) consists of minimal P^--proof schemes of tree($c(\sigma)$) followed by the triple $\langle c(\sigma), (9)^*, \emptyset \rangle$ where $(9)^*$ is the ground instance of (9) where X is replaced by $c(\sigma)$. Thus all the minimal $Q_{a,e}$- proof schemes of notpath($c(\sigma)$), where σ does not start with 0, have empty support. Similarly, In($c(\sigma)$) can be derived only using clause (10) if σ has length 1 or 2 so that all minimal $Q_{a,e}$-proof schemes of $In(c(\sigma))$, where σ has length 1 or 2, have empty support. Clause (12) is the only way to derive control(0) so that the only minimal $Q_{a,e}$-proof scheme of control(0) uses clause (12) and has empty support.

The only way to derive $\text{In}(\sigma)$ for σ of length 3 is via an instance of clause (11). Such clauses will allow us to derive $\text{In}(c((m,n,0)))$ for any $m > 0$ and $n \geq 0$ with a proof scheme whose support is of the form $\{\text{In}(c((m,n,p)))\}$ for some $p > 0$ where $(m,n,p) \in T_{\psi(a,e)}$. Since we always put $(m,n,1) \in T_{\psi(a,e)}$, there is at least one such proof scheme but there could be infinitely many proof schemes if $(m,n,p) \in T_{\psi(a,e)}$ for infinitely many $p > 0$. It then follows that from our definition of $T_{\psi(a,e)}$ that there will be finitely many $m > 0$ and $n \geq 0$ such that $\text{In}(c((m,n,0)))$ has infinitely many proof schemes if and only if $\text{NotFirst0}(T_{\psi(a,e)})$ is *a.a.* bounded, which is if and only if $a \in S$. Now if T_e is bounded, then we can use the same argument that we used in Theorem 4.1 to show that there are only finitely many minimal $Q_{a,e}$-proofs schemes for the ground instances of predicates in the heads of such clauses for $\sigma \in T_{a,e}$ that start with 0. It follows that if T_e is bounded, then $a \in S$ if and only if $Q_{a,e}$ has the *a.a.* FS property.

We can use the same arguments that we used in Theorem 4.1 to show that the clauses (1)–(7) force that the only stable models correspond to M_β where $\beta = (0, \beta(1), \beta(2), \ldots) \in \omega^\omega$ and $(\beta(1), \beta(2), \ldots) \in [T_e]$. The only difference is that clause (12) allows us to derive control(0) directly. Thus if T_e is bounded, then there will be an effective one-to-one degree-preserving correspondence between $\text{Stab}(Q_{a,e})$ and $[T_{\psi)a,e)}]$ and $Q_{a,e}$ has the *a.a.* FSP if and only if $a \in S$.

The Σ_4^0-completeness results for the remaining parts of our theorem can all be proved by the following type argument. Suppose, for example, that we want to prove that

$$A = \{e : Q_e \text{ has the } a.a. \text{ FSP and } Stab(Q_e) \text{ is nonempty}$$

$$\text{and rec. empty}\}$$

is Σ_4^0-complete. Then we know that there exists an *r.b.* tree T which is nonempty but which has no recursive elements by Jockusch and Soare [23]. Thus fix e such that T_e is *r.b.* and $[T_e]$ is nonempty and has no recursive elements. Then for our Σ_4^0 predicate S, we have the property that $a \in S$ if and only if $T_{\psi(a,e)}$ is *a.a.* bounded and $[T_{\psi(a,e)}]$ is nonempty and has no recursive elements. But then $T_{\psi(a,e)}$ is *a.a.* bounded and $[T_{\psi(a,e)}]$ is nonempty and has no recursive elements if and only if $Q_{a,e}$ is *a.a.* bounded and $\mathrm{Stab}(Q_{a,e})$ is nonempty and has no recursive elements. Now if g is the recursive function such that $Q_{g(a)} = Q_{a,e}$, then $a \in S$ if and only if $g(a) \in A$. Thus A is complete for Σ_4^0 sets. □

Acknowledgments

During the work on this paper D. Cenzer was partially supported by NSF grant DMS-65372. V. W. Marek was partially supported by the following grants and contracts: Image-Net: Discriminatory Imaging and Network Advancement for Missiles, Aviation, and Space, United States Army SMDC contract, NASA-JPL Contract, Kentucky Science and Engineering Foundation grant, and NSF ITR: Decision-Theoretic Planning with Constraints grant. J. B. Remmel was partially supported by NSF grant DMS 0654060.

References

[1]   H. Andreka and I. Nemeti, The generalized completeness of Horn predicate logic as a programming language, *Acta Cybernetica* **4** (1978) 3–10.

[2]   K. R. Apt and H. A. Blair, Arithmetic classification of perfect models of stratified programs, *Fund. Inf.* **14** (1991) 339–343. *Fund. Inf.* **14** (1991) 339–343.

[3]   K. R. Apt, H. A. Blair and A. Walker, Towards a theory of declarative knowledge, in *Foundations of Deductive Databases and Logic Programming*, (Elsevier, 1988), pp. 89–148.

[4]   Asparagus, Benchmarks for answer set programming, http://asparagus.cs.uni-potsdam.de, Accessed on August 2014.

[5]   P. Bonatti, Reasoning with infinite stable models, *Artificial Intell. J.* **156** (2004) 75–111.

[6]   M. Calautti, S. Greco, F. Spezzano and I. Trubitsyna, Checking termination of bottom-up evaluation of logic programs with function

symbols, to appear in *Theory and Practice of logic Programming* Vol. 15 (2015), pp. 854–889.

[7]   F. Calimeri, S. Cozza, G. Ianni and N. Leone, Computable functions in ASP: Theory and implementation, in *Proc. of ICLP 2008* (2008), pp. 407–424.

[8]   D. Cenzer, V. W. Marek and J. B. Remmel, Index sets for finite predicate logic programs, in T. Eiter and G. Gottlob (eds.), *FLOC '99 Workshop on Complexity-Theoretic and Recursion-Theoretic Methods in Databases, Artificial Intelligence and Finite Model Theory*, (IRST, Trento, 1999), pp. 72–80.

[9]   D. Cenzer and J. B. Remmel, Index sets for Π_1^0 classes, *Ann. Pure Appl. Logic* **93** (1998) 3–61.

[10] D. Cenzer and J. B. Remmel, Π_1^0 classes in mathematics, in Yu. L. Ershov, S. S. Goncharov, A. Nerode and J. B. Remmel (eds.) *Handbook of Recursive Mathematics: Volume 2*, Studies in Logic and the Foundations of Mathematics, Vol. 139 (Elsevier, 1998), pp. 623–822.

[11] D. Cenzer and J. B. Remmel, *Effectively Closed Sets* (Cambridge University Press, 2020).

[12] D. Cenzer and J. B. Remmel, A connection between Cantor–Bendixson derivatives and the well-founded semantics of finite logic programs, *Ann. Math. Artificial Intell.* **65**(1) (2012) 1–24.

[13] K. Clark, Negation as failure, in H. Gallaire and J. Minker (eds.), *Logic and Data Bases* (Plenum Press, 1978), pp. 293–322.

[14] M. Denecker and E. Ternovska, A logic of nonmonotone inductive definitions. *ACM Trans. Comput. Log.* **9** (2008) 14.

[15] P. M. Dung and K. Kanchanasut, A fixpoint approach to declarative semantics of logic programs, in E. L. Lusk and R. A. Overbeek (eds.), *Proc. North American Conf. Logic Programming* (MIT Press, 1989), pp. 604–625.

[16] D. East and M. Truszczyński, The aspps system, in S. Flesca, S. Greco, N. Leone and G. Ianni (eds.), *Proc. Logics in Artificial Intelligence, European Conf. (JELIA 2002)*, Lecture Notes in Computer Science, Vol. 2424 (Springer, 2002), pp. 533–536.

[17] The Fifth Answer Set Programming Competition, https://www.mat.unical.it/aspcomp2014, 2014.

[18] A. Ehrenfeucht, Separable theories, *Bull. Acad. Polon.Sci. Sér. Sci. Math. Astronom. Phys.* **9** (1961) 17–19.

[19] M. Gebser, B. Kaufmann, A. Neumann and T. Schaub, Conflict-driven answer set solving, in M. Veloso (ed.), *Proc. Joint Int. Conf. Artificial Intelligence* (MIT Press, 2007), p. 386.

[20] M. Gelfond and N. Leone, Logic programming and knowledge representation: The A-prolog perspective, *Artificial Intell.* **138** (2002) 3–38.

[21] M. Gelfond and V. Lifschitz, The stable semantics for logic programs, in *Proc. fifth Int. Symp. Logic Programming* (MIT Press, 1988), pp. 1070–1080.

[22] C. Jockusch and R. Soare, A minimal pair of Π_1^0 classes, *J. Symb. Logic* **36** (1971) 66–78.

[23] C. Jockusch and R. Soare, Π_1^0 classes and degrees of theories, *Trans. Amer. Math. Soc.* **173** (1972) 33–56.

[24] C. Jockusch and R. Soare, Degrees of members of Π_1^0 classes, *Pacific J. Math.* **40** (1972) 605–616.

[25] Y. Lierer and V. Lifschitz, One more decidable class of finitely ground programs, *Proc. ICLP 09* (Springer LNCS 5649, 2009), pp. 489–493.

[26] J. Lloyd, *Foundations of Logic Programming* (Springer-Verlag, 1989).

[27] V. W. Marek, A. Nerode and J. B. Remmel, How complicated is the set of stable models of a recursive logic program, *Ann. Pure Appl. Logic* **56** (1992) 119–136.

[28] V. W. Marek, A. Nerode and J. B. Remmel, The stable models of predicate logic programs. *J. Logic Program.* **21** (1994) 129–153.

[29] I. Niemelä and P. Simons, Smodels — an implementation of the stable model and well-founded semantics for normal logic programs, in J. Dix, U. Furbach and A. Nerode (eds.), *Proc. 4th Int. Conf. Logic Programming and Nonmonotonic Reasoning* Lecture Notes in Computer Science Vol. 1265 (Springer, 1997), pp. 420–429.

[30] A. Nerode and R. Shore, *Logic for Applications* (Springer, 1993).

[31] R. Soare, *Recursively Enumerable Sets and Degrees* (Springer, 1987).

[32] T. Syrjänen, Omega-restricted logic programs, in *Proc. LPNMR 2001* (Springer, 2001), pp. 267–279.

[33] A. Van Gelder, The alternating fixpoint of logic programs with negation, in *J. Computer and System Sciences* **47** (1993), 185–221.

[34] J. C. Shepherdson, Unsolvable problems for SLDNF-resolution, *J. Logic Program.* **10** (1991) 19–22.

[35] J. Shoenfield, Degrees of models, *J. Symbolic Logic* **25** (1960) 233–237.

[36] R. M. Smullyan, *First-Order Logic* (Springer, 1968).

Chapter 7

Computability and Definability

Trang Ha*,‡, Valentina Harizanov*,§, Leah Marshall*,¶
and Hakim Walker†,‖

*Department of Mathematics, George Washington University,
Washington, DC 20052, USA
†Department of Mathematics, Harvard University,
Cambridge, MA 02138, USA
‡trangtha09@gmail.com
§harizanv@email.gwu.edu
¶lbm@email.gwu.edu
‖hjwalker@math.harvard.edu

The connection between computability and definability is one of the main themes in computable model theory. A decidable theory has a *decidable model*, that is, a model with a decidable elementary diagram. On the other hand, in a *computable model* only the atomic diagram must be decidable. For some nonisomorphic structures that are elementarily equivalent, we can use computable (infinitary) sentences to describe different structures. In general, if two computable structures satisfy the same computable sentences, then they are isomorphic. Roughly speaking, computable formulas are $L_{\omega_1\omega}$-formulas with disjunctions and conjunctions over computably enumerable index sets. Let $\alpha > 0$ be a computable ordinal. A computable Σ_α (Π_α, respectively) formula is a computably enumerable disjunction (conjunction, respectively) of formulas $\exists \overline{u}\, \psi(\overline{x}, \overline{u})$ ($\forall \overline{u}\, \psi(\overline{x}, \overline{u})$, respectively) where ψ is computable Π_β (Σ_β, respectively) for some $\beta < \alpha$. A computable Σ_0 or Π_0 formula is a finitary quantifier-free formula. To show that our descriptions of structures by computable infinitary formulas are optimal we often consider index sets and analyze their complexity. The index set of a structure $\mathcal{A}$ is the set of all Gödel indices for computable isomorphic copies of $\mathcal{A}$.

For a complexity class $\mathcal{P}$, a computable structure $\mathcal{A}$ is *$\mathcal{P}$-categorical* if for all computable $\mathcal{B}$ isomorphic to $\mathcal{A}$, there is an isomorphism in $\mathcal{P}$. A computable structure $\mathcal{A}$ is *relatively $\mathcal{P}$-categorical* if for all $\mathcal{B}$ isomorphic to $\mathcal{A}$, there is an isomorphism that is $\mathcal{P}$ relative to the atomic diagram of $\mathcal{B}$. There is a powerful syntactic condition that is equivalent to relative Δ^0_α-categoricity. The condition is that there is a computably enumerable Scott family of computable Σ_α formulas. For every computable α, there are Δ^0_α-categorical structures that do not have corresponding effective Scott families.

A relation R on a computable structure $\mathcal{A}$, which is not named in the language of $\mathcal{A}$, is called *intrinsically $\mathcal{P}$* on $\mathcal{A}$ if the image of R under every isomorphism from $\mathcal{A}$ onto another computable structure belongs to $\mathcal{P}$. A relation R is *relatively intrinsically $\mathcal{P}$* on $\mathcal{A}$ if the image of R under every isomorphism from $\mathcal{A}$ to any structure $\mathcal{B}$ is $\mathcal{P}$ relative to the atomic diagram of $\mathcal{B}$. A relation R is relatively intrinsically Σ^0_α if and only if R is definable by a computable Σ_α formula with finitely many parameters. For every computable α, there are intrinsically Σ^0_α relations on computable structures, which are not definable by computable Σ_α formulas. On the other hand, intrinsically Δ^1_1 relations on computable structures coincide with relatively intrinsically Δ^1_1 relations, and are exactly the relations definable by computable infinitary formulas with finitely many parameters.

1. Introduction and Preliminaries: Theories, Diagrams, and Models

In the 1970s, Metakides and Nerode, together with other researchers in the United States, initiated a systematic study of computability in mathematical structures and constructions by using modern computability-theoretic tools, such as the priority method and various coding techniques. At the same time and independently, computable model theory was developed in the Siberian school of constructive mathematics by Goncharov, Nurtazin and Peretyat'kin. While in classical mathematics we can replace some constructions by effective ones, for others such replacement is impossible in principle. For example, from the point of view of computability theory, isomorphic structures may have very different properties.

We will assume that all structures are at most countable and their languages are computable. A computable language is a countable language with algorithmically presented set of symbols and their arities. The universe A of an infinite countable structure $\mathcal{A}$ can be identified with ω. If L is the language of $\mathcal{A}$, then L_A is the language L expanded by adding a constant symbol for every $a \in A$, and $\mathcal{A}_A = (\mathcal{A}, a)_{a \in A}$ is the corresponding expansion of $\mathcal{A}$ to L_A. The *atomic (open) diagram* of a structure $\mathcal{A}$, $D(\mathcal{A})$, is the set of all quantifier-free sentences of L_A true in $\mathcal{A}_A$. A structure is *computable* if its atomic diagram is computable. The *Turing degree of* $\mathcal{A}$, $\deg(\mathcal{A})$, is the Turing degree of the atomic diagram of $\mathcal{A}$. The *elementary* (complete, full) *diagram* of $\mathcal{A}$, denoted by $D^c(\mathcal{A})$, is the set of all sentences of L_A that are true in $\mathcal{A}_A$.

We will assume that our theories are countable and consistent. Henkin's construction of a model for a complete decidable theory is effective and produces a structure $\mathcal{A}$ with a computable domain such that the elementary diagram of $\mathcal{A}$ is computable. A structure $\mathcal{A}$ is called *decidable* if its elementary diagram $D^c(\mathcal{A})$ is computable. Thus, in the case of a computable structure, our starting point is semantic, while in the case of a decidable structure, the starting point is syntactic. It is easy to see that not every computable structure is decidable. For example, the *standard model of arithmetic*, $\mathcal{N} = (\omega, +, \cdot, S, 0)$, is computable but not decidable. On the other hand, Tennenbaum showed that if $\mathcal{A}$ is a nonstandard model of *Peano arithmetic* (PA), then $\mathcal{A}$ is not computable. Harrison-Trainor [87] has recently established that characterizing those computable structures that have a decidable copy is a Σ_1^1-complete problem. There are familiar structures $\mathcal{A}$ such that for all $\mathcal{B} \cong \mathcal{A}$, we have $D^c(\mathcal{B}) \equiv_T D(\mathcal{B})$. In particular, this is true for algebraically closed fields, and for other structures for which we have effective elimination of quantifiers.

In computable model theory, we investigate structures, their theories, fragments of diagrams, relations, and isomorphisms within various computability-theoretic hierarchies, such as arithmetic or, more generally, hyperarithmetical hierarchy, or within Turing degree and other degree-theoretic hierarchies. Computability-theoretic notation in this paper is standard and as in [145]. We review some basic notions and notation. For $X \subseteq \omega$, let $\varphi_0^X, \varphi_1^X, \varphi_2^X, \ldots$ be a fixed effective enumeration of all unary X-computable functions. For a structure $\mathcal{B}$, $\varphi_e^{\mathcal{B}}$ stands for $\varphi_e^{D(\mathcal{B})}$. If X is computable, we omit the superscript X. For $e \in \omega$, let $W_e^X = \text{dom}(\varphi_e^X)$. Hence $W_0, W_1, W_2, \ldots$ is an effective enumeration of all computably enumerable (c.e.) sets. By $X \leq_T Y$ ($X \equiv_T Y$, respectively) we denote that X is Turing reducible to Y (X is Turing equivalent to Y, respectively). By $X <_T Y$, we denote that $X \leq_T Y$ but $Y \not\leq_T X$. We write $\mathbf{x} = \deg(X)$ for the Turing degree of X. Thus, $\mathbf{0} = \deg(\emptyset)$. Let $n \geq 1$. Then $\mathbf{x}^{(n)} = \deg(X^{(n)})$, where $X^{(n)}$ is the nth Turing jump of X. For a set X, we define the ω-*jump* of X by $X^{(\omega)} = \{\langle x, n \rangle : x \in X^{(n)}\}$, and let $\mathbf{x}^{(\omega)} = \deg(X^{(\omega)})$. The degree $\mathbf{0}^{(\omega)}$ is a natural upper bound for the sequence $(\mathbf{0}^{(n)})_{n \in \omega}$, although no ascending sequence of Turing degrees has a least upper bound. By $\oplus$ we denote the join of Turing degrees. A set $X \leq_T \emptyset'$ and its Turing degree $\mathbf{x}$ are called *low* if $\mathbf{x}' \leq 0'$, and *low*$_n$ if $\mathbf{x}^{(n)} \leq 0^{(n)}$.

An important question is when a theory has a decidable or a computable model, or a model of certain Turing degree. It is easy to see that the theory of a structure $\mathcal{A}$ is computable in $D^c(\mathcal{A})$, and that $D^c(\mathcal{A})$ is computable in $(D(\mathcal{A}))^{(\omega)}$. The *low basis theorem* of Jockusch and Soare can be used to obtain for a theory S, a model $\mathcal{A}$ with

$$(D^c(\mathcal{A}))' \leq_T S'.$$

The atomic diagram of a model of a theory may be of much lower Turing degree than the theory itself. For example, while the standard model of arithmetic $\mathcal{N}$ is computable, its theory, *true arithmetic*, $Th(\mathcal{N}) = TA$, is of Turing degree $\mathbf{0}^{(\omega)}$. Harrington and Knight [11] proved that there is a nonstandard model $\mathcal{M}$ of PA such that $\mathcal{M}$ is *low* and $Th(\mathcal{M}) \equiv_T \emptyset^{(\omega)}$. Knight [103] proved that if $\mathcal{A}$ is a nonstandard model of PA, then there exists $\mathcal{B}$ isomorphic to $\mathcal{A}$ such that $D(\mathcal{B}) <_T D(\mathcal{A})$.

A set is Σ_n^0 if it is c.e. relative to $\mathbf{0}^{(n-1)}$. A set is Π_n^0 if its negation is Σ_n^0, and a set is Δ_n^0 if it is both Σ_n^0 and Π_n^0. Let $\Delta_0^0 =_{\text{def}} \Delta_1^0$. A set X is *arithmetical* if $X \leq \emptyset^{(k)}$ for some $k \geq 0$. Gödel established that a set X, or a relation R, is arithmetical if and only if it is definable in the standard model of arithmetic $\mathcal{N}$. Moreover, R is Σ_n^0 (Π_n^0, respectively) if and only if it is definable in $\mathcal{N}$ by a Σ_n^0 (Π_n^0, respectively) formula. Σ_1^0 sets are exactly c.e. sets. One of the main results in computable mathematics, due to Matiyasevich, which implies the undecidability of the Hilbert Tenth Problem is that the Diophantine sets of natural numbers coincide with the c.e. sets. A set $X \subseteq \omega$ is *Diophantine* if there is a polynomial $p(y, x_1, \ldots, x_m)$ with integer coefficients such that for every natural number n, we have that $n \in X$ if and only if there are natural numbers $c_1, \ldots, c_m$ such that $p(n, c_1, \ldots, c_m) = 0$.

The set of all computable types of a complete decidable theory is a Π_2^0 set. Every principal type of such a theory is computable, and the set of all its principal types is Π_1^0. A countable structure $\mathcal{A}$ is *homogeneous* if for every two finite sequences $\overrightarrow{a}$ and $\overrightarrow{b}$ of the same length n, if $\overrightarrow{a}$ and $\overrightarrow{b}$ realize the same n-type in $\mathcal{A}$, then there is an automorphism of $\mathcal{A}$ taking $\overrightarrow{a}$ to $\overrightarrow{b}$. Prime models and countable saturated models are examples of homogeneous models. The study of the computable content of these models was initiated in the 1970s. A model $\mathcal{A}$ of a theory T is *prime* if for all models $\mathcal{B}$ of T, $\mathcal{A}$ elementarily embeds into $\mathcal{B}$. It is well known that all prime models of a given theory are isomorphic, and that every complete atomic theory has a prime model. Goncharov and Nurtazin [75], and independently Harrington [83] established that a complete decidable theory T with a prime model has a decidable prime model if and only if the set of all principal types of T is uniformly computable.

Khisamiev [100] showed that there is a complete theory of abelian groups with both a computable model and a prime model, but no computable prime model. More recently, Hirschfeldt [93] showed that there is a complete theory of linear orderings having a computable model and a prime model, but no computable prime model.

A countable model is *saturated* if it realizes every type of its language augmented by any finite tuple of constants for its elements. Morley [131] and Millar [125] independently proved that a complete decidable theory T has a decidable saturated model if and only if the set of all types of T is uniformly computable. For example, it was shown recently in [26] that the theory of differentially closed fields of characteristic 0, in symbols DCF_0, has a decidable saturated model. If the types are not uniformly computable, then the existence of a decidable saturated model is not guaranteed, as shown by counterexamples constructed independently by Goncharov and Nurtazin [75], Morley [131] and Millar [125].

For a structure $\mathcal{A}$, the *type spectrum* of $\mathcal{A}$ is the set of all types realized in $\mathcal{A}$. Goncharov [62], Peretyat'kin [136] and Millar [124] independently showed that there exists a complete decidable theory T having a homogeneous model $\mathcal{M}$ without a decidable copy, such that the type spectrum of $\mathcal{M}$ consists only of computable types and is computable. In fact, Goncharov [62] and Peretyat'kin [136] provided a criterion for a homogeneous model to be decidable. Their criterion can be stated in terms of the effective extension property. A computable set of computable types of a theory has the *effective extension property* if there is a partial computable function f that given a type Γ_n of arity k and a formula θ_i of arity $k + 1$ (identified with their indices), outputs the index for a type containing Γ_n and θ_i, if there exists such a type.

A theory is called *Ehrenfeucht* if it has finitely many but at least two countable models, up to isomorphism. By Vaught's theorem, if a theory has two nonisomorphic models, then it has at least three nonisomorphic models. An example of a theory with exactly three countable models was given by Ehrenfeucht. His result can be easily generalized to obtain a theory with exactly n countable models, for any finite $n \geq 3$. Gavryushkin constructed examples of computable Ehrenfeucht models of arbitrarily high arithmetical and nonarithmetical complexity.

Theorem 1 ([58]). *For every $n \geq 3$, there exists an Ehrenfeucht theory T of arbitrary arithmetical complexity such that it has n countable models, up to isomorphism, and it has a computable model among them. There also exists such a theory that is Turing equivalent to $\emptyset^{(\omega)}$.*

A theory is called κ-*categorical*, where κ is an infinite cardinal, if it has exactly one model of cardinality κ, up to isomorphism. The theories that are $\aleph_0$-categorical are also called *countably categorical*. Morley's categoricity theorem states that if a theory T is κ-categorical for some uncountable cardinal κ, then T is λ-categorical for all uncountable cardinals λ. Hence, theories categorical in an uncountable cardinal are also called *uncountably categorical*. A theory that is both countably and uncountably categorical is simply called *totally categorical*. For the case of countably categorical theories, Lerman and Schmerl [113] gave sufficient conditions for the existence of a computable model, which were later extended by Knight in [104]. Knight proved that if T is a countably categorical theory such that $T \cap \Sigma_{n+2}$ is Σ_{n+1}^0 uniformly in n, then T has a computable model. The natural question is whether there exist such countably categorical theories of higher complexity. Fokina established the following result using the method of Marker's extension. (For Marker's extension, see [71].)

Theorem 2 ([53]). *There exists a countably categorical theory of arbitrary arithmetical complexity, which has a computable model.*

The problem of the existence of a countably categorical theory of non-arithmetical complexity was resolved by Khoussainov and Montalbán [101]. They showed that there exists a countably categorical theory S with a computable model such that $S \equiv_T 0^{(\omega)}$. The unique model of their theory in an infinite language is, up to isomorphism, a modification of the random graph. Andrews [3] later established the following result about truth-table (tt-) degrees. Truth-table reducibility, a version of strong Turing reducibility where the reduction presents a single list of questions to the oracle simultaneously (depending only on the input) and then after seeing the answers produces the output.

Theorem 3 ([3]). *In every tt-degree that is $\leq 0^{(\omega)}$, there is a countably categorical theory in a finite language with a computable model.*

A complete theory T is *strongly minimal* if any definable (with parameters) subset of any model $\mathcal{M}$ of T is finite or cofinite. We call a structure strongly minimal if it has a strongly minimal theory. Andrews and Knight [5] showed that if T is a strongly minimal theory and for $n \geq 1$, $T \cap \Sigma_{n+2}$ is Δ_n^0, uniformly in n, then every model has a computable copy. This result relativized to $\emptyset'''$ gives the following corollary.

Theorem 4 ([5]). *If a strongly minimal theory has a computable model, then every model has a Δ_4^0 copy.*

Andrews and Miller [7] defined the (*Turing degree*) *spectrum of a theory* T to be the set of Turing degrees of models of T. The idea behind this notion is to better understand the relationship between the model-theoretic properties of a theory and the computability-theoretic complexity of its models. On the other hand, the (*Turing*) *degree spectrum* of a structure $\mathcal{A}$ is

$$\mathrm{DgSp}(\mathcal{A}) = \{\deg(D(\mathcal{B})) : \mathcal{B} \cong \mathcal{A}\}.$$

A structure $\mathcal{A}$ is called *automorphically trivial* if there exists a finite subset $\{a_1, \ldots, a_n\}$ of the domain of $\mathcal{A}$ such that every permutation f of the domain with $f(a_i) = a_i$ for $i \in \{1, \ldots, n\}$ is an automorphism of $\mathcal{A}$. These structures include all finite structures and also some infinite structures such as the complete graph on countably many vertices. A structure $\mathcal{A}$ in a finite language is automorphically trivial if and only if its spectrum is $\{\mathbf{0}\}$. The spectrum of an automorphically trivial structure always contains exactly one Turing degree, but if the language is infinite, that degree can be noncomputable. Knight proved the following fundamental result. If $\mathcal{A}$ is not automorphically trivial, then for any two Turing degrees $\mathbf{c} \leq \mathbf{d}$, if $\mathbf{c} \in \mathrm{DegSp}(\mathcal{A})$, then also $\mathbf{d} \in \mathrm{DegSp}(\mathcal{A})$. Moreover, we have the following result.

Theorem 5 ([82]). (a) *For every automorphically nontrivial structure* $\mathcal{A}$, *and every set* $X \geq_T D^c(\mathcal{A})$, *there exists* $\mathcal{B} \cong \mathcal{A}$ *such that*

$$D^c(\mathcal{B}) \equiv_T D(\mathcal{B}) \equiv_T X.$$

(b) *For every automorphically trivial structure* $\mathcal{A}$, *we have* $D^c(\mathcal{A}) \equiv_T D(\mathcal{A})$.

Theory spectra may coincide with degree spectra of structures, e.g., the cones above arbitrary Turing degrees are both theory spectra and degree spectra, as well as the set of all noncomputable degrees. On the other hand, there are examples of theory spectra that are not degree spectra for any structure, such as the degrees of complete extensions of Peano arithmetic, and the union of the cones above two incomparable Turing degrees [7]. On the other hand, by [76], there is a structure the degree spectrum of which consists of exactly the nonhyperarithmetical degrees, while as shown in [7], the set of nonhyperarithmetical degrees is not the spectrum of a theory. Further interesting examples of theory spectra can be found in [7], and for the case of atomic theories in [6].

A structure $\mathcal{A}$ is called *n-decidable* for $n \geq 1$ if the Σ_n-diagram of $\mathcal{A}$ is decidable. We will denote the Σ_n-diagram of $\mathcal{A}$ by $D_n(\mathcal{A})$. For sets X and Y, we say that Y is *c.e. in and above (c.e.a. in)* X if Y is c.e. relative to X, and $X \leq_T Y$. For any structure $\mathcal{A}$, $D_{n+1}(\mathcal{A})$ is c.e.a. in $D_n(\mathcal{A})$, uniformly in n, where $D_0(\mathcal{A}) = D(\mathcal{A})$. For all n, there are n-decidable structures that do not have $(n+1)$-decidable copies. Chisholm and Moses [39] established that there is a linear ordering that is n-decidable for every $n \in \omega$, but has no decidable copy. Goncharov [64] obtained a similar result for Boolean algebras. Harrison-Trainor [87] has recently established that for $n \geq 1$, characterizing those computable structures that have an n-decidable copy is a Σ_1^1-complete problem.

Turing jump of a structure and different forms of the jump inversion for a structure have been studied independently by Baleva, Soskov, and Soskova in Bulgaria, by Morozov, Stukachev, and Puzarenko in Russia, and by Montalbán in the United States. We say that a structure $\mathcal{A}$ admits *strong jump inversion* if for every oracle X, if X' computes $D(\mathcal{C})'$ for some $\mathcal{C} \cong \mathcal{A}$, then X computes $D(\mathcal{B})$ for some $\mathcal{B} \cong \mathcal{A}$. Equivalently, for every oracle X, if $\mathcal{A}$ has a copy that is low over X, then it has a copy that is computable in X. Here, when we say that $\mathcal{C}$ is low over X, we mean that $D(\mathcal{C})' \leq_T X'$. For example, if $\mathcal{A}$ is an equivalence structure with infinitely many infinite classes, then $\mathcal{A}$ admits strong jump inversion. That is because if $\mathcal{A}$ is low over X, then the character of $\mathcal{A}$, consisting of pairs (n, k) such that there are at least k classes of size n, is Σ_2^0 relative to $\mathcal{A}$, so it is Σ_2^0 relative to X. Then $\mathcal{A}$ has an X-computable copy. Downey and Jockusch [44] established that every Boolean algebra admits strong jump inversion. More recently, Marker and Miller [117] have shown that all countable models of the theory of differentially closed fields of characteristic 0, in symbols DCF_0, admit strong jump inversion.

Not all countable structures admit strong jump inversion. Jockusch and Soare [97] proved that there are *low* linear orderings without computable copies. If T is a low completion of PA, then there is a model $\mathcal{A}$ such that its atomic diagram $D(\mathcal{A})$ is computable in T, hence $D(\mathcal{A})'$ is Δ_2^0. However, since $\mathcal{A}$ is necessarily nonstandard, it does not have a computable copy.

The authors of [26] established a general result with sufficient conditions for a structure $\mathcal{A}$ to admit strong jump inversion. The conditions involve saturation properties and an enumeration of B_1-types, where these are made up of formulas that are Boolean combinations of existential formulas. The general result applies to some familiar kinds of structures, including some classes of linear orderings and trees, as well as DCF_0. It also applies to Boolean algebras with no 1-atom, with some extra information on the complexity of the isomorphism. Our general result gives the result of Marker and Miller. In order to apply our general result, we produce a computable enumeration of the types realized in models of DCF_0. This also yields the fact that the saturated model of DCF_0 has a decidable copy.

The general result applies to structures from familiar algebraic classes, including certain classes of linear orderings, abelian p-groups, equivalence structures, and trees. When a structure $\mathcal{A}$ admits strong jump inversion, and $\mathcal{A}$ is low relative to an oracle X, the authors of [26] also considered the complexity of the isomorphisms between $\mathcal{A}$ and its X-computable copies. In the case of an infinite Boolean algebra with no 1-atom, such an isomorphism can be chosen to be Δ_3^0 relative to X. This is interesting because Knight and Stob [108] established that any low Boolean algebra has a computable copy and a corresponding Δ_4^0 isomorphism, and this bound has been proven to be sharp.

2. Computable Infinitary Formulas: Scott Rank

Several important notions of computability on effective structures have syntactic characterizations, which involve computable infinitary formulas introduced by Ash. Formulas of $L_{\omega_1\omega}$ are infinitary formulas with countable disjunctions and conjunctions, but only finite strings of quantifiers. If we restrict the disjunctions and conjunctions to c.e. sets, then we have the computable infinitary formulas. Ash defined computable Σ_α and Π_α formulas of $L_{\omega_1\omega}$, where α is a computable ordinal, recursively and simultaneously and together with their Gödel numbers.

An ordinal is *computable* if it is finite or is the order type of a computable well ordering on ω. The computable ordinals form a countable initial segment of the ordinals. *Kleene's $\mathcal{O}$* is the set of notations for computable ordinals, together with a partial ordering $<_\mathcal{O}$ (see [140, 141]). The ordinal 0 gets notation 1. If a is a notation for α, then 2^a is a notation for $\alpha+1$. Then $a <_\mathcal{O} 2^a$, and also, if $b <_\mathcal{O} a$, then $b <_\mathcal{O} 2^a$. Suppose α is a limit ordinal. If φ_e is a total function, giving notations for an increasing sequence of ordinals with limit α, then $3 \cdot 5^e$ is a notation for α. For all n, we have $\varphi_e(n) <_\mathcal{O} 3 \cdot 5^e$, and if $b <_\mathcal{O} \varphi_e(n)$, then $b <_\mathcal{O} 3 \cdot 5^e$. Let $|a|$ denote the ordinal with notation a. If $a \in \mathcal{O}$, then the restriction of $<_\mathcal{O}$ to the set $\mathrm{pred}(a) = \{b \in \mathcal{O} : b <_\mathcal{O} a\}$ is a well ordering of type $|a|$. For $a \in \mathcal{O}$, $\mathrm{pred}(a)$ is c.e., uniformly in a. The set $\mathcal{O}$ is Π_1^1-complete. A Π_1^1 subset of $\mathcal{O}$ is Δ_1^1 if and only it is contained in a set of the form $\mathcal{O}_\alpha = \{b \in \mathcal{O} : |b| < \alpha\}$, where α is a computable ordinal.

For computable infinitary formulas, we cannot bring the quantifiers outside, but we can bring negations inside. We have a resemblance to normal form, and we can classify formulas according to the number of alternations of infinite disjunction/$\exists$ with infinite conjunction/$\forall$. The computable Σ_0 and Π_0 formulas are the finitary quantifier-free formulas. The computable $\Sigma_{\alpha+1}$ formulas are of the form

$$\bigvee_{n \in W_e} \exists \vec{y}_n \psi_n(\vec{x}, \vec{y}_n),$$

where for $n \in W_e$, ψ_n is a Π_α formula indexed by its Gödel number, and $\exists \vec{y}_n$ is a finite block of existential quantifiers. That is, $\Sigma_{\alpha+1}$ formulas are c.e. disjunctions of $\exists\Pi_\alpha$ formulas. Similarly, $\Pi_{\alpha+1}$ formulas are c.e. conjunctions of $\forall\Sigma_\alpha$ formulas. It can be shown that a computable Σ_1 formula is of the form

$$\bigvee_{n\in\omega} \exists \overrightarrow{y}_n \theta_n(\overrightarrow{x}, \overrightarrow{y}_n),$$

where $(\theta_n(\overrightarrow{x}, \overrightarrow{y}_n))_{n\in\omega}$ is a computable sequence of quantifier-free formulas. If α is a limit ordinal, then Σ_α (Π_α, respectively) formulas are of the form $\bigvee_{n\in W_e} \psi_n$ ($\bigwedge_{n\in W_e} \psi_n$, respectively), such that there is a sequence $(\alpha_n)_{n\in W_e}$ of ordinals having limit α, given by the ordinal notation for α, and every ψ_n is a Σ_{α_n} (Π_{α_n}, respectively) formula. For a more precise definition of computable Σ_α and Π_α formulas, see [11].

The least noncomputable ordinal is denoted by ω_1^{CK}, where CK stands for Church–Kleene. To extend the arithmetical hierarchy, we define the representative sets in the hyperarithmetical hierarchy, H_a for $a \in \mathcal{O}$. The definition is recursive, and is based on iterating Turing jump: $H_1 = \emptyset$, $H_{2^a} = (H_a)'$, and $H_{3.5^e} = \{2^x 3^n : x \in H_{\varphi_e(n)}\}$. Let β be an infinite computable ordinal. Then a set is Σ_β^0 if it is c.e. relative to some H_a such that β is represented by notation a. A set is Π_β^0 if its negation is Σ_β^0, and a set is Δ_β^0 if it is both Σ_β^0 and Π_β^0. A set is *hyperarithmetical* if it is Δ_α^0 for some computable α. Hence, a set X is hyperarithmetical if $(\exists a \in \mathcal{O})[X \leq_T H_a]$. The hyperarithmetical sets coincide with Δ_1^1 sets.

The important property of these formulas is given in the following theorem due to Ash.

Theorem 6. *For a structure* $\mathcal{A}$, *if* $\theta(\overrightarrow{x})$ *is a computable* Σ_α (Π_α) *formula, then the set* $\{\overrightarrow{a} : \mathcal{A}_A \models \theta(\overrightarrow{a})\}$ *is* Σ_α^0 (Π_α^0) *relative to the atomic diagram of* $\mathcal{A}$.

As an example, we will consider some definable properties of a countable reduced abelian p-group, where p is a prime number. Recall that a *p-group* is a group in which every element has order p^n for some n. Countable reduced Abelian p-groups are of particular interest because of their classification up to isomorphism by Ulm. We define a sequence of subgroups G_α, letting $G_0 = G$, $G_{\alpha+1} = pG_\alpha$, and for limit α, $G_\alpha = \bigcap_{\beta<\alpha} G_\beta$. There is a countable ordinal α such that $G_\alpha = G_{\alpha+1}$. The least such α is the *length* of G, denoted by $\lambda(G)$. The group is *reduced* if $G_{\lambda(G)} = \{0\}$. An element $x \neq 0$ has *height* β if $x \in G_\beta - G_{\beta+1}$. Let $P(G)$ be the set of element of G of order p. Let $P_\alpha = G_\alpha \cap P(G)$. For each $\beta < \lambda(G)$, $P_\beta/P_{\beta+1}$ is a vector space over $\mathbb{Z}_p$ of dimension $\leq \aleph_0$, and this dimension is denoted by $u_\beta(G)$. The *Ulm sequence* for G is the sequence $(u_\beta(G))_{\beta<\lambda(G)}$.

For any computable ordinal α, it is fairly straightforward to write a computable infinitary sentence stating that G is a reduced Abelian p-group of length at most α, and describing its Ulm invariants. In particular, Barker established the following results.

Proposition 1 ([16]). *Let G be a computable Abelian p-group.*

(1) $G_{\omega\cdot\alpha}$ *is* $\Pi^0_{2\alpha}$.
(2) $G_{\omega\cdot\alpha+m}$ *is* $\Sigma^0_{2\alpha+1}$.
(3) $P_{\omega\cdot\alpha}$ *is* $\Pi^0_{2\alpha}$.
(4) $P_{\omega\cdot\alpha+m}$ *is* $\Sigma^0_{2\alpha+1}$.

Proof. It is easy to see that (3) and (4) follow from (1) and (2), respectively. Toward (1) and (2), note the following:

$$x \in G_m \iff \exists y[p^m y = x];$$

$$x \in G_\omega \iff \bigvee_{m\in\omega} \exists y[p^m y = x];$$

$$x \in G_{\omega\cdot\alpha+m} \iff \exists y[p^m y = x \ \& \ G_{\omega\cdot\alpha}(y)];$$

$$x \in G_{\omega\cdot\alpha+\omega} \iff \bigvee_{m\in\omega} \exists y[p^m y = x \ \& \ G_{\omega\cdot\alpha}(y)];$$

$$x \in G_{\omega\cdot\alpha} \iff \bigvee_{\gamma<\alpha} G_{\omega\cdot\gamma}(x) \text{ for limit } \alpha. \qquad \square$$

Using these results, it is easy to write, for any computable ordinal β, a computable $\Pi_{2\beta+1}$ sentence the models of which are exactly the reduced abelian p-groups of length $\omega\beta$.

Harizanov, Knight, and Morozov gave conditions for intrinsic collapse of the complete diagram to the n-diagram using infinitary formulas.

Theorem 7 ([82]). *For any structure $\mathcal{A}$ and any n, the following conditions are equivalent.*

(i) *For all $\mathcal{B} \cong \mathcal{A}$, $D^c(\mathcal{B}) \equiv_T D_n(\mathcal{B})$.*
(ii) *For some tuple $\overrightarrow{c}$, there is a computable function d taking each (finitary) formula $\theta(\overrightarrow{x})$ to a formula $d_\theta(\overline{c}, \overrightarrow{x})$, a c.e. disjunction of (finitary) Σ^0_{n+1} formulas with parameters $\overrightarrow{c}$, such that*

$$\mathcal{A} \models \forall \overrightarrow{x} \, [\theta(\overrightarrow{x}) \Leftrightarrow d_\theta(\overrightarrow{c}, \overrightarrow{x})].$$

As an application of this theorem, let $\mathcal{B}$ be a linear ordering of type $\omega^n \cdot \eta$ for $n \geq 1$. Then

$$D^c(\mathcal{B}) \equiv_T D_{2n}(\mathcal{B})$$

and we obtain formulas d_θ as follows. Let $\mathcal{A}$ be an ordering of type $\omega^n \cdot \eta$ such that $D^c(\mathcal{A})$ is computable. For each tuple $\overrightarrow{a}$, we can find a (finitary) Σ^0_{2n+1} formula $\psi_{\overrightarrow{a}}(\overrightarrow{x})$ defining the orbit of $\overrightarrow{a}$. Then for each formula $\theta(\overrightarrow{x})$, $d_\theta(\overrightarrow{x})$ is the disjunction of these $\psi_{\overrightarrow{a}}(\overrightarrow{x})$ for $\overrightarrow{a}$ satisfying $\theta(\overrightarrow{x})$. It can be shown that there is a formula θ such that d_θ cannot be made finitary even for $n = 1$.

Ehrenfeucht gave an example of a complete theory with exactly three models, up to isomorphism. The language of the theory has a binary relation symbol $<$ and constants c_n for $n \in \omega$. The axioms say that $<$ is a dense linear ordering without endpoints, and the constants are strictly increasing. The theory T has the following three countable models, up to isomorphism. There is the *prime* model, in which there is no upper bound for the constants. There is the *saturated* model, in which the constants have an upper bound but no least upper bound. There is the *middle* model, in which there is a least upper bound for the constants. Let $\mathcal{A}^1$ be the prime model, let $\mathcal{A}^2$ be the middle model, and let $\mathcal{A}^3$ be the saturated model. The following examples of Scott sentences are due to S. Quinn [29].

A computable Π_2 sentence characterizing the models of T such that

$$(\forall x) \bigvee_{n \in \omega} x < c_n$$

is a Scott sentence for $\mathcal{A}^1$.

We have a computable Σ_3 Scott sentence for $\mathcal{A}^2$ describing a model of T such that

$$(\exists x) \left[\bigwedge_{n \in \omega} x > c_n \ \& \ (\forall y) \left[\left(\bigwedge_{n \in \omega} y > c_n \right) \to y \geq x \right] \right].$$

We have a computable Π_3 Scott sentence for $\mathcal{A}^3$, describing a model of T such that

$$(\exists x) \left[\bigwedge_{n \in \omega} x > c_n \right] \ \& \ (\forall y)$$

$$\left[\bigwedge_{n \in \omega} y > c_n \implies (\exists z) \left[\bigwedge_{n \in \omega} z > c_n \ \& \ z < y \right] \right].$$

A *d-Σ_α^0 formula* is the conjunction of a Σ_α^0 formula and a Σ_α^0 formula. Knight and Saraph [107] proved that a finitely generated abelian group has a computable d-Σ_2^0 Scott sentence. The infinite dihedral group is given by the presentation $\langle a, b \mid a^2, b^2 \rangle$. Knight and Saraph [107] also proved that a computable infinite dihedral group has a computable d-Σ_2^0 Scott sentence. Scott sentences for many other classes of groups were investigated. For example, Ho [96] proved that every computable polycyclic group has a computable d-Σ_2^0 Scott sentence.

We can measure the complexity of a countable structure by looking for a Scott sentence of minimal complexity. A. Miller [126], using a result of D. Miller [127], proved that for a countable ordinal $\gamma \geq 1$, if $\mathcal{A}$ has a $\Pi^0_{\gamma+1}$ Scott sentence and a $\Sigma^0_{\gamma+1}$ Scott sentence, then it must have a d-Σ^0_γ Scott sentence. Hence for $\alpha \geq 2$, if $\mathcal{A}$ has a Π^0_β Scott sentence and a Σ^0_β Scott sentence, then it must have a d-Σ^0_α Scott sentence for some $\alpha < \beta$. Therefore, the optimal Scott sentence for a given structure is Σ^0_α, Π^0_α, or d-Σ^0_α for some α. Alvir, Knight, and McCoy [2] established an effective version of A. Miller's result.

Theorem 8 ([2]). *Let $\alpha \geq 2$ be a computable ordinal. If a structure $\mathcal{A}$ has a computable Π^0_β Scott sentence and a computable Σ^0_β Scott sentence, then it must have a computable d-Σ^0_α Scott sentence for some $\alpha < \beta$.*

Knight and Saraph [107] showed that a finitely generated structure always has a Σ^0_3 Scott sentence. However, many finitely generated groups have a simpler description, which is d-Σ^0_2. Harrison-Trainor and Ho [88] gave a characterization of finitely generated structures for which the Σ^0_3 Scott sentence is optimal. A substructure $\mathcal{B}$ of a structure $\mathcal{A}$ is a Σ^0_1-elementary substructure of $\mathcal{M}$ if for every existential formula $\theta(\overrightarrow{x})$ and $\overrightarrow{b} \in B^{lh(\overrightarrow{x})}$, we have $\mathcal{A} \models \theta(\overrightarrow{b})$ if and only if $\mathcal{A}_B \models \theta(\overrightarrow{b})$.

Theorem 9 ([2, 88]). *For a finitely generated structure $\mathcal{A}$, the following conditions are equivalent.*

(i) *$\mathcal{A}$ has a d-Σ^0_2 Scott sentence.*
(ii) *$\mathcal{A}$ does not contain a proper Σ^0_1-elementary substructure isomorphic to itself.*
(iii) *For all (or some) generating tuples of $\mathcal{A}$, the orbit is defined by a Π^0_1 formula.*

The equivalence of (i) and (iii) has been established by Alvir, Knight, and McCoy [2]. Since every finitely generated field satisfies condition (ii) of the previous theorem, it follows that a finitely generated field has a d-Σ^0_2 Scott sentence.

The compactness theorem of Kreisel and Barwise states that if Γ is a Π^1_1 set of computable infinitary sentences such that every Δ^1_1 subset of Γ has a model, then Γ has a model. As a corollary, we obtain the following result (see [11]).

Theorem 10. *Let Γ be a Π^1_1 set of computable infinitary sentences. If every Δ^1_1 set $\Sigma \subseteq \Gamma$ has a computable model, then Γ has a computable model.*

The following result is a special case of a result of Ressayre in [139].

Theorem 11. *Let $\mathcal{A}$ be a hyperarithmetical structure. If $\vec{a}$ and $\vec{b}$ are tuples in $\mathcal{A}$ satisfying the same computable infinitary formulas, then there is an automorphism of $\mathcal{A}$ taking $\vec{a}$ to $\vec{b}$.*

Similarly, if $\mathcal{A}$ and $\mathcal{B}$ are hyperarithmetical structures satisfying the same computable sentences, then $\mathcal{A} \cong \mathcal{B}$ (see [72]).

The Scott isomorphism theorem says that for any countable structure $\mathcal{A}$ for a computable language, there is an $L_{\omega_1\omega}$ sentence σ such that the countable models of σ are exactly the isomorphic copies of $\mathcal{A}$. For a proof of the Scott isomorphism theorem, see [11]. The proof leads to an assignment of ordinals to countable structures, which we call Scott rank. By a result of Nadel [133], for any hyperarithmetical structure, there is a computable Scott sentence if and only if the Scott rank is computable.

There are several different definitions of Scott rank. The one used by Sacks [142] involves a sequence of expansions of $\mathcal{A}$. Let $\mathcal{A}_0 = \mathcal{A}$, let $\mathcal{A}_{\alpha+1}$ be the result of adding to $\mathcal{A}_\alpha$ predicates for the types realized in $\mathcal{A}_\alpha$, and for limit α, let $\mathcal{A}_\alpha$ be the limit of the expansions $\mathcal{A}_\beta$, for $\beta < \alpha$. For some countable ordinal α, $\mathcal{A}_\alpha$ is atomic. The least such α is the *rank*. However, we will use the definition of the Scott rank given in [11] (also see [30]). First we define a family of equivalence relations on finite tuples $\overrightarrow{a}$ and $\overrightarrow{b}$ of elements in $\mathcal{A}$, of the same length.

(1) We say that $\overrightarrow{a} \equiv^0 \overrightarrow{b}$ if $\overrightarrow{a}$ and $\overrightarrow{b}$ satisfy the same quantifier-free formulas.
(2) For $\alpha > 0$, we say that $\overrightarrow{a} \equiv^\alpha \overrightarrow{b}$ if for all $\beta < \alpha$, for every $\overrightarrow{c}$, there exists $\overrightarrow{d}$, and for every $\overrightarrow{d}$, there exists $\overrightarrow{c}$, such that $\overrightarrow{a}, \overrightarrow{c} \equiv^\beta \overrightarrow{b}, \overrightarrow{d}$.

The *Scott rank of a tuple* $\overrightarrow{a}$ in $\mathcal{A}$ is the least β such that for all $\overrightarrow{b}$, the relation $\overrightarrow{a} \equiv^\beta \overrightarrow{b}$ implies $(\mathcal{A}, \overrightarrow{a}) \cong (\mathcal{A}, \overrightarrow{b})$. The *Scott rank of* $\mathcal{A}$, $\mathrm{SR}(\mathcal{A})$, is the least ordinal α greater than the ranks of all tuples in $\mathcal{A}$. For example, if $\mathcal{L}$ is a linear order of type ω, then $\mathrm{SR}(\mathcal{L}) = 2$. For a hyperarithmetical structure, the Scott rank is at most $\omega_1^{\mathrm{CK}} + 1$. A *Harrison ordering* is a computable ordering of type $\omega_1^{\mathrm{CK}}(1 + \eta)$, where η is the order type of the rationals. Here, for orderings $\mathcal{L}_1$ and $\mathcal{L}_2$, $\mathcal{L}_1\mathcal{L}_2$ is the result of replacing each element of $\mathcal{L}_2$ by a copy of $\mathcal{L}_1$. Harrison [85] showed such an ordering exists. It can be shown that its Scott rank is $\omega_1^{\mathrm{CK}}+1$.

In general, it can be shown (see [11, 30]) that for a computable structure $\mathcal{A}$, we have the following.

(1) $\mathrm{SR}(\mathcal{A}) < \omega_1^{\mathrm{CK}}$ if there is a computable ordinal β such that the orbits of all tuples are defined by computable Π_β formulas.
(2) $\mathrm{SR}(\mathcal{A}) = \omega_1^{\mathrm{CK}}$ if the orbits of all tuples are defined by computable infinitary formulas, but there is no bound on the complexity of these formulas.
(3) $\mathrm{SR}(\mathcal{A}) = \omega_1^{\mathrm{CK}} + 1$ if, there is some tuple, the orbit of which is not defined by any computable infinitary formula.

There are structures in natural classes, for example, abelian p-groups, where p is a prime number, with arbitrarily large computable ranks, and of rank $\omega_1^{\mathrm{CK}} + 1$, but none of rank ω_1^{CK} (see [18]). Makkai was the first to prove the existence of a structure of Scott rank ω_1^{CK}.

Theorem 12 ([115]). *There is an arithmetical structure of Scott rank ω_1^{CK}.*

In [106], Millar and Knight showed that such structure can be made computable. Through the recent work of Calvert, Knight, and Millar [31], Calvert, Goncharov, and Knight [27], and Freer [56], we started to better understand the structures of Scott rank ω_1^{CK}.

Theorem 13 ([27, 31]). *There are computable structures of Scott rank ω_1^{CK} in the following classes: trees, undirected graphs, fields of any fixed characteristic, and linear orders.*

Sacks asked whether for known examples of computable structures of Scott rank ω_1^{CK}, the computable infinitary theories are $\aleph_0$-categorical. In [28], Calvert, Goncharov, Millar, and Knight gave an affirmative answer for known examples. In [122], Millar and Sacks introduced an innovative technique that produced a countable structure $\mathcal{A}$ of Scott rank ω_1^{CK} such that $\omega_1^{\mathcal{A}} = \omega_1^{CK}$ and the $L_{\omega_1^{CK},\omega}$-theory of $\mathcal{A}$ is *not* $\aleph_0$-categorical. Finally, Harrison-Trainor, Igusa and Knight gave a negative answer to Sacks's question.

Theorem 14 ([89]). *There is a computable structure $\mathcal{M}$ of Scott rank ω_1^{CK} such that the computable infinitary theory of $\mathcal{M}$ is not $\aleph_0$-categorical.*

3. Index Sets of Structures and Classes of Structures

In order to measure computability-theoretic complexity of countable structures, one of the main goals is to find an optimal definition of the class of structures under investigation. This often requires the use of various internal properties of the structures in the class. After a reasonable definition is found, it is necessary to prove its sharpness. Usually, this is done by proving completeness in some complexity class.

We may state our goal as follows. Let K be a class of structures. We denote by K^c the set of computable structures in K. A *computable characterization* of K should separate computable structures in K from all other structures (those not in K, or noncomputable ones). A *computable classification* for K up to an equivalence relation E (isomorphism, computable isomorphism, etc.) should determine each computable element, up to the equivalence E, in terms of relatively simple invariants. In [72], Goncharov and Knight presented three possible approaches to the study of computable characterizations of classes of structures.

Within the framework of the first approach, we say that K has a *computable characterization* if K^c is the set of computable models of a computable sentence. The class of linear orderings can be characterized by a single first-order sentence. The class of abelian p-groups is characterized by a single computable Π_2 sentence. The classes of well orderings and reduced abelian p-groups cannot be characterized by single computable sentences.

Furthermore, we say that there is a *computable classification* for K if there is a computable bound on the ranks of elements of K^c. By a *computable rank* $R^c(\mathcal{A})$ of a structure $\mathcal{A}$ we mean the least ordinal α such that for all tuples $\overrightarrow{a}$ and $\overrightarrow{b}$ in A, of the same length, if for all $\beta < \alpha$, all computable Π_β formulas that true of $\overrightarrow{a}$ are also true of $\overrightarrow{b}$, then there is an automorphism of $\mathcal{A}$ taking $\overrightarrow{a}$ to $\overrightarrow{b}$. For example, the computable rank of a vector space over $\mathbb{Q}$ is 1. There is no computable bound on computable ranks of linear orders and abelian p-groups. The computable rank is not the same as the Scott rank. However, for a hyperarithmetical structure, its computable rank is a computable ordinal just in case its Scott rank is computable (see [72]). If $\mathcal{A}$ is hyperarithmetical, then $R^c(\mathcal{A}) \leq \omega_1^{\mathrm{CK}}$.

The second approach involves the notion of an index set. For a computable structure $\mathcal{A}$, an *index* is a number e such that $\varphi_e = \chi_{D(\mathcal{A})}$, where $(\varphi_e)_{e \in \omega}$ is a computable enumeration of all unary partial computable functions. We denote the structure with index e by $\mathcal{A}_e$. The *index set* for $\mathcal{A}$ is the set $I(\mathcal{A})$ of all indices for computable (isomorphic) copies of $\mathcal{A}$. For a class K of structures, closed under isomorphism, the *index set* is the set $I(K)$ of all indices for computable members of K. For an equivalence relation E on a class K, we define

$$I(E, K) = \{(m, n) : m, n \in I(K) \ \& \ \mathcal{A}_m E \mathcal{A}_n\}.$$

Within this approach, we say that K has a *computable characterization* if $I(K)$ is hyperarithmetical. The class K has a *computable classification* up to E if $I(E, K)$ is hyperarithmetical.

The first and the second approach are known to be equivalent [72]. In fact, we do not know a better way to estimate the complexity of an index set than by giving a description by a computable formula. (The third approach of Goncharov and Knight to computable characterization of classes of structures, equivalent to the other two approaches, involves the notion of an *enumeration*.)

Theorem 15 ([72]). *For the following classes K, the index set $I(K)$ is Π_2^0:*

(1) *linear orderings,*
(2) *Boolean algebras,*
(3) *abelian p-groups,*
(4) *vector spaces over $\mathbb{Q}$.*

The results in the following theorem are well-known and can be attributed to Kleene and Spector.

Theorem 16. *For the following classes K, the index set $I(K)$ is not hyperarithmetical:*

(1) *well-orderings,*
(2) *superatomic Boolean algebras,*
(3) *reduced abelian p-groups.*

In the next theorem, the complexity of index sets for classes of structures with important model-theoretic properties are given by White [152], Pavlovskii [135], and Fokina [51, 52]. By $\Sigma_3^0 - \Sigma_3^0$ we denote the difference of two Σ_3^0 sets. This difference is also denoted by $d\text{-}\Sigma_3^0$.

Theorem 17.

(a) ([52]) *The index set of structures with decidable countably categorical theories is an m-complete $\Sigma_3^0 - \Sigma_3^0$ set.*

(b) ([51]) *The index set of decidable structures is Σ_3^0-complete.*

(c) ([135, 152]) *The index set of computable prime models is an m-complete $\Pi_{\omega+2}^0$ set.*

(d) ([152]) *The index set of computable homogeneous models is an m-complete $\Pi_{\omega+2}^0$ set.*

Calvert, Fokina, Goncharov, Knight, Kudinov, Morozov, and Puzarenko [25] investigated index set complexity of structures of certain Scott ranks.

Theorem 18 ([25]).

(a) *The index set of computable structures with noncomputable Scott ranks is m-complete Σ_1^1.*

(b) *The index set of structures with the Scott rank ω_1^{CK} is m-complete Π_2^0 relative to Kleene's $\mathcal{O}$.*

(c) *The index set of structures with the Scott rank $\omega_1^{CK} + 1$ is m-complete Σ_2^0 relative to Kleene's $\mathcal{O}$.*

A computable structure $\mathcal{A}$ may not have a computable Scott sentence. If it does have a computable Scott sentence σ, then the complexity of the index set $I(\mathcal{A})$ is bounded by the complexity of σ. For many structures from familiar classes, it is often the case that the complexity of the index set matches that of an optimal Scott sentence. However, Knight and McCoy demonstrated in [105] that this is not always the case. Namely, they found a subgroup of $\mathbb{Q}$, which does not have a $d\text{-}\Sigma_2^0$ Scott sentence but its index set is $d\text{-}\Sigma_2^0$.

For some structures, we obtain more meaningful results by locating the given computable structure $\mathcal{A}$ within some natural class K closed under isomorphism.

Definition 1. A sentence σ is a *Scott sentence for $\mathcal{A}$ within K* if the countable models of σ in K are exactly the isomorphic copies of $\mathcal{A}$.

We say how to describe $\mathcal{A}$ within K, and also how to calculate the complexity of $I(\mathcal{A})$ within $I(K)$.

Definition 2. Let Γ be a complexity class.

(1) $I(\mathcal{A})$ is Γ *within* K if $I(\mathcal{A}) = R \cap I(K)$ for some $R \in \Gamma$.
(2) $I(\mathcal{A})$ is *m-complete* Γ *within* K if $I(\mathcal{A})$ is Γ within K and for any $S \in \Gamma$, there is a computable function $f : \omega \to I(K)$ such that there is a uniformly computable sequence $(\mathcal{C}_n)_{n \in \omega}$ for which

$$n \in S \text{ if and only if } \mathcal{C}_n \cong \mathcal{A},$$

that is,

$$n \in S \text{ if and only if } f(n) \in I(\mathcal{A}).$$

Recall Ehrenfeucht's example of a complete theory with exactly three countable models, up to isomorphism, in the language with a binary relation symbol $<$ and constants c_n for $n \in \omega$.

Theorem 19 ([29]). *Let K be the class of models of the original Ehrenfeucht theory T. Let $\mathcal{A}^1$ be the prime model, let $\mathcal{A}^2$ be the middle model, and let $\mathcal{A}^3$ be the saturated model.*

(a) *$I(\mathcal{A}^1)$ is m-complete Π_2^0 within K.*
(b) *$I(\mathcal{A}^2)$ is m-complete Σ_3^0 within K.*
(c) *$I(\mathcal{A}^3)$ is m-complete Π_3^0 within K.*

Finite structures are easy to describe. By $\Sigma_1^0 - \Sigma_1^0$ we denote the difference of two c.e. sets. Such a set is also called *d-c.e.* where d stands for difference.

Theorem 20. *Let L be a finite relational language. Let K be the class of finite L-structures, and let $\mathcal{A} \in K$.*

If $\mathcal{A}$ has size $n \geq 1$, then $I(\mathcal{A})$ is m-complete $\Sigma_1^0 - \Sigma_1^0$ within K.

Proof. We have a finitary existential sentence θ stating that there is a substructure isomorphic to $\mathcal{A}$, and another finitary existential sentence ψ stating that there are at least $n + 1$ elements. Then $\theta \wedge \neg\psi$ is a Scott sentence for $\mathcal{A}$. It follows that $I(\mathcal{A})$ is d-c.e. within K. For completeness, let $S = S_1 - S_2$, where S_1 and S_2 are c.e. We have the usual finite approximations $S_{1,s}$, $S_{2,s}$.

Let $\mathcal{A}^-$ be a proper substructure of $\mathcal{A}$, and let $\mathcal{A}^+$ be a finite proper superstructure of $\mathcal{A}$. We will build a uniformly computable sequence $(\mathcal{A}_n)_{n \in \omega}$ such that

$$
\mathcal{A}_n \cong \begin{cases} \mathcal{A}^- & \text{if } n \notin S_1, \\ \mathcal{A} & \text{if } n \in S_1 - S_2, \\ \mathcal{A}^+ & \text{if } n \notin S_1 \cap S_2. \end{cases}
$$

To accomplish this, let $D_0 = D(\mathcal{A}^-)$. At stage s, if $n \notin S_{1,s}$, we let D_s be the atomic diagram of $\mathcal{A}^-$. If $n \in S_{1,s} - S_{2,s}$, we let D_s be the atomic diagram of $\mathcal{A}$. If $n \in S_{1,s} \cap S_{2,s}$, we let D_s be the atomic diagram of $\mathcal{A}^+$. There is some s_0 such that for all $s \geq s_0$, $n \in S_1$ if and only if $n \in S_{1,s}$, and $n \in S_2$ if and only if $n \in S_{2,s}$. Let $\mathcal{A}_n$ be the structure with diagram D_s for $s \geq s_0$. It is clear that $\mathcal{A}_n \cong \mathcal{A}$ if and only if $n \in S$. $\square$

The finite-dimensional vector spaces over a fixed infinite computable field are completely determined by a finite a basis. For concreteness, we assume that vector spaces are over $\mathbb{Q}$. Let K be the class of $\mathbb{Q}$-vector spaces, and let $\mathcal{A}$ be a one-dimensional member of K. First, we show that $\mathcal{A}$ has a computable Π_2 Scott sentence. We have a computable Π_2 sentence characterizing the class K. We take the conjunction of this with the sentence saying

$$(\exists x)\ x \neq 0\ \&\ (\forall x)\,(\forall y)\ \bigvee_{\lambda \in \Lambda} \lambda(x, y) = 0,$$

where Λ is the set of all nontrivial linear combinations $q_1 x + q_2 y$, for $q_i \in \mathbb{Q}$. Now, $I(\mathcal{A})$ is Π_2^0. We do not need to locate $\mathcal{A}$ within K, since the set of indices for members of K is Π_2^0. We can also show Π_2^0-hardness. Suppose $\mathcal{B}$ in K has dimension k, where $k > 1$. Then $\mathcal{B}$ has a d-Σ_2 Scott sentence. We take the conjunction of the axioms for $\mathbb{Q}$-vector spaces, and we add a sentence saying that there are at least k independent elements, and that there are not at least $k+1$. Then $I(\mathcal{B})$ is d-Σ_2^0. For $\mathcal{C}$, we have a computable Π_3 Scott sentence, obtained by taking the conjunction of the axioms for $\mathbb{Q}$-vector spaces and the conjunction over all $k \in \omega$ of computable Σ_2 sentences saying that the dimension is at least k. Therefore, $I(\mathcal{C})$ is Π_3^0. We can also establish the corresponding hardness results.

Theorem 21 ([29]). *Let K be the class of computable vector spaces over $\mathbb{Q}$, and let $\mathcal{A}, \mathcal{B}, \mathcal{C} \in K$.*

(a) *If* $\dim(\mathcal{A}) = 1$, *then* $I(\mathcal{A})$ *is m-complete* Π_2^0 *within* K.
(b) *If* $\dim(\mathcal{B}) > 1$, *then* $I(\mathcal{B})$ *is m-complete* d-Σ_2^0 *within* K.
(c) *Let* $\mathcal{C}$ *be of infinite dimension. Then* $I(\mathcal{A})$ *is m-complete* Π_3^0 *within* K.

Archimedean ordered fields are isomorphic to subfields of the reals. They are determined by the Dedekind cuts that are filled. It follows that for any computable Archimedean ordered field $\mathcal{A}$, the index set $I(\mathcal{A})$ is Π_3^0. It is enough to show that $\mathcal{A}$ has a computable Π_3 Scott sentence. We have a computable Π_2 sentence σ_0 characterizing the Archimedean ordered fields. For each $a \in \mathcal{A}$, we have a computable Π_1 formula $c_a(x)$ saying that x is in the cut corresponding to a — we take the conjunction of a c.e. set of formulas saying $q < x < r$, for rationals q, r such that $\mathcal{A} \models q < a < r$. Let σ_1 be $\bigwedge_a (\exists x)\, c_a(x)$, and let σ_2 be $(\forall x)\ \bigvee_a c_a(x)$. The conjunction of σ_0, σ_1, and σ_2 is a Scott sentence, which we may take to be computable Π_3.

Theorem 22 ([29]). *Let K be the class of Archimedean real closed ordered fields, and let $\mathcal{A}$ be a computable member of K.*

(a) *If the transcendence degree of $\mathcal{A}$ is 0 (i.e., $\mathcal{A}$ is isomorphic to the ordered field of algebraic reals), then $I(\mathcal{A})$ is m-complete Π_2^0 within K.*

(b) *If the transcendence degree of $\mathcal{A}$ is finite but not 0, then $I(\mathcal{A})$ is m-complete d-Σ_2^0 within K.*

(c) *If the transcendence degree of $\mathcal{A}$ is infinite, then $I(\mathcal{A})$ is m-complete Π_3^0 within K.*

For a computable member $\mathcal{A}$ of K, to show that $I(\mathcal{A})$ is Π_2^0, we show that there is a computable Π_2 Scott sentence. We take the conjunction of a sentence characterizing the real closed ordered fields, and a sentence saying that each element is a root of some polynomial.

For reduced abelian p-groups of length $< \omega^2$, we have the following result.

Theorem 23 ([29]). *Let K be the class of reduced Abelian p-groups of length $\omega M + N$ for some $M, N \in \omega$. Let $\mathcal{A} \in K$.*

(a) *If $\mathcal{A}_{\omega M}$ is minimal for the given length (of the form $\mathbb{Z}_{p^N}$), then $I(\mathcal{A})$ is m-complete Π_{2M+1}^0 within K.*

(b) *If $\mathcal{A}_{\omega M}$ is finite but not minimal for the given length, then $I(\mathcal{A})$ is m-complete d-Σ_{2M+1}^0 within K.*

(c) *If there is a unique $k < N$ such that $u_{\omega M+k}(\mathcal{A}) = \infty$, and for all $m < k$, $u_{\omega M+m}(\mathcal{A}) = 0$, then $I(\mathcal{A})$ is m-complete Π_{2M+2}^0 within K.*

(d) *If there is a unique $k < N$ such that $u_{\omega M+k}(\mathcal{A}) = \infty$ and for some $m < k$ we have $0 < u_{\omega M+m}(\mathcal{A}) < \infty$, then $I(\mathcal{A})$ is m-complete d-Σ_{2M+2}^0 within K.*

(e) *If there exist $m < k < N$ such that $u_{\omega M+m}(\mathcal{A}) = u_{\omega M+k}(\mathcal{A}) = \infty$, then $I(\mathcal{A})$ is m-complete Π_{2M+3}^0 within K.*

The case when the length is ω was proved in [23].

Proof. Let K be the class of reduced Abelian p-groups of length ωM, and let $\mathcal{A} \in K$. Then $I(\mathcal{A})$ is m-complete Π^0_{2M+1} within K. Let $\mathcal{A} \in K$. First, we show that $\mathcal{A}$ has a computable Π_{2M+1} Scott sentence. There is a computable Π_2 sentence θ characterizing the Abelian p-groups. Next, there is a computable Π_{2M+1} sentence λ characterizing the groups which are reduced and have length at most ωM. For each $\alpha < \omega M$, we can find a computable Σ_{2M} sentence $\varphi_{\alpha,k}$ saying that $u_\alpha(\mathcal{A}) \geq k$. The set of these Σ_{2M} sentences true in $\mathcal{A}$ is Σ^0_{2M}. For each $\varphi_{\alpha,k}$, we can find a computable Π_{2M} sentence equivalent to the negation, and the set of these sentences true in $\mathcal{A}$ is Π^0_{2M}. We have a computable Π_{2M+1} sentence v equivalent to the conjunction of the sentences $\pm\varphi_{\alpha,k}$ true in $\mathcal{A}$. Then we have a computable Π_{2M+1} Scott sentence equivalent to $\theta \wedge \lambda \wedge v$. It follows that $I(\mathcal{A})$ is Π^0_{2M+1}.

For completeness, let S be a Π^0_{2M+1} set. We will produce a uniformly computable sequence $(\mathcal{A}_n)_{n\in\omega}$ of elements of K, such that $n \in S$ if and only if $\mathcal{A}_n$ is isomorphic to $\mathcal{A}$. $\square$

A group G is *free* if there is a set B of elements such that B generates G and there are no nontrivial relations on elements of B. We call B a *basis* for G. If B and U are two bases for a free group G, then B and U have the same cardinality. For a free group G, the cardinality of a basis is called the *rank*. We write F_n for the free group of rank n, and F_∞ for the free group of rank $\aleph_0$. The groups F_n and F_∞ all have computable copies. Sela in a series of seven papers 2001–06 gave a positive solution to the problem of elementary equivalence of free groups of different finite ranks greater than 1, posed by Tarski in the 1940s. (Also, see work of Kharlampovich and Myasnikov [99].) That is, all non-abelian free groups with finitely many generators have the same elementary first-order theory. Inspired by this result, we investigated free groups in the context of computable model theory [32, 121].

Theorem 24 ([32]). *Within the class of free groups:*

(a) $I(F_2)$ *is m-complete* Π_2^0.
(b) *For* $n > 2$, *the set* $I(F_n)$ *is m-complete* d-Σ_2^0.
(c) $I(F_\infty)$ *is m-complete* Π_3^0.

Theorem 25 ([32, 121]). *Within the class of all groups:*

(a) *For* $n \geq 1$, *the set* $I(F_n)$ *is m-complete* d-Σ_2^0.
(b) *The set* $I(F_\infty)$ *is m-complete* Π_4^0.

We also define complexity of one class "within" a larger class. This definition allows us to analyze situations where, for instance, determining whether an index is in B is harder than Γ, but *once we know* that the index is in B, the problem of determining whether it is also in A not harder than Γ.

Definition 3. Let Γ be a complexity class and let $A \subseteq B$.

(1) We say that A is Γ *within* B if there is some $C \in \Gamma$ such that $A = C \cap B$.
(2) We say that A is Γ-*hard within* B if for any set S in Γ, there is a computable function $f : \omega \to B$ such that $f(n) \in A$ if and only if $n \in S$.
(3) We say that A is *m-complete* Γ *within* B if A is Γ within B and A is Γ-hard within B.

Let *FinGen* denote the class of all finitely generated groups.

Theorem 26. (a) *The set* $I(FinGen)$ *is m-complete* Σ_3^0 *within the class of free groups.*

(b) *The set* $I(FinGen)$ *is m-complete* Σ_3^0 *within the class of all groups.*

(c) *The set* $I(LocFr)$ *is m-complete* Π_2^0 *within the class of all groups.*

4. Relatively Δ_α^0-categorical Structures

The complexity of isomorphisms between a computable structure and its isomorphic copies can be of various complexity.

The main notion in this area of investigation is that of computable categoricity. A computable structure $\mathcal{M}$ is *computably categorical* if for every computable structure $\mathcal{A}$ isomorphic to $\mathcal{M}$, there exists a computable isomorphism from $\mathcal{M}$ onto $\mathcal{A}$. In [114], Mal'cev considered the notion of a recursively (computably) stable structure. A computable structure $\mathcal{M}$ is *computably stable* if every isomorphism from $\mathcal{M}$ to another computable structure is computable. In Ershov [60], Mal'cev investigated the notion of *autostability* of structures, which is equivalent to that of computably categoricity. Since then computable categoricity has been studied extensively. It has been extended to arbitrary levels of hyperarithmetical hierarchy, and more precisely to Turing degrees $\mathbf{d}$. Computable categoricity of a computable structure $\mathcal{M}$ can also be relativized to all (including noncomputable) structures $\mathcal{A}$ isomorphic to $\mathcal{M}$.

Definition 4. A computable structure $\mathcal{M}$ is $\mathbf{d}$-*computably categorical* if for every computable structure $\mathcal{A}$ isomorphic to $\mathcal{M}$, there exists a $\mathbf{d}$-computable isomorphism from $\mathcal{M}$ onto $\mathcal{A}$.

In the case when $\mathbf{d} = \mathbf{0}^{(n-1)}$, $n \geq 1$, we also say that $\mathcal{M}$ is Δ_n^0-*categorical*. Thus, computably categorical is the same as $\mathbf{0}$-computably categorical or Δ_1^0-categorical. We can similarly define Δ_α^0-categorical structures for any computable ordinal α.

Computably categorical structures tend to be very particular. For a structure in a typical algebraic class, being computably categorical is usually equivalent to having a finite basis or a finite generating set (such as in the case of a vector space), or to being highly homogeneous (such as in the case of a random graph). For example, Ershov established that a computable algebraically closed field is computably categorical if and only if it has a finite transcendence degree over its prime subfield. Goncharov, Lempp, and Solomon [73] proved that a computable, ordered, abelian group is computably categorical if and only if it has finite rank. Similarly, they showed that a computable, ordered, Archimedean group is computably categorical if and only if it has finite rank.

An *injection structure* $\mathcal{A} = (A, f)$ consists of a nonempty set A and an one-to-one function $f : A \to A$. Given $a \in A$, the *orbit* $O_f(a)$ of a under f is $\{b \in A : (\exists n \in \mathbb{N})[f^n(a) = b \vee f^n(b) = a]\}$. An injection structure (A, f) may have two types of infinite orbits: $\mathbb{Z}$-orbits, which are isomorphic to $(\mathbb{Z}, S)$, and ω-orbits, which are isomorphic to (ω, S). Cenzer, Harizanov, and Remmel [33] characterized computably categorical injection structures as those that have finitely many infinite orbits.

On the other hand, Miller and Schoutens [130] constructed a computable field that has *infinite* transcendence degree over the rationals, yet is computably categorical. Their idea uses a computable set of rational polynomials called Fermat polynomials to "tag" elements of a transcendence basis. Hence their field has an infinite computable transcendence basis that is computable in every isomorphic computable copy of the field, and with each single element effectively distinguishable from the others.

We can relativize the notion of Δ^0_α-categoricity by studying the complexity of isomorphisms from a computable structure to any countable isomorphic structure.

Definition 5. A computable structure $\mathcal{M}$ is *relatively Δ^0_α-categorical* if for every $\mathcal{A}$ isomorphic to $\mathcal{M}$, there is an isomorphism from $\mathcal{M}$ to $\mathcal{A}$, which is Δ^0_α relative to the atomic diagram of $\mathcal{A}$.

Clearly, a relatively Δ^0_α-categorical structure is Δ^0_α-categorical.

A remarkable feature of relative Δ^0_α-categoricity is that it admits a syntactic characterization. This characterization involves the existence of certain effective Scott families. Scott families come from the Scott Isomorphism Theorem. A *Scott family* for a structure $\mathcal{A}$ is a countable family Φ of $L_{\omega_1\omega}$-formulas with finitely many fixed parameters from A such that:

(1) each finite tuple in $\mathcal{A}$ satisfies some $\psi \in \Phi$;
(2) if $\overrightarrow{a}$, $\overrightarrow{b}$ are tuples in $\mathcal{A}$, of the same length, satisfying the same formula in Φ, then there is an automorphism of $\mathcal{A}$, which maps $\overrightarrow{a}$ to $\overrightarrow{b}$.

If we strengthen condition (1) to require that the formulas in Φ define each tuple in $\mathcal{A}$, then Φ is called a *defining family* for $\mathcal{A}$. A *formally* Σ_α^0 *Scott family* is a Σ_α^0 Scott family consisting of computable Σ_α formulas. In particular, it follows that a formally c.e. Scott family is a c.e. Scott family consisting of finitary existential formulas. The following equivalence was established by Goncharov [65] for $\alpha = 1$, and by Ash, Knight, Manasse, and Slaman [12] and independently by Chisholm [37] for any computable ordinal α.

Theorem 27 ([12, 37]). *The following conditions are equivalent for a computable structure $\mathcal{A}$.*

(i)   *The structure $\mathcal{A}$ is relatively Δ_α^0-categorical.*
(ii) *The structure $\mathcal{A}$ has a formally Σ_α^0 Scott family Φ with finitely many fixed parameters.*
(iii) *The structure $\mathcal{A}$ has a c.e. Scott family consisting of computable Σ_α formulas with finitely many fixed parameters.*

For example, consider a computable equivalence structure $\mathcal{A}$, a computable set with a single equivalence relation. If $\mathcal{A}$ has a bound on the size of its finite equivalence classes, then we can show that $\mathcal{A}$ is relatively Δ_2^0-categorical. Let k be the maximum size of any finite equivalence class. Then $[a]$ is infinite if and only if $[a]$ contains at least $k + 1$ elements, which is a Σ_1^0 condition. There is a Δ_2^0 formula that characterizes the elements a with a finite equivalence class of size m. Then a Scott formula for the tuple $(a_1, \ldots, a_m)$ includes a formula $\psi_i(x_i)$ for each a_i, giving the cardinality of $[a_i]$, together with formulas $\psi_{i,j}(x_i, x_j)$ for each i, j, which express whether $a_i E^\mathcal{A} a_j$ and whether $a_i = a_j$. Moreover, every computable equivalence structure is relatively Δ_3^0-categorical since every element with an infinite equivalence class has a Π_2 Scott formula, while the other elements even have Δ_2 Scott formulas. Thus, every finite tuple has a Σ_3 Scott formula.

A structure is *rigid* if it does not have nontrivial automorphisms. A computable structure is Δ_α^0-*stable* if every isomorphism from $\mathcal{A}$ onto a computable structure is Δ^0_α. If a computable structure is rigid and Δ^0_α-categorical, then it is Δ^0_α-stable. A *defining family* for a structure $\mathcal{A}$ is a set Φ of formulas with one free variable and a fixed finite tuple of parameters from A such that:

(1) every element of A satisfies some formula $\psi \in \Phi$;
(2) no formula of Φ is satisfied by more than one element of A.

For a rigid computable structure $\mathcal{A}$, there is a formally Σ^0_α Scott family if and only if there is a formally Σ^0_α defining family.

Let us recall the definition of a Fraïssé limit. The *age* of a structure $\mathcal{M}$ is the class of all finitely generated structures that can be embedded in $\mathcal{M}$. Fraïssé showed that a (nonempty) finite or countable class $\mathbb{K}$ of finitely generated structures is the age of a finite or a countable structure if and only if $\mathbb{K}$ has the hereditary property and the joint embedding property. A class $\mathbb{K}$ has the *hereditary property* if whenever $\mathcal{C} \in \mathbb{K}$ and $\mathcal{S}$ is a finitely generated substructure of $\mathcal{C}$, then $\mathcal{S}$ is isomorphic to some structure in $\mathbb{K}$. A class $\mathbb{K}$ has the *joint embedding property* if for every $\mathcal{B}, \mathcal{C} \in \mathbb{K}$ there is $\mathcal{D} \in \mathbb{K}$ such that $\mathcal{B}$ and $\mathcal{C}$ embed into $\mathcal{D}$. A structure $\mathcal{U}$ is *ultrahomogeneous* if every isomorphism between finitely generated substructures of $\mathcal{U}$ extends to an automorphism of $\mathcal{U}$. A structure $\mathcal{A}$ is a *Fraïssé limit* of a class of finitely generated structures $\mathbb{K}$ if $\mathcal{A}$ is countable, ultrahomogeneous, and has age $\mathbb{K}$. Fraïssé proved that the Fraïssé limit of a class of finitely generated structures is unique up to isomorphism. We say that a structure $\mathcal{A}$ is a Fraïssé limit if for some class $\mathbb{K}$, $\mathcal{A}$ is the Fraïssé limit of $\mathbb{K}$.

Theorem 28 ([1, 55]). *Let $\mathcal{A}$ be a computable structure, which is a Fraïssé limit. Then $\mathcal{A}$ is relatively Δ^0_2-categorical.*

Proof. Because of ultrahomogeneity, we can construct isomorphisms between $\mathcal{A}$ and an isomorphic structure $\mathcal{B}$ using a back-and-forth argument, as long as we can determine for every two sequences $\overrightarrow{a}$ and $\overrightarrow{b}$ of the same length of elements in $\mathcal{A}$ and $\mathcal{B}$, respectively, whether there is an isomorphism from the structure generated by $\overrightarrow{a}$ to the structure generated by $\overrightarrow{b}$, which maps $\overrightarrow{a}$ to $\overrightarrow{b}$ in order. This can be determined by $(D(\mathcal{B}))'$, since there is such an isomorphism precisely if there is no atomic formula ϕ with $\mathcal{A} \models \phi(\overrightarrow{a})$ and $\mathcal{B} \not\models \phi(\overrightarrow{b})$. This is a Π_1^0 condition relative to $\mathcal{A} \oplus \mathcal{B} \equiv_T \mathcal{B}$.

Therefore, we can use $(D(\mathcal{B}))'$ as an oracle to perform the back-and-forth construction of an isomorphism, and so there is an isomorphism that is Δ_2^0 relative to $\mathcal{B}$. $\qquad\square$

Adams and Cenzer [1] defined a structure $\mathcal{A}$ to be *weakly ultrahomogeneous* if there is a finite sequence of elements $\overrightarrow{a}$ from its domain such that $(\mathcal{A}, \overrightarrow{a})$ becomes ultrahomogeneous in the language extended by constants representing these elements. Adams and Cenzer [1] proved that every computable weakly ultrahomogeneous structure is relatively Δ_2^0-categorical. They also proved that every computable, relational, weakly ultrahomogeneous structure is relatively computably categorical. Hence every computable weakly ultrahomogeneous graph is computably categorical, but there are computably categorical graphs that are not weakly ultrahomogeneous.

Theorem 29 ([1]). (a) *A computable linear ordering is weakly homogeneous if and only if it is relatively computably categorical.*

(b) *A computable equivalence structure is weakly homogeneous if and only if it is relatively computably categorical.*

(c) *For a computable injection structure, computable categoricity implies weak ultrahomogeneity, which implies relative Δ_2^0-categoricity, but neither implication can be reversed.*

Goncharov and Dzgoev, and independently Remmel characterized computably categorical linear orderings in terms of the number of successor pairs (also called adjacencies). Similarly, they and also LaRoche (independently) characterized computably categorical Boolean algebras.

Theorem 30. (a) ([66, 137]) *A computable linear ordering is computably categorical if and only if it has only finitely many successor pairs.*

Every computably categorical linear ordering is relatively computably categorical.

(b) ([66, 111, 138]) *A computable Boolean algebra is computably categorical if and only if it has finitely many atoms.*

Every computably categorical Boolean algebra is relatively computably categorical.

Goncharov [61] and Smith [144] independently characterized computably categorical abelian p-groups. By $\mathbb{Z}(p^n)$ we denote the cyclic group of order p^n, and by $\mathbb{Z}(p^\infty)$ the quasicyclic (Prüfer) abelian p-group. Related to abelian p-groups are equivalence structures. Calvert, Cenzer, Harizanov, and Morozov [24] characterized computably categorical equivalence structures.

Theorem 31. (a) ([61, 144]) *An abelian p-group is computably categorical if and only if it can be written in one of the following forms: $(\mathbb{Z}(p^\infty))^l \oplus F$ for $l \in \omega \cup \{\infty\}$ and F is a finite group, or $(\mathbb{Z}(p^\infty))^n \oplus H \oplus (\mathbb{Z}(p^k))^\infty$, where $n, k \in \omega$ and H is a finite group.*

Every computably categorical abelian p-group is relatively computably categorical.

(b) [24] *A computable equivalence structure $\mathcal{A}$ is computably categorical if and only if either $\mathcal{A}$ has finitely many finite equivalence classes, or $\mathcal{A}$ has finitely many infinite classes, upper bound on the size of finite classes, and exactly one finite k with infinitely many classes of size k.*

Every computably categorical equivalence structure is relatively computably categorical.

Lempp, McCoy, Miller, and Solomon [112] characterized computably categorical trees of finite height and showed that they are relatively computably categorical. R. Miller [128] previously established that no computable well-founded tree of infinite height is computably categorical. Equivalence structures can be generalized to allow for more than one equivalence relation on the universe. For finite $n \geq 2$, an n-equivalence structure is a structure $\mathcal{A} = (A, E_1, \ldots, E_n)$ where each E_i is an equivalence relation on A. An n-equivalence structure is nested if for $i < j \leq n$ we have $x E_j y \Rightarrow x E_i y$, i.e., $E_j \subseteq E_i$ as subsets of $A \times A$. For $a \in A$, we let $[a]_i$ denote the equivalence class of a under E_i. Thus for a nested equivalence structure, $i < j \leq n$ implies that $[a]_j \subseteq [a]_i$, so that the E_i classes are partitioned by E_j. There is also an equivalence relation $E_0 = A \times A$, so that $[a]_0 = A$ for all a.

In [118], Marshall described an effective correspondence between nested n-equivalence structures and certain trees of finite height where the branching of the tree reflects the containment of equivalence classes. This correspondence allows many effective properties to be transferred between nested n-equivalence structures and trees of finite height. More precisely, for any nested n-equivalence structure $\mathcal{A} = (A, E_1, \ldots, E_n)$, let E_{n+1} be the equality, and define the tree $T_{\mathcal{A}}$ as follows. The universe of $T_{\mathcal{A}}$ is the set $\{[a]_i : a \in A \wedge i = 1, \ldots, n\}$ and the partial ordering is inclusion. This means that for each a and $i \leq n$, $[a]_i$ is the predecessor of $[a]_{i+1}$. Marshall shows that a presentation of $T_{\mathcal{A}}$ can be computed from $\mathcal{A}$ so that the mapping from a to $[a]$ is also computable from $\mathcal{A}$.

Theorem 32 ([118]). *Let $\mathcal{A}$ be a computable nested n-equivalence structure and $T_{\mathcal{A}}$ its corresponding tree of finite height. Then the following conditions are equivalent.*

(i) *$\mathcal{A}$ is computably categorical.*
(ii) *$\mathcal{A}$ is relatively computably categorical.*
(iii) *$(T_{\mathcal{A}}, \prec)$ is computably categorical.*
(iv) *$(T_{\mathcal{A}}, \prec)$ is relatively computably categorical.*

Now, let us denote the predecessor function in $T_{\mathcal{A}}$ by f.

Theorem 33 ([1]). *The following conditions are equivalent.*

(i) $\mathcal{A}$ *is weakly ultrahomogeneous.*

(ii) $(T_{\mathcal{A}}, f)$ *is weakly ultrahomogeneous.*

Cenzer, Harizanov, and Remmel [33] established that computably categorical injection structures are also relatively computably categorical. Miller and Shlapentokh [129] proved that a computable algebraic field F with a splitting algorithm is computably categorical if and only if it is decidable which pairs of elements of F belong to the same orbit under automorphisms. They also showed that this criterion is equivalent to relative computable categoricity of F.

Goncharov [63] was the first to show that computable categoricity of a computable structure does not imply relative computable categoricity. The main idea of his proof was to code a special kind of family of sets into a computable structure. Such families were constructed independently by Badaev [15] and Selivanov [143]. Hirschfeldt, Khoussainov, Shore, and Slinko [94] established a general result that implies that there are computably categorical but not relatively computably categorical structures in the following classes: partial orderings, lattices, 2-step nilpotent groups, commutative semigroups, and integral domains of arbitrary characteristic. Hirschfeldt, Kramer, Miller, and Shlapentokh [95] characterized relative computable categoricity for computable algebraic fields and used their characterization to construct a field with the following property.

Theorem 34 ([95]). *There is a computably categorical algebraic field, which is not relatively computably categorical.*

Infinitary language is essential for Scott families. Cholak, Shore, and Solomon [41] proved the existence of a computably categorical graph that does not have a Scott family of finitary formulas. It follows that this structure is not relatively computably categorical.

The result about the existence of computably categorical structures that are not relatively computably categorical was lifted to higher levels in the hyperarithmetical hierarchy by Goncharov, Harizanov, Knight, McCoy, Miller, and Solomon for successor ordinals [69], and by Chisholm, Fokina, Goncharov, Harizanov, Knight, and Quinn for limit ordinals [38]. It is an open question whether every Δ_1^1-categorical structure must be relatively Δ_1^1-categorical.

Theorem 35 ([38, 69]). *For every computable ordinal α, there is a Δ_α^0-categorical but not relatively Δ_α^0-categorical structure.*

There is a complete description of higher levels of categoricity (in fact, stability) for well-orderings due to Ash [9]. Harris [84] has a description of Δ_n^0-categorical Boolean algebras for any $n < \omega$. However, not enough is known about Δ_n^0-categoricity for $n \geq 2$ for structures from many natural classes of algebraic structures. The study of higher level categoricity often leads to the study of algebraic properties of a family of relations specific for a given class (such the independence relations or back-and-forth relations). Obtaining classification of categoricity is usually a difficult task. The reason is either the absence of algebraic invariants (such as for the linear orderings, and abelian and nilpotent groups), or the lack of suitable computability-theoretic notions that would capture the property of being Δ_n^0-categorical (such as in the case of Δ_2^0-categoricity of equivalence structures). Even for $n = 2$, the following problems remain open. Describe Δ_2^0-categorical linear orderings. Describe Δ_2^0-categorical equivalence relations. Describe Δ_2^0-categorical abelian p-groups. Describe Δ_2^0-categorical trees of finite height.

In [119], McCoy characterized relatively Δ_2^0-categorical linear orderings and Boolean algebras. In [120], McCoy gave a complete description of relatively Δ_3^0-categorical Boolean algebras. Frolov [57] found a Δ_3^0-categorical linear ordering that is not relatively Δ_3^0-categorical. In the following theorem, we state McCoy's characterizations of relatively Δ_2^0-categorical linear orderings and

Boolean algebras. As usual, by ω^* we denote the reverse order type of ω, and by η the order type of rationals.

Theorem 36 ([119]). (a) *A computable linear ordering is relatively Δ_2^0-categorical if and only if it is a sum of finitely many intervals, each of type $m, \omega, \omega^*, \mathbb{Z}$, or $n \cdot \eta$, so that each interval of type $n \cdot \eta$ has a supremum and infimum.*

(b) *A computable Boolean algebra is relatively Δ_2^0-categorical if and only if it can be expressed as a finite direct sum $c_1 \vee \cdots \vee c_n$, where each c_i is either atomless, an atom, or a 1-atom.*

Bazhenov [19] and Harris [84] independently showed that for Boolean algebras the notions of Δ_2^0-categoricity and relative Δ_2^0-categoricity coincide. It is not known whether every Δ_2^0-categorical linear ordering is relatively Δ_2^0-categorical.

Calvert, Cenzer, Harizanov, and Morozov characterized relative Δ_2^0-categoricity for equivalence structures [24] and abelian p-groups [23]. Recall that the length of an abelian p-group G, $\lambda(G)$, is the least ordinal α such that $p^{\alpha+1}G = p^\alpha G$. The divisible part of G is $\mathrm{Div}(G) = p^{\lambda(G)}G$ and is a direct summand of G. The group G is said to be reduced if $\mathrm{Div}(G) = \{0\}$. For a group G, the *period* of G is $\max\{\mathrm{order}(g) : g \in G\}$ if this quantity is finite, and ∞ otherwise.

Theorem 37. (a) ([24]) *A computable equivalence structure is relatively Δ_2^0-categorical if and only if it either has finitely many infinite equivalence classes, or there is an upper bound on the size of its finite equivalence classes.*

(b) ([23]) *A computable abelian p-group G is relatively Δ_2^0-categorical if and only if G is reduced and $\lambda(G) \leq \omega$, or G is isomorphic to $\bigoplus_\alpha \mathbb{Z}(p^\infty) \oplus H$, where $\alpha \leq \omega$ and H has finite period.*

Kach and Turetsky [98] showed that there exists a Δ_2^0-categorical equivalence structure, which is not relatively Δ_2^0-categorical. Downey, Melnikov, and Ng [48] built examples of abelian p-groups that show that the notions of

Δ_2^0-categoricity and relative Δ_2^0-categoricity do not coincide for these groups. Every computable equivalence structure is relatively Δ_3^0-categorical. There is no such bound for a computable abelian p-group G. For example, it follows from the index set results in [29] that if $\lambda(G) = \omega \cdot n$ and $m \leq 2n - 1$, or if $\lambda(G) > \omega \cdot n$ and $m \leq 2n - 2$, then G is not Δ_m^0-categorical. Barker [16] proved that for every computable ordinal α, there are $\Delta_{2\alpha+2}^0$-categorical but not $\Delta_{2\alpha+1}^0$-categorical abelian p-groups.

Cenzer, Harizanov, and Remmel [33] characterized relative Δ_2^0-categoricity for injection structures.

Theorem 38 ([33]). *A computable injection structure is relatively Δ_2^0-categorical if and only if it has finitely many orbits of type ω, or finitely many orbits of type $\mathbb{Z}$. Every Δ_2^0-categorical injection structure is relatively Δ_2^0-categorical.*

Every computable injection structure is relatively Δ_3^0-categorical.

In [34], Cenzer, Harizanov, and Remmel investigated computability-theoretic properties of a computable structure (A, f) with a single unary function f such that for every x in the preimage, $f^{-1}(x)$ has exactly two elements, which is called a 2:1 structure. Every computable 2:1 structure is Δ_2^0-categorical.

Theorem 39 ([34]). *A 2:1 structure is computably categorical if and only if it has finitely many $\mathbb{Z}$-chains.*

Structures for which $f^{-1}(x)$ has either exactly two or zero elements are called $(2,0){:}1$ structures. We can identify (A, f) with its directed graph $G(A, f)$, which has vertex set A and where the edge set consists of all pairs $(i, f(i))$ for $i \in A$. Given $a \in A$, we let the orbit $\mathcal{O}_A(a)$ consist of the set of all points in A, which lie in the connected component of $G(A, f)$ containing a. We let $\mathrm{tree}_A(a) =_{\mathrm{def}} \{y \in A : (\exists n)(f^n(y) = x)\}$. We say that a $(2,0){:}1$ structure (A, f) is *locally finite* if $\mathrm{tree}_A(a)$ is finite for all $a \in A$. Every computable locally finite $(2,0){:}1$ structure is Δ_3^0-categorical. Cenzer, Harizanov, and Remmel [34] proved that every computable locally finite $(2,0)$:1 structure with finitely many ω-chains is Δ_2^0-categorical. Walker [150, 151] extended this investigation to $(2,1)$:1 structures where for every x, the preimage $f^{-1}(x)$ has either two or one element.

There is no known characterization of Δ_2^0-categoricity or of higher level categoricity for trees of finite height. Lempp, McCoy, Miller, and Solomon [112] proved that for every $n \geq 1$, there is a computable tree of finite height, which is Δ_{n+1}^0-categorical but not Δ_n^0-categorical. Fokina, Harizanov, and Turetsky established the following result, which also holds when a tree is presented as a directed graph.

Theorem 40 ([55]). *There is a Δ_2^0-categorical tree of finite height, which is not relatively Δ_2^0-categorical. There is also such a tree of infinite height.*

It follows from [32, 121] that every computable, free, non-abelian group is Δ_4^0-categorical, and the result cannot be improved to Δ_3^0. It was shown in [47] that every computable, free, abelian group is Δ_2^0-categorical, and the result cannot be improved to computable categoricity.

A *homogeneous, completely decomposable, abelian group* is a group of the form $\bigoplus_{i \in \kappa} H$, where H is a subgroup of the additive group of the rationals, $(\mathbb{Q}, +)$. Note that we have only a single H in the sum — any two summands are isomorphic. It is well known that such a group is computably categorical if and only if κ is finite; the proof is similar to the analogous result that a computable vector space is computably categorical if and only if it has finite dimension.

For P a set of primes, define $Q^{(P)}$ to be the subgroup of $(\mathbb{Q}, +)$ generated by $\{\frac{1}{p^k} : p \in P \wedge k \in \omega\}$. Downey and Melnikov [47] showed that a computable, homogeneous, completely decomposable, abelian group of infinite rank is Δ_2^0-categorical if and only if it is isomorphic to $\bigoplus_\omega Q^{(P)}$, where P is c.e. and the set $(\text{Primes} - P)$ is semi-low. Recall that a set $S \subseteq \omega$ is *semi-low* if the set $H_S = \{e : W_e \cap S \neq \emptyset\}$ is computable from $\emptyset'$. In [55], we

proved that a computable, homogeneous, completely decompos-
able, abelian group of infinite rank is relatively Δ_2^0-categorical if
and only if it is isomorphic to $\bigoplus_\omega Q^{(P)}$, where P is a computable
set of primes. Since there exist co-c.e. sets that are semi-low and
noncomputable, we obtained the following result.

Theorem 41. *There is a homogeneous, completely decomposable, abelian group, which is Δ_2^0-categorical but not relatively
Δ_2^0-categorical.*

The notions of computable categoricity and relative computable categoricity coincide if we add more effectiveness
requirements on the structure. Goncharov [65] proved that in
the case of 2-decidable structures, computable categoricity and
relative computable categoricity coincide. Kudinov showed that
the assumption of 2-decidability cannot be weakened, by giving in [110] an example of 1-decidable and computably categorical structure, which is not relatively computably categorical.
Recently, Fokina, Harizanov and Turetsky obtained such an
example of a Fraïssé limit.

Theorem 42 ([55]). *There is a 1-decidable structure $\mathcal{F}$ that
is a Fraïssé limit and computably categorical, but not relatively
computably categorical. Moreover, the language for such $\mathcal{F}$ can
be finite or it can be relational.*

Ash [8] established that for every computable ordinal α,
under certain decidability conditions on $\mathcal{A}$, if $\mathcal{A}$ is Δ_α^0-
categorical, then $\mathcal{A}$ is relatively Δ_α^0-categorical.

Millar [123] proved that if a structure $\mathcal{A}$ is 1-decidable,
then any expansion of $\mathcal{A}$ by finitely many constants remains
computably categorical. Cholak, Goncharov, Khoussainov, and
Shore showed that the assumption of 1-decidability is important.
They showed that there is a computable structure, which is computably categorical, but ceases to be after naming any element
of the structure. It follows that this structure is not relatively
computably categorical since it cannot have a formally c.e. Scott

family. Furthermore, Khoussainov and Shore [102] proved that there is a computably categorical structure $\mathcal{A}$ without a formally c.e. Scott family such that the expansion of $\mathcal{A}$ by any finite number of constants is computably categorical.

Downey, Kach, Lempp, and Turetsky have obtained the following result.

Theorem 43 ([46]). *Any 1-decidable computably categorical structure is relatively Δ_2^0-categorical.*

Based on Theorem 43, we could conjecture that every computable structure that is computably categorical should be relatively Δ_3^0-categorical. However, this is not the case, as proved by Downey, Kach, Lempp, Lewis, Montalbán, and Turetsky.

Theorem 44 ([45]). *For every computable ordinal α, there is a computably categorical structure that is not relatively Δ_α^0-categorical.*

Thus, a natural question arises whether there is a computably categorical structure that is not relatively hyperarithmetically categorical.

Downey, Kach, Lempp, and Turetsky [46] established the following index set complexity result for relatively computably categorical structures.

Theorem 45 ([46]). *The index set of relatively computably categorical structures is Σ_3^0-complete.*

On the other hand, in [45], Downey, Kach, Lempp, Lewis, Montalbán, and Turetsky established that there is no simple syntactic characterization of computable categoricity, thus answering a long-standing open question.

Theorem 46 ([45]). *The index set of computably categorical structures is Π_1^1-complete.*

Goncharov also investigated categoricity restricted to decidable structures (for example, see [59]).

Definition 6. A decidable structure $\mathcal{A}$ is called *decidably categorical* if every two decidable copies of $\mathcal{A}$ are computably isomorphic.

Nurtazin gave the following characterization of decidably categorical structures. Recall that for a complete theory T, a formula $\theta(\overrightarrow{x})$ is called *complete* if for every formula $\psi(\overrightarrow{x})$, either $T \vdash (\theta(\overrightarrow{x}) \Rightarrow \psi(\overrightarrow{x}))$ or $T \vdash (\theta(\overrightarrow{x}) \Rightarrow \neg\psi(\overrightarrow{x}))$.

Theorem 47 ([134]). *Let $\mathcal{A}$ be a decidable structure. Then $\mathcal{A}$ is decidably categorical if and only if there is a finite tuple $\overrightarrow{c}$ of elements in A such that $(\mathcal{A}, \overrightarrow{c})$ is a prime model of the theory $Th(\mathcal{A}, \overrightarrow{c})$ and the set of complete formulas of this theory is computable.*

Moreover, Nurtazin proved that if there is no such $\overrightarrow{c}$, then there are infinitely many decidable copies of $\mathcal{A}$, no two of which are computably isomorphic.

Goncharov and Marchuk [74] showed that the index set of computable structures with decidably categorical copies is $\Sigma^0_{\omega+3}$-complete, while for decidably categorical structures, the index set is Σ^0_3-complete. Index sets for decidably categorical structures with particular properties were further investigated by Bazhenov, Goncharov, and Marchuk.

We say that a structure $\mathcal{A}$ is *categorical relative to n-decidable presentations* if any two n-decidable copies of $\mathcal{A}$ are computably isomorphic. For $n = 0$, we have a computably categorical structure. Fokina, Goncharov, Harizanov, Kudinov, and Turetsky investigated for various $m, n \in \omega$, the index sets $I_{n,m}$ for n-decidable structures categorical relative to m-decidable presentations.

Theorem 48 ([54]). (a) *In the case when $m \geq n \geq 0$, the index set $I_{n,m}$ is Π^1_1-complete.*

(b) *In the case when $m = n - 1 \geq 0$, the index set $I_{n,m}$ is Π^0_4-complete.*

(c) *In the case when $0 \leq m \leq n - 2$, the index set $I_{n,m}$ is Σ^0_3-complete.*

5. Definability and Complexity of Relations on Structures

One of the important questions in computable model theory is how a specific property of a computable structure may change if the structure is isomorphically transformed so that it remains computable. A computable property of a computable structure $\mathcal{A}$, which Ash and Nerode [13] considered, is given by an additional (new) computable relation R on the domain A of $\mathcal{A}$. (That is, R is not named in the language of $\mathcal{A}$.) Ash and Nerode investigated syntactic conditions on $\mathcal{A}$ and R under which for every isomorphism f from $\mathcal{A}$ onto a computable structure $\mathcal{B}$, $f(R)$ is c.e. Such relations are called *intrinsically c.e.* on $\mathcal{A}$. In general, we have the following definition. Let $\mathcal{P}$ be a certain complexity class.

Definition 7 ([13]). An additional relation R on the domain of a computable structure $\mathcal{A}$ is called *intrinsically $\mathcal{P}$* on $\mathcal{A}$ if the image of R under every isomorphism from $\mathcal{A}$ to a computable structure belongs to $\mathcal{P}$.

For example, the successor relation, and being an even number are not intrinsically computable relations on $(\omega, <)$. Clearly, if $\mathcal{A}$ is a computably stable structure, then every computable relation on its domain is intrinsically computable.

If R is definable in $\mathcal{A}$ by a computable Σ_1 formula with finitely many parameters, then R is intrinsically c.e. Ash and Nerode [13] proved that, under a certain extra decidability condition on $\mathcal{A}$ and R, the relation R is intrinsically c.e. on $\mathcal{A}$ if and only if R is definable by a computable Σ_1 formula with finitely many parameters. The *Ash–Nerode decidability condition* says that for an m-ary relation R, there is an algorithm that determines for every existential formula $\psi(x_0, \ldots, x_{m-1}, \overrightarrow{y})$ and every $\overrightarrow{c} \in A^{lh(\overrightarrow{y})}$, whether the following implication holds for every $\overrightarrow{a} \in A^m$:

$$(\mathcal{A} \vDash \psi(\overrightarrow{a}, \overrightarrow{c})) \Rightarrow R(\overrightarrow{a}).$$

Barker [17] lifted the Ash–Nerode theorem to arbitrary levels of the hyperarithmetical hierarchy. He proved that for a structure $\mathcal{A}$ and an additional relation R on $\mathcal{A}$, under some effectiveness conditions, R is definable by a computable Σ_α formula with finitely many parameters.

For the relative notions, the effectiveness conditions are not needed. Let $\mathcal{P}$ be a certain complexity class, which can be relativized, such as the class of all Σ_α^0 sets.

Definition 8. An additional relation R on the domain of a computable structure $\mathcal{A}$ is called *relatively intrinsically $\mathcal{P}$ on $\mathcal{A}$* if the image of R under every isomorphism from $\mathcal{A}$ to any structure $\mathcal{B}$ is $\mathcal{P}$ relative to the atomic diagram of $\mathcal{B}$.

The following equivalence is due to Ash, Knight, Manasse, and Slaman [12], and independently Chisholm [37].

Theorem 49 ([12, 37]). *Let $\mathcal{A}$ be a computable structure. An additional relation R on $\mathcal{A}$ is relatively intrinsically Σ_α^0 if and only if R is definable by a computable Σ_α formula with finitely many parameters.*

A relation R on a structure $\mathcal{A}$ that is definable by a computable Σ_α formula with finitely many parameters is also called *formally Σ_α^0 on $\mathcal{A}$*.

Goncharov [63] and Manasse [116] gave examples of intrinsically c.e. relations on computable structures, which are not relatively intrinsically c.e. This result was lifted to higher levels of the hyperarithmetical hierarchy by Goncharov, Harizanov, Knight, McCoy, Miller, and Solomon for successor ordinals [69], and by Chisholm, Fokina, Goncharov, Harizanov, Knight, and Quinn for limit ordinals [38].

Theorem 50 ([38, 69]). *For every computable ordinal α, there is a computable structure $\mathcal{A}$ with an intrinsically Σ_α^0 relation R such that R is not definable by a computable Σ_α formula with finitely many parameters.*

We will assume that $\mathcal{A}$ is an infinite computable structure, and that R is an additional infinite co-infinite relation on $\mathcal{A}$. Without loss of generality, we assume that R is unary. We are interested in syntactic conditions under which there is a computable copy of $\mathcal{A}$ in which the image of R is simple. We may also ask when the image of $\neg R$ is only immune.

A subset of ω is called *immune* if it is infinite and contains no infinite c.e. subset. A set is *simple* if it is c.e. and its complement is immune. It is established in computability theory that a unary relation C on ω is *hyperimmune*, abbreviated by *h-immune*, if and only if it is infinite and no computable function majorizes its principal function p_C, where $p_C(n) =_{\text{def}} c_n$ provided that $C = \{c_0 < c_1 < c_2 < \cdots\}$. A set is called *hypersimple*, abbreviated by *h-simple*, if it is c.e. and its complement is hyperimmune. The following (canonical) indexing of finite sets is standard. Let $D_0 =_{\text{def}} \emptyset$. For $m > 0$, let $D_m = \{d_0, \ldots, d_{k-1}\}$, where $d_0 < \cdots < d_{k-1}$ and $m = 2^{d_0} + \cdots + 2^{d_{k-1}}$. A sequence $(U_i)_{i \in \omega}$ of finite sets is a *strong array* if there is a unary computable function f such that for every $i \in \omega$, $U_i = D_{f(i)}$. A strong array is *disjoint* if its members are pairwise disjoint. Let $S \subseteq \omega$. The relation $\neg S$ is *h-immune* (on ω) if it is infinite and there is no disjoint strong array $(U_i)_{i \in \omega}$ such that for every $n \in \omega$, we have $U_i \cap \overline{S} \neq \emptyset$. We can similarly define h-immune relations on any computable set. Every h-immune set is *immune*, that is, infinite but without any infinite c.e. subset. Not every immune set is h-immune.

Results establishing various equivalences of syntactic and corresponding semantic conditions in computable copies of $\mathcal{A}$ usually involve additional effectiveness conditions, expressed in terms of $\mathcal{A}$ and R. To discover syntactic conditions governing the algorithmic properties of images of R in computable copies of $\mathcal{A}$, it is sometimes helpful to consider arbitrary copies of $\mathcal{A}$ and relative versions of the algorithmic properties. One advantage is that we may use the forcing method instead of the priority method—the latter is more complicated. In addition, the relative results should require no additional effectiveness conditions, which often mask the syntactic conditions.

A new relation on a countable structure $\mathcal{B}$ is *immune relative to $\mathcal{B}$* if it is infinite and contains no infinite subset that is c.e. relative to $\mathcal{B}$. A new relation on a countable structure $\mathcal{B}$ is *simple relative to $\mathcal{B}$* if it is c.e. relative to $\mathcal{B}$ and its complement is immune relative to $\mathcal{B}$. If we are to construct an isomorphic copy of $\mathcal{A}$ in which the image of $\neg R$ is relatively immune, there must be no infinite subset D of $\neg R$ definable in $\mathcal{A}$ by a computable Σ_1 formula $\varphi(\vec{c}, x)$ (with a finite tuple of parameters $\vec{c}$). This obvious necessary condition turns out to be sufficient.

Theorem 51 ([68]). *Let $\mathcal{A}$ be a computable L-structure, and let R be a unary infinite and co-infinite relation on A. Then the following conditions are equivalent.*

(i) *For all copies $\mathcal{B}$ of $\mathcal{A}$ and all isomorphisms F from $\mathcal{A}$ onto $\mathcal{B}$, $\neg F(R)$ is not immune relative to $\mathcal{B}$.*

(ii) *There are an infinite set D and a finite tuple $\vec{c}$ such that $D \subseteq \neg R$ and D is definable in $\mathcal{A}$ by a computable Σ_1 formula $\varphi(\vec{c}, x)$.*

To prove (i) $\Rightarrow$ (ii) we build a "generic" copy $(\mathcal{B}, S)$ of $(\mathcal{A}, R)$. Under the assumption that $\neg S$, the image of $\neg R$, is not immune relative to $\mathcal{B}$, we produce the set D and a tuple $\vec{c}$ as in (ii). Let B be an infinite computable set, the universe of $\mathcal{B}$. The forcing conditions are the finite one-to-one partial functions from B to A.

Let $\mathcal{A}$ be an L-structure, and $\mathbf{R}$ be an additional unary relation symbol. If we are interested in c.e. relations, computable Σ_1 formulas with positive occurrences of $\mathbf{R}$ in the expanded language $L \cup \{\mathbf{R}\}$ play an important role. Assume that there is an infinite set $D \subseteq \neg R$ such that D is definable in $(\mathcal{A}, R)$ by a computable Σ_1 formula with finitely many parameters and with only positive occurrences of $\mathbf{R}$. In any copy $\mathcal{B}$ of $\mathcal{A}$, if the image of R is c.e. relative to $\mathcal{B}$, then so is the image of D. Therefore, under this definability assumption, the image of R cannot be made simple relative to $\mathcal{B}$. It turns out that this is the only obstacle.

Theorem 52 ([68]). *Let $\mathcal{A}$ be an infinite computable structure in a relational language L, and let R be a computable unary infinite and co-infinite relation on A. Then the following conditions are equivalent.*

(i) *For all copies $\mathcal{B}$ of $\mathcal{A}$ and all isomorphisms F from $\mathcal{A}$ onto $\mathcal{B}$, $F(R)$ is not simple relative to $\mathcal{B}$.*

(ii) *There are an infinite set D and a finite tuple of parameters $\vec{c}$ such that $D \subseteq \neg R$, and D is definable in $(\mathcal{A}, R)$ by a computable Σ_1 formula $\varphi(\vec{c}, x)$ of $L \cup \{\mathbf{R}\}$ with only positive occurrences of $\mathbf{R}$.*

The following results give syntactic conditions that allow the existence of an isomorphism F from $\mathcal{A}$ onto a *computable* copy such that $\neg F(R)$ is immune (or simple). The results involve extra decidability conditions, which imply that both $\mathcal{A}$ and R are computable.

Theorem 53 ([68]). *Let $\mathcal{A}$ be an infinite (computable) L-structure, and let R be a unary (computable) infinite and co-infinite relation on A. Assume that we have an effective procedure for deciding whether*

$$(\mathcal{A}_A, R) \models (\exists x \in \mathbf{R})\, \theta(\vec{c}, x),$$

where $\theta(\vec{c}, x)$ is a finitary existential formula of L with finitely many parameters. If there is no infinite set D such that $D \subseteq \neg R$ and D is definable in $\mathcal{A}$ by a computable Σ_1 formula of L with finitely many parameters, then there is an isomorphism F from $\mathcal{A}$ onto a computable copy $\mathcal{B}$ such that the relation $\neg F(R)$ is immune.

The proof uses the finite injury priority method.

Theorem 54 ([68]). *Let $\mathcal{A}$ be an infinite (computable) L-structure, and let R be a unary (computable) infinite and co-infinite relation on A. Assume that we have an effective procedure for deciding whether*

$$(\mathcal{A}_A, R) \models (\exists x \in \mathbf{R})\, \varphi(\vec{c}, x),$$

where φ is a finitary existential formula in $L \cup \{\mathbf{R}\}$ with finitely many parameters and with positive occurrences of $\mathbf{R}$. If there is no infinite $D \subseteq \neg R$ definable by such a formula, then there is an isomorphism F from $\mathcal{A}$ onto a computable copy $\mathcal{B}$ such that $F(R)$ is simple.

For a finite sequence (tuple) of elements $\vec{c}$, we write $a \in \vec{c}$ to say that $a \in \text{ran}(\vec{c})$, and $\vec{c} \cap \vec{d} = \emptyset$ to denote that $\text{ran}(\vec{c}) \cap \text{ran}(\vec{d}) = \emptyset$.

Example 1. Let $\mathcal{A} = (\omega, <_\omega)$ and let R be the set of all even numbers. First, we show that no infinite subset of the odds is definable by a computable Σ_1 formula (in the language $\{<, \mathbf{R}\}$) with finitely many parameters $\vec{c}$ and positive occurrences of $\mathbf{R}$. Otherwise, we can assume, without loss of generality, that a disjunct of such a formula is a finitary formula $\exists \vec{u} \psi(\vec{c}, x, \vec{u})$ so that the following conditions are true:

(i) the formula $\psi(\vec{c}, x, \vec{u})$ is a conjunct which gives the complete ordering of $\vec{c}, x, \vec{u}$ and expresses that certain elements of $\vec{c}, \vec{u}$ are in R;

(ii) there exist a tuple $\vec{d}$ and an odd number a bigger than every element in $\vec{c}$ so that $(\mathcal{A}_A, R) \models \psi(\vec{c}, a, \vec{d})$.

Define a' and a tuple $\vec{d'}$ as follows:

(i) $a' = a + 1$;
(ii) if $d_i \in \vec{d}$ and d_i is less than a, set $d'_i =_{def} d_i$;
(iii) if $d_i \in \vec{d}$ and d_i is greater than a, set $d'_i =_{def} d_i + 2$.

Clearly, $(\mathcal{A}_A, R) \models \psi(\vec{c}, a', \vec{d'})$. Hence $(\mathcal{A}_A, R) \models \exists \vec{u} \psi(\vec{c}, a', \vec{u})$, but a' is even, which is a contradiction.

Next, the structure $(\mathcal{A}, R)$ satisfies the decidability condition of Theorem 54. Therefore, there is a computable copy $\mathcal{B}$ of $\mathcal{A}$ and $F : \mathcal{A} \cong \mathcal{B}$ so that $F(R)$ is simple.

Example 2. Let $\mathcal{A}$ be an equivalence structure with infinitely many equivalence classes, all of size 2. Let R be a relation containing exactly one element from each class so that the pair $(\mathcal{A}, R)$ satisfies the decidability condition of Theorem 53. No infinite subset of $\neg R$ is definable by a computable Σ_1 formula (in the language $\{E\}$) with only finitely many parameters: if an element a and its equivalent are both outside the parameters, then any formula satisfied by a is also satisfied by its equivalent element. Therefore, there is a computable copy $\mathcal{B}$, and $F : \mathcal{A} \cong \mathcal{B}$ so that $\neg F(R)$ is immune.

However, $\neg R$ is definable by a computable Σ_1 formula $\varphi(x)$ in $\{E, \mathbf{R}\}$ with only positive occurrences of $\mathbf{R}$. Namely, $\varphi(x)$ is the following finitary formula: $\exists y (\mathbf{R}(y) \wedge yEx \wedge y \neq x)$. Therefore, in any copy $\mathcal{B}$ in which $F(R)$ is c.e. relative to $\mathcal{B}$, $F(R)$ is, in fact, computable relative to $\mathcal{B}$.

Example 3. Let $\mathcal{A}$ be an equivalence structure with infinitely many equivalence classes, all of size 2. Let R be a relation such that the following conditions are satisfied:

(i) there are infinitely many equivalence classes from which R contains exactly one element;

(ii) there are no equivalence classes from which R contains both elements;

(iii) there are infinitely many equivalence classes from which R contains neither element;

(iv) the pair $(\mathcal{A}, R)$ satisfies the decidability condition of Theorem 53.

No infinite subset of $\neg R$ is definable by a computable Σ_1 formula (in the language $\{E\}$) with only finitely many parameters, so there is a computable copy $\mathcal{B}$, and $F : \mathcal{A} \cong \mathcal{B}$ in which $\neg F(R)$ is immune.

Furthermore, there is a computable copy $\mathcal{B}$ in which the image of R is c.e., but not computable. However, the formula $\varphi(x)$ in the language $\{E, \mathbf{R}\}$: $\exists y(\mathbf{R}(y) \wedge yEx \wedge y \neq x)$ defines an infinite subset of $\neg R$. Consequently, there is no $F : \mathcal{A} \cong \mathcal{B}$ such that $F(R)$ is simple relative to $\mathcal{B}$.

Example 4. Let $\mathcal{A}$ be the structure $(\mathcal{Q}, <_{\mathcal{Q}})$, and let R be the set of all rationals less than π. There is no computable formula (in the language $\{<\}$) with finitely many parameters which defines $\neg R$. However, the formula "$5 < x$" does define an infinite subset of $\neg R$. Consequently, there is no $F : \mathcal{A} \cong \mathcal{B}$ in which $\neg F(R)$ is immune relative to $\mathcal{B}$.

Example 5. Let $\mathcal{A}$ be an $\aleph_0$-dimensional vector space over a finite field, say over a field with three elements. Let R be the domain of a subspace of $\mathcal{A}$ of infinite dimension and infinite co-dimension. There is a computable copy of $\mathcal{A}$ in which the image of R is immune, since the only sets definable in $\mathcal{A}$ are finite and co-finite, and there is a copy also satisfying the effectiveness condition of Theorem 53.

For $a \notin R$, the formula $\varphi(a, x) = (\exists y)[x = a + y]$ defines an infinite subset of $\neg R$ that is c.e. (relative to $\mathcal{B}$) if the image of R is. It follows that the image of R can never be relatively simple.

The following definition of Hird introduces a syntactic property corresponding to h-immunity. We will term it "being formally h-immune on $\mathcal{A}$".

Definition 9 (Hird [91]).

(1) A *formal strong array* on $\mathcal{A}$ is a computable sequence of existential formulas in L with finitely many parameters $\vec{c}$, $(\psi_i(\vec{c}, \vec{x}_i))_{i \in \omega}$, such that for every finite set $G \subseteq A$ there exist $i \in \omega$ and a sequence $\vec{a}_i \in A^{lh(\vec{x}_i)}$ with

$$(\mathcal{A}_A \models \psi_i(\vec{c}, \vec{a}_i)) \wedge (\vec{a}_i \cap G = \emptyset).$$

(2) We say that the relation $\neg R$ is *formally h-immune on* $\mathcal{A}$ if there is no formal strong array $(\psi_i(\vec{c}, \vec{x}_i))_{i \in \omega}$ on $\mathcal{A}$ such that

for every $i \in \omega$,

$$(\forall \vec{a}_i \in A^{lh(\vec{x}_i)})[(\mathcal{A}_A \models \psi_i(\vec{c}, \vec{a}_i)) \Rightarrow (\vec{a}_i \cap \neg R \neq \emptyset)].$$

Being formally h-immune on $\mathcal{A}$ turns out to be a necessary condition for the existence of a computable copy of $\mathcal{A}$ such that the corresponding image of R is h-immune (see [91]). Assume that $\mathcal{B}$ is a computable copy of $\mathcal{A}$ and that F is an isomorphism from $\mathcal{A}$ onto $\mathcal{B}$. The following result establishes that, under some extra decidability conditions for $(\mathcal{A}, R)$, the existence of a computable copy $\mathcal{B}$ of $\mathcal{A}$ such that the image of $\neg R$ is h-immune relative to $\mathcal{B}$ is equivalent to $\neg R$ being formally h-immune on $\mathcal{A}$.

Theorem 55 ([91]). (a) *Assume that $\mathcal{B}$ is a computable copy of $\mathcal{A}$ and that F is an isomorphism from $\mathcal{A}$ onto $\mathcal{B}$. If $F(\neg R)$ is h-immune on B, then $\neg R$ is formally h-immune on $\mathcal{A}$.*

(b) *Assume that there is an algorithm which decides for a given sequence $\vec{c} \in A^{<\omega}$ and an existential formula $\psi(\vec{u}, \vec{x})$ in L, $lh(\vec{u}) = lh(\vec{c})$, whether*

$$(\forall \vec{a} \in A^{lh(\vec{x})})[(\mathcal{A}_A \models \psi(\vec{c}, \vec{a})) \Rightarrow (\vec{a} \cap \neg R \neq \emptyset)].$$

If $\neg R$ is formally h-immune on $\mathcal{A}$, then there is a computable structure $\mathcal{B}$ and an isomorphism F from $\mathcal{A}$ onto $\mathcal{B}$ such that the relation $F(\neg R)$ is h-immune on B.

We now introduce a relative version of h-immunity.

Definition 10. Let S be an additional (unary) relation on the domain B of a countable structure $\mathcal{B}$.

(1) A sequence $(U_i)_{i \in \omega}$ of finite sets is a *strong array relative to $\mathcal{B}$* if there is a unary $\mathcal{B}$-computable function f such that for every $i \in \omega$, $U_i = D_{f(i)}$.

(2) The relation $\neg S$ is *h-immune relative to $\mathcal{B}$* if it is infinite and there is no disjoint strong array relative to $\mathcal{B}$, $(U_i)_{i \in \omega}$, such that for every $n \in \omega$, we have $U_i \cap \overline{S} \neq \emptyset$.

If there is an isomorphic copy of $\mathcal{A}$ on which the image of $\neg R$ is relatively h-immune, then $\neg R$ must be formally h-immune on $\mathcal{A}$. This necessary syntactic condition turns out to be sufficient.

Theorem 56 ([67]). *Let $\mathcal{A}$ be a computable L-structure, and let R be a unary infinite and co-infinite relation on A. Then the following conditions are equivalent.*

(i) *For all copies $\mathcal{B}$ of $\mathcal{A}$ and all isomorphisms F from $\mathcal{A}$ onto $\mathcal{B}$, $\neg F(R)$ is not h-immune relative to $\mathcal{B}$.*

(ii) *The relation $\neg R$ is not formally h-immune on $\mathcal{A}$.*

Hird [91] established that, under a suitable decidability condition, R is formally h-simple on $\mathcal{A}$ if and only if there is a computable copy $\mathcal{B}$ of $\mathcal{A}$ such that the image of R under an isomorphism from $\mathcal{A}$ onto $\mathcal{B}$ is h-simple on $\mathcal{B}$. In [67] we give a relative analogue of Hird's result for h-simple relations on computable copies. Harizanov [77] gave a sufficient general conditions for the existence of an h-simple relation on a computable copy of $\mathcal{A}$, in arbitrary nonzero c.e. Turing degree. For more on syntactic characterizations of relations having Post-type and similar properties on structures, or their degree-theoretic complexity see [14, 67, 68, 77, 78, 90, 91].

In addition to considering the complexity of relations on computable structures within hyperarithmetical hierarchy, we can also consider their degrees, such as Turing degrees or strong degrees. Harizanov introduced the following notion.

Definition 11 ([81]). The *Turing degree spectrum* of R on $\mathcal{A}$, in symbols $DgSp_{\mathcal{A}}(R)$, is the set of all Turing degrees of the images of R under all isomorphisms from $\mathcal{A}$ onto computable structures.

Let $\mathcal{L} = (\omega, \prec)$ be the following computable linear order of ordering type $\omega + \omega^*$:

$$0 \prec 2 \prec 4 \prec \cdots \prec 5 \prec 3 \prec 1.$$

We define a computable relation R to be the initial segment of type ω; that is, $R = 2\omega$. An early result, obtained independently by Tennenbaum and Denisov, is that there is an isomorphic computable copy of $\mathcal{L}$ such that its initial segment of type ω is not computable. It is easy to see that the relation R is intrinsically Δ_2^0 on $\mathcal{L}$, because of the following definability of R and $\neg R$:

$$x \in R \Leftrightarrow \bigvee_{n \in \omega} \exists x_0 \ldots \exists x_n [x_0 \prec x_1 \prec \cdots \prec x_n \wedge x = x_n \wedge$$

$$\forall y [\neg(y \prec x_0) \wedge \neg(x_0 \prec y \prec x_1) \wedge \cdots \wedge \neg(x_{n-1} \prec y \prec x_n)],$$

and

$$x \notin R \Leftrightarrow \bigvee_{n \in \omega} \exists x_0 \ldots \exists x_n [x_0 \succ x_1 \succ \cdots \succ x_n \wedge x = x_n \wedge$$

$$\forall y [\neg(y \succ x_0) \wedge \neg(x_0 \succ y \succ x_1) \wedge \cdots \wedge \neg(x_{n-1} \succ y \succ x_n)].$$

It can further be shown that the degree spectrum $Dg_{\mathcal{L}}(R)$ consists of all Δ_2^0 degrees (see [79]).

Ershov classified Δ_2^0 sets as follows. Let α be a computable ordinal. A set $C \subseteq \omega$ is α-c.e. if there are a computable function $f : \omega^2 \to \{0,1\}$ and a computable function $o : \omega \times \omega \to \alpha + 1$ with the following properties:

$$(\forall x)[f(x,0) = 0 \wedge \lim_{s \to \infty} f(x,s) = C(x)],$$

$$(\forall x)(\forall s)[o(x,0) = \alpha \wedge o(x, s+1) \leq o(x,s)], \text{ and}$$

$$(\forall x)(\forall s)[f(x, s+1) \neq f(x, s) \Rightarrow o(x, s+1) < o(x,s)].$$

In particular, 1-c.e. sets are c.e. sets, and 2-c.e. sets are d.c.e. sets.

Let R be an additional unary computable relation on the domain of a computable structure $\mathcal{A}$. We are interested in syntactic conditions such that for an α-c.e. degree $\mathbf{c}$, there is an isomorphism f of degree $\mathbf{c}$ to a computable structure $\mathcal{B}$ for which $f(R)$ is also of degree $\mathbf{c}$. First, we need the following definition.

The complement of R with respect to A is denoted $\overline{R}$. Let $\mathbf{R}$ be a symbol for R. If we are interested in the c.e. images of R, certain first-order formulas with positive occurrences of $\mathbf{R}$ in the expanded language $L(\mathcal{A}) \cup \{\mathbf{R}\}$ play a special role. A Σ_1 formula in $L(\mathcal{A}) \cup \{\mathbf{R}\}$ in which $\mathbf{R}$ occurs only positively is also called a Σ_1^Γ formula. This notation was introduced by Ash and Knight, who defined a hierarchy of infinitary formulas in a general setting in which Γ is a function assigning computable ordinals to relation symbols.

For an ordinal γ, define a binary relation $\leq_\gamma$ on finite sequences of elements from A, of equal length, by: $\vec{l} \leq_\gamma \vec{r}$ if and only if every Π_γ formula true of $\vec{l}$ is also true of $\vec{r}$ (equivalently, every Σ_γ formula true of $\vec{r}$ is also true of $\vec{l}$).

Definition 12. Let $\vec{c} \in A^{<\omega}$ and $a \in A$.

(1) (Harizanov) We say that a is *free over* $\vec{c}$ (also called 1-*free* over $\vec{c}$) if $a \in \overline{R}$ and for every finitary Σ_1^Γ formula $\psi(\vec{z}, x)$, $lh(\vec{z}) = lh(\vec{c})$, if

$$(\mathcal{A}_A, R) \models \psi(\vec{c}, a)$$

then $(\exists a' \in R)[(\mathcal{A}_A, R) \models \psi(\vec{c}, a')]$.

(2) (Ash–Knight) Let β be a computable ordinal such that $\beta > 1$. The element a is β-*free over* $\vec{c}$ if $a \in \overline{R}$ and for every ordinal γ such that $1 \leq \gamma < \beta$,

$$(\forall \vec{u})(\exists a' \in R)(\exists \vec{v})[\vec{c}\,\hat{}\,a\,\hat{}\,\vec{u} \leq_\gamma \vec{c}\,\hat{}\,a'\,\hat{}\,\vec{v}].$$

Let the set of all free elements over $\vec{c}$ be denoted by $\mathrm{fr}(\vec{c})$. Note that $\mathrm{fr}(\vec{c}) \subseteq \overline{R}$. Clearly, if $a \in \mathrm{fr}(\vec{c})$ and $\vec{d}$ is a subsequence of $\vec{c}$, then $a \in \mathrm{fr}(\vec{d})$. Let

$$bd(\vec{c}) =_{\mathrm{def}} \{a \in \overline{R} : a \text{ is not free over } \vec{c}\}.$$

Thus, if $a \in bd(\vec{c})$ and $\vec{c}$ is a subsequence of $\vec{d}$, then $a \in bd(\vec{d})$. A maximal relation on $\overline{R}$ (with respect to the set-theoretic inclusion) that is definable by a computable Σ_1^Γ formula with parameters $\overrightarrow{c}$ is of the form $bd(\overrightarrow{c})$. Conversely, if $bd(\overrightarrow{c})$ is definable by a computable Σ_1^Γ formula with parameters $\overrightarrow{c}$, then $bd(\overrightarrow{c})$ is a maximal relation on $\overline{R}$ definable by such a formula. For example, if $(\mathcal{A}, R) = (\mathcal{A}_0, R_0)$ then

$$a \in \mathrm{fr}(\overrightarrow{c}) \Leftrightarrow a \prec c_{i_0},$$

where c_{i_0} is the $\prec$-least element in $\overline{R} \cap \mathrm{ran}(\overrightarrow{c})$. If $\mathcal{A}$ is $(\omega, =)$ and R is a computable infinite co-infinite subset of ω, then

$$a \in \mathrm{fr}(\overrightarrow{c}) \Leftrightarrow a \notin \mathrm{ran}(\overrightarrow{c}).$$

Theorem 57 (Harizanov [80]). *Assume that there is an algorithm which for every $\overrightarrow{c} \in A^{<\omega}$ outputs an element $a \in A$ such that a is free over $\overrightarrow{c}$. Let $C \subseteq \omega$ be a c.e. set. Then there exist a computable structure $\mathcal{B}$ and an isomorphism $f : \mathcal{A} \to \mathcal{B}$ such that*

$$f(R) \equiv_T f \equiv_T C \& (f(R) \text{ is c.e.}).$$

Proof. Let $\{C_s\}_{s \in \omega}$ be a computable enumeration of C such that at every stage s, C receives at most one new element, and that element is $\leq s$. Let $B = \omega$ and let $(\theta_e)_{e \in \omega}$ be an effective list of all atomic sentences in $L(\mathcal{A})_B$. For every e, either θ_e or $\neg\theta_e$ will be enumerated in the diagram of $\mathcal{B}$. At every stage s of the construction, we define a finite isomorphism f_s from $\mathcal{B}$ to $\mathcal{A}$. Let $X_s = f_s^{-1}(R)$ and

$$\overline{X}_s = \{d_0^s < d_1^s < d_2^s < \cdots\}.$$

We set $X =_{def} \bigcup_{s \in \omega} X_s$. During the construction, we define a partially computable function $h(n, s)$ such that for every s,

$$h(n, s) \downarrow \Leftrightarrow n \in \{0, \ldots, s+1\}.$$

For every n, there exists $\lim_{s \to \infty} h(n, s)$, and for every c,

$$c \in C_s - C_{s-1} \Rightarrow d_{\gamma(c,s-1)}^{s-1} \in X_s. \qquad \square$$

Let α be a nonzero ordinal. By Cantor's normal form theorem, there is a unique representation

$$\alpha = \omega^{\alpha_1} \cdot n_1 + \omega^{\alpha_2} \cdot n_2 + \cdots + \omega^{\alpha_k} \cdot n_k,$$

where $\alpha_1 > \alpha_2 > \cdots > \alpha_k$ and $0 < n_1, n_2, \ldots, n_k < \omega$. Let

$$c_n(\alpha) =_{\mathrm{def}} \omega^{\alpha_1} \cdot (nn_1) + \omega^{\alpha_2} \cdot (nn_2) + \cdots + \omega^{\alpha_k} \cdot (nn_k),$$

and

$$c^*(\alpha) =_{def} \sup\{c_n(\alpha) : n \in \omega\}.$$

Hence, if β is the greatest ordinal such that $\omega^\beta \leq \alpha$, then $c^*(\alpha) = \omega^{\beta+1} \leq \alpha \cdot \omega$.

Theorem 58 (Ash–Cholak–Knight [10]). *Let $C \subseteq \omega$ be an α-c.e. set where $\alpha \geq 2$. Assume that the relations $(\leq_\gamma)_{1 \leq \gamma < c^*(\alpha)}$ are uniformly c.e. Assume that for every $n \in \omega$, for every sequence $\vec{c} \in A^{<\omega}$, there is $a \in \overline{R}$ such that a is $c_n(\alpha)$-free over $\vec{c}$. Then there exist a computable model $\mathcal{B}$ and an isomorphism f from $\mathcal{A}$ to $\mathcal{B}$ such that*

$$f(R) \equiv_T C \equiv_T f.$$

Montalbán [86] introduced the following definition of a degree spectrum of a relation on a cone. The intuition is that we have some fixed set of information we can access, and we must use the same information to view all copies of the structure.

Definition 13. Let R be an additional relation on a structure $\mathcal{A}$, and S be an additional relation on a structure $\mathcal{B}$. We say that R and S have the *same Turing degree spectrum on a cone* if there is a Turing degree $\mathbf{d}$ such that for every Turing degree $\mathbf{c} \geq \mathbf{d}$, we have

$$\{\deg(R^{\mathcal{A}_1}) \oplus \mathbf{c} : \mathcal{A}_1 \cong \mathcal{A} \,\&\, \deg(\mathcal{A}_1) \leq \mathbf{c}\}$$
$$= \{\deg(S^{\mathcal{B}_1}) \oplus \mathbf{c} : \mathcal{B}_1 \cong \mathcal{B} \,\&\, \deg(\mathcal{B}_1) \leq \mathbf{c}\}.$$

A set X and its Turing degree are called n-CEA for $n \in \omega$, if there is a sequence $X_0, X_1, \ldots, X_n$ such that $X = X_n$, X_0 is c.e., and X_{i+1} is c.e. in and above X_i for $0 \leq i \leq n - 1$.

Theorem 59. *Let $\mathcal{A}$ be a structure and R be an additional relation on $\mathcal{A}$.*

(a) (Harizanov [80]) *Relative to a cone, every degree spectrum of a relation is either the singleton consisting of a computable degree or contains all c.e. degrees.*

(b) (Harrison-Trainor [86]) *Relative to a cone, every degree spectrum of a relation is either intrinsically Δ_2^0 or contains all 2-CEA degrees.*

Harrison-Trainor also showed that there is a computable structure $\mathcal{A}$ with relatively intrinsically d-c.e. relations that have incomparable degree spectra relative to every oracle.

For some familiar relations on computable structures, their Turing degree spectra exhibit the dichotomy: either singletons or infinite. Harizanov [80] established that if for a non-intrinsically c.e. relation R on $\mathcal{A}$, the Ash-Nerode decidability condition holds, then $\mathrm{DgSp}_{\mathcal{A}}(R)$ must be infinite. Moses [132] proved that a computable relation on a computable linear ordering is either definable by a quantifier-free formula with finitely many constants, or is not intrinsically computable. Hirschfeldt [92] gave a sufficient condition for a relation to have infinite degree spectrum. That is, if R is a computable relation on the domain of a computable structure $\mathcal{A}$ such that there is a Δ_2^0 function f such that $f(\mathcal{A})$ is a computable structure but $f(R)$ is not a computable relation, then $\mathrm{DgSp}_{\mathcal{A}}(R)$ must be infinite. Applying this condition to linear orderings and using the proof of Moses's result, Hirschfeldt obtained the following result.

Theorem 60 ([92]). *A computable relation on a computable linear ordering is either definable by a quantifier-free formula (in which case it is obviously intrinsically computable) or has an infinite Turing degree spectrum.*

Downey, Goncharov, and Hirschfeldt [43] proved that a computable relation on a computable Boolean algebra is either intrinsically computable or has infinite Turing degree spectrum.

Theorem 61 ([43]). *A computable relation on a computable Boolean algebra is either definable by a quantifier-free formula with finitely many constants (hence intrinsically computable) or has infinite degree spectrum.*

A similar question about Turing degree spectra dichotomy can be asked for computable relations on other classes of structures such as computable abelian groups. Another interesting question from [43] is whether the degree spectrum of an intrinsically Δ_2^0 relation on a computable linear order is always a singleton or infinite.

Csima, Harizanov, Miller, and Montalbán studied computable Fraïssé limits and relations on their domains. A class K of structures is *computably locally finite* if there exists a computable function $g : \omega \to \omega$ such that every structure in K that is generated by n elements contains at most $g(n)$ elements.

Theorem 62 ([42]). *Let $\mathcal{A}$ be a 1-decidable structure for a finite language L, which is homogeneous and computably locally finite. Let R be a unary relation on $\mathcal{A}$. Then the following conditions are equivalent:*

(i) *The degree spectrum of R on $\mathcal{A}$ is not upward closed under Turing reducibility.*

(ii) *The relation R is definable by a quantifier-free formula with parameters in $\mathcal{A}$.*

(iii) *The relation R is intrinsically computable.*

As a corollary of this theorem, we obtain that if K is a class of finite structures for a finite language L, such that $Th_L(K)$ is computably axiomatizable and locally finite, and with computable Fraïssé limit $\mathcal{A}$, then $\mathcal{A}$ is as in Theorem 62.

For more complicated relations on computable structures, Soskov [147] established the following definability result.

Theorem 63. *Let $\mathcal{A}$ be a computable structure and let R be a Δ_1^1 relation on its domain, which is invariant under automorphisms of $\mathcal{A}$. Then R is definable in $\mathcal{A}$ by a computable infinitary formula (without parameters).*

This led to the following characterization of intrinsically Δ_1^1 relations.

Theorem 64 ([147]). *For a computable structure $\mathcal{A}$, and a relation R on $\mathcal{A}$, the following conditions are equivalent:*

(i) *R is intrinsically Δ_1^1 on $\mathcal{A}$;*
(ii) *R is relatively intrinsically Δ_1^1 on $\mathcal{A}$;*
(iii) *R is definable in $\mathcal{A}$ by a computable infinitary formula with finitely many parameters.*

In the following theorem characterizing intrinsically Π_1^1 relations, Soskov [146] established the equivalence (ii) $\Leftrightarrow$ (iii), while (i) $\Leftrightarrow$ (ii) was established in [70].

Theorem 65 ([70, 146]). *For a computable structure $\mathcal{A}$ and relation R on $\mathcal{A}$, the following conditions are equivalent:*

(i) *R is intrinsically Π_1^1 on $\mathcal{A}$;*
(ii) *R is relatively intrinsically Π_1^1 on $\mathcal{A}$;*
(iii) *R is definable in $\mathcal{A}$ by a Π_1^1 disjunction of computable infinitary formulas with finitely many parameters.*

A relation R on $\mathcal{A}$ defined in $\mathcal{A}$ by a Π_1^1 disjunction of computable infinitary formulas with finitely many parameters is also called *formally Π_1^1 on $\mathcal{A}$*. In [70], we showed that if $\mathcal{A}$ is a computable structure and let R be a relation on $\mathcal{A}$, which is Π_1^1 and invariant under automorphisms of $\mathcal{A}$, then R is definable in $\mathcal{A}$ by a Π_1^1 disjunction of computable infinitary formulas without parameters.

Here are some examples of computable structures with intrinsically Π_1^1 relations. Let $\mathcal{A}$ be a Harrison ordering, that is, a computable linear ordering of type $\omega_1^{\text{CK}}(1 + \eta)$, where η is the order type of the rationals. Let R be the initial segment of type ω_1^{CK}.

Structure and Randomness in Computability and Set Theory

This set R is intrinsically Π_1^1, since it is defined by the disjunction of computable infinitary formulas saying that the interval to the left of x has order type β, for computable ordinals β.

For an ordering $\mathcal{L}$, the interval algebra $I(\mathcal{L})$ is the algebra generated, under finite union, by the half-open intervals $[a, b)$, $(-\infty, b)$, $[a, \infty)$, with endpoints in $\mathcal{L}$. A *Harrison Boolean algebra* is a computable Boolean algebra of type $I(\omega_1^{CK}(1 + \eta))$. Let $\mathcal{A}$ be a Harrison Boolean algebra, and let R be the set of superatomic elements. This R is intrinsically Π_1^1, since it is defined by the disjunction of computable infinitary formulas saying that x is a finite join of α-atoms, for computable ordinals α.

A *Harrison p-group* is a computable abelian p-group $\mathcal{G}$ such that its length $\lambda(\mathcal{G}) = \omega_1^{CK}$, every element in its Ulm sequence $(u_\alpha(G))_{\alpha < \omega_1^{CK}}$ is ∞, and the divisible part has infinite dimension. A *Harrison group* is a Harrison p-group for some p. Recall that the Ulm subgroups G_α are defined by $G_\alpha = p^{\omega\alpha}G$, and $u_\alpha(G) =_{\text{def}} \dim_{\mathbb{Z}_p} P_\alpha(G)/P_{\alpha+1}(G)$, where $P_\alpha(G) = G_\alpha \cap \{x \in G : px = 0\}$. Let $\mathcal{A}$ be a Harrison group, and let R be the set of elements that have computable ordinal heights, that is, the complement of the divisible part. Then R is intrinsically Π_1^1 on $\mathcal{A}$, since it is defined by the disjunction of computable infinitary formulas saying that x has height α, for computable ordinals α. The divisible part of $\mathcal{G}$ has the same degree as its complement R.

Theorem 66 ([70]). *The following sets of Turing degrees are equal:*

(i) *the set of Turing degrees of maximal well-ordered initial segments of Harrison orderings;*

(ii) *the set of Turing degrees of superatomic parts of Harrison Boolean algebras;*

(iii) *the set of Turing degrees of divisible parts of Harrison p-groups;*

(iv) *the set of Turing degrees of left-most paths of computable trees $\mathcal{T} \subseteq \omega^{<\omega}$ such that T has a path, but no hyperarithmetical path;*

(v) *the set of Turing degrees of Π_1^1 paths through Kleene's $\mathcal{O}$.*

Acknowledgments

Harizanov was partially supported by the Simons Foundation-Collaboration Grant and CCAS Dean's Research Chair award of the George Washington University.

References

[1] F. Adams and D. Cenzer, Computability and categoricity of weakly untrahomogeneous structures, *Computability* **6** (2017) 365–389.

[2] R. Alvir, J. F. Knight and C. F. D. McCoy, Complexity of Scott sentences, Preprint (2018), https://arxiv.org/abs/1807.02715.

[3] U. Andrews, The degrees of categorical theories with recursive models, *Proc. Amer. Math. Soc.* **131** (2013) 2501–2514.

[4] U. Andrews, New spectra of strongly minimal theories in finite languages, *Ann. Pure Appl. Logic* **162** (2011) 367–372.

[5] U. Andrews and J. F. Knight, Strongly minimal theories with recursive models, *J. European Math. Soc.* **20** (2018) 1561–1594.

[6] U. Andrews and J. F. Knight, Spectra of atomic theories, *J. Symbolic Logic* **78** (2013) 1189–1198.

[7] U. Andrews and J. S. Miller, Spectra of theories and structures, *Proc. Amer. Math. Soc.* **143** (2015) 1283–1298.

[8] C. J. Ash, Categoricity in hyperarithmetical degrees, *Ann. Pure Appl. Logic* **34** (1987) 1–14.

[9] C. J. Ash, Recursive labeling systems and stability of recursive structures in hyperarithmetical degrees, *Trans. Amer. Math. Soc.* **298** (1986) 497–514.

[10] C. J. Ash, P. Cholak and J. F. Knight, Permitting, forcing, and copying of a given recursive relation, *Ann. Pure Appl. Logic* **86** (1997) 219–236.

[11] C. Ash and J. Knight, *Computable Structures and the Hyperarithmetical Hierarchy* (Elsevier, Amsterdam, 2000).

[12] C. Ash, J. Knight, M. Manasse and T. Slaman, Generic copies of countable structures, *Ann. Pure Appl. Logic* **42** (1989) 195–205.

[13] C. J. Ash and A. Nerode, Intrinsically recursive relations, in J. N. Crossley (ed.), *Aspects of Effective Algebra* (U. D.A. Book Co., Steel's Creek, Australia, 1981), pp. 26–41.

[14] C. J. Ash, J. F. Knight and J. B. Remmel, Quasi-simple relations in copies of a given recursive structure, *Ann. Pure Appl. Logic* **86** (1997) 203–218.

[15] S. A. Badaev, Computable enumerations of families of general recursive functions, *Algebra and Logic* **16** (1977) 129–148 (in Russian); (1978) 83–98 (English translation).

[16] E. Barker, Back and forth relations for reduced abelian *p*-groups, *Ann. Pure Appl. Logic* **75** (1995) 223–249.

[17] E. Barker, Intrinsically Σ_α^0 relations, *Ann. Pure Appl. Logic* **39** (1988) 105–130.

[18] J. Barwise, Infinitary logic and admissible sets, *J. Symbolic Logic* **34** (1969) 226–252.

[19] N. A. Bazhenov, Δ_2^0-categoricity of Boolean algebras, *J. Math. Sci.* **203** (2014) 444–454.

[20] W. Calvert, Algebraic structure and computable structure, Ph.D. dissertation, University of Notre Dame (2005).

[21] W. Calvert, The isomorphism problem for computable abelian *p*-groups of bounded length, *J. Symbolic Logic* **70** (2005) 331–345.

[22] W. Calvert, The isomorphism problem for classes of computable fields, *Arch. Math. Logic* **43** (2004) 327–336.

[23] W. Calvert, D. Cenzer, V. Harizanov and A. Morozov, Effective categoricity of Abelian *p*-groups, *Ann. Pure Appl. Logic* **159** (2009) 187–197.

[24] W. Calvert, D. Cenzer, V. Harizanov and A. Morozov, Effective categoricity of equivalence structures, *Ann. Pure Appl. Logic* **141** (2006) 61–78.

[25] W. Calvert, E. Fokina, S. Goncharov, J. Knight, O. Kudinov, A. Morozov and V. Puzarenko, Index sets for classes of high rank structures, *J. Symbolic Logic* **72** (2007) 1418–1432.

[26] W. Calvert, A. Frolov, V. Harizanov, J. Knight, C. McCoy, A. Soskova and S. Vatev, Strong jump inversion, *J. Logic Comput.* **7** (2018) 1499–1522.

[27] W. Calvert, S. S. Goncharov and J. F. Knight, Computable structures of Scott rank ω_1^{CK} in familiar classes, in S. Gao, S. Jackson and Y. Zhang, (eds.), *Advances in logic*, Contemporary Mathematics, Vol. 425 (American Mathematical Society, Providence, RI, 2007), pp. 49–66.

[28] W. Calvert, S. Goncharov, J. Millar and J. Knight, Categoricity of computable infinitary theories, *Arch. Math. Logic* **48** (2009) 25–38.

[29] W. Calvert, V. S. Harizanov, J. F. Knight and S. Miller, Index sets of computable structures, *Algebra Logic* **45** (2006) 306–325.

[30] W. Calvert and J. F. Knight, Classification from a computable point of view, *Bull. Symbolic Logic* **12** (2006) 191–218.

[31] W. Calvert, J. F. Knight and J. Millar, Computable trees of Scott rank ω_1^{CK}, and computable approximability, *J. Symbolic Logic* **71** (2006) 283–298.

[32] J. Carson, V. Harizanov, J. Knight, K. Lange, C. Safranski, C. McCoy, A. Morozov, S. Quinn and J. Wallbaum, Describing free groups, *Trans. Amer. Math. Soc.* **364** (2012) 5715–5728.

[33] D. Cenzer, V. Harizanov and J. Remmel, Effective categoricity of injection structures, *Algebra Logic* **53** (2014) 39–69 (English translation).

[34] D. Cenzer, V. Harizanov and J. Remmel, Two-to-one structures, *J. Logic Comput.* **23** (2013) 1195–1223.

[35] D. Cenzer and J. B. Remmel, Complexity-theoretic model theory and algebra, in Yu. L. Ershov, S. S. Goncharov, A. Nerode and J. B. Remmel, eds., *Handbook of Recursive Mathematics*, Vol. 1, Studies in Logic and the Foundations of Mathematics, Vol. 139 (North-Holland, Amsterdam, 1998), pp. 381–513.

[36] C. C. Chang and H. J. Keisler, *Model Theory* (North-Holland, Amsterdam, 1973).

[37] J. Chisholm, Effective model theory vs. recursive model theory, *J. Symbolic Logic* **55** (1990) 1168–1191.

[38] J. Chisholm, E. Fokina, S. Goncharov, V. Harizanov, J. Knight and S. Quinn, Intrinsic bounds on complexity and definability at limit levels, *J. Symbolic Logic* **74** (2009) 1047–1060.

[39] J. Chisholm and M. Moses, An undecidable linear order that is n-decidable for all n, *Notre Dame J. Formal Logic* **39** (1998) 519–526.

[40] P. Cholak, S. Goncharov, B. Khoussainov and R. A. Shore, Computably categorical structures and expansions by constants, *J. Symbolic Logic* **64** (1999) 13–37.

[41] P. Cholak, R. A. Shore and R. Solomon, A computably stable structure with no Scott family of finitary formulas. *Arch. Math. Logic* **645** (2006) 519–538.

[42] B. Csima, V. Harizanov, R. Miller and A. Montalbán, Computability of Fraïssé limits, *J. Symbolic Logic* **76** (2011) 66–93.

[43] R. G. Downey, S. S. Goncharov and D. R. Hirschfeldt, Degree spectra of relations on Boolean algebras, *Algebra and Logic* **42** (2003) 105–111.

[44] R. Downey and C. G. Jockusch, Every low Boolean algebra is isomorphic to a recursive one, *Proc. Amer. Math. Soc.* **122** (1994) 871–880.

[45] R. G. Downey, A. M. Kach, S. Lempp, A. E. M. Lewis-Pye, A. Montalbán and D. D. Turetsky, The complexity of computable categoricity, *Adv. Math.* **268** (2015) 423–466.

[46] R. G. Downey, A. M. Kach, S. Lempp and D. D. Turetsky, Computable categoricity versus relative computable categoricity, *Fund. Math.* **221** (2013) 129–159.

[47] R. G. Downey and A. G. Melnikov, Effectively categorical abelian groups, *J. Algebra* **373** (2013) 223–248.

[48] R. G. Downey, A. G. Melnikov and K. M. Ng, Abelian p-groups and the halting problem, *Ann. Pure Appl. Logic* **167** (2016) 1123–1138.

[49] Yu. L. Ershov, *Decidability Problems and Constructive Models* (Nauka, Moscow, 1980) (in Russian).

[50] Yu. L. Ershov and S. S. Goncharov, *Constructive Models*, Siberian School of Algebra and Logic (Kluwer Academic/Plenum Publishers, 2000) (English translation).

[51] E. B. Fokina, Index sets for some classes of structures, *Ann. Pure Appl. Logic* **157** (2009) 139–147.

[52] E. B. Fokina, Index sets of decidable models, *Siberian Math. J.* **48** (2007) 939–948 (English translation).

[53] E. Fokina, On complexity of categorical theories with computable models, *Vestnik NGU* **5** (2005) 78–86 (in Russian).

[54] E. Fokina, S. Goncharov, V. Harizanov, O. Kudinov and D. Turetsky, Index sets for n-decidable structures categorical relative to m-decidable presentations, *Algebra Logic* **54** (2015) 336–341 (English translation).

[55] E. Fokina, V. Harizanov and D. Turetsky, Computability-theoretic categoricity and Scott families, *Ann. Pure Appl. Logic* **170** (2019) 699–717.

[56] C. Freer, Models with high scott rank, Ph.D. dissertation, Harvard University (2008).

[57] A. N. Frolov, Effective categoricity of computable linear orderings, *Algebra Logic* **54** (2015) 415–417 (English translation).

[58] A. N. Gavryushkin, Spectra of computable models for Ehrenfeucht theories, *Algebra Logic* **46** (2007) 149–157 (English translation).

[59] S. S. Goncharov, Autostability of prime models with respect to strong constructivizations, *Algebra Logic* **48** (2009) 410–417 (English translation).

[60] S. S. Goncharov, Autostable models and algorithmic dimensions, in Yu. L. Ershov, S. S. Goncharov, A. Nerode and J. B. Remmel (eds.), *Handbook of Recursive Mathematics*, Vol. 1, (North-Holland, Amsterdam, 1998), pp. 261–287.

[61] S. S. Goncharov, Autostability of models and abelian groups, *Algebra Logic* **19** (1980) 13–27 (English translation).

[62] S. S. Goncharov, Strong constructivizability of homogeneous models, *Algebra Logic* **17** (1978) 247–263 (English translation).

[63] S. S. Goncharov, The quantity of nonautoequivalent constructivizations, *Algebra Logic* **16** (1977) 169–185 (English translation).

[64] S. S. Goncharov, Restricted theories of constructive Boolean algebras, *Siberian Math. J.* **17** (1976) 601–611 (English translation).

[65] S. S. Goncharov, Selfstability and computable families of constructivizations, *Algebra Logic* **14** (1975) 647–680 (in Russian).

[66] S. S. Goncharov and V. D. Dzgoev, Autostability of models, *Algebra Logic* **19** (1980) 28–37 (English translation).

[67] S. S. Goncharov, V. S. Harizanov, J. F. Knight and C. F.D. McCoy, Relatively hyperimmune relations on structures, *Algebra Logic* **43** (2004) 94–101 (English translation).

[68] S. S. Goncharov, V. S. Harizanov, J. F. Knight and C. F.D. McCoy, Simple and immune relations on countable structures, *Arch. Math. Logic* **42** (2003) 279–291.

[69] S. Goncharov, V. Harizanov, J. Knight, C. McCoy, R. Miller and R. Solomon, Enumerations in computable structure theory, *Ann. Pure Appl. Logic* **136** (2005) 219–246.

[70] S. S. Goncharov, V. S. Harizanov, J. F. Knight and R. A. Shore, Π^1_1 relations and paths through $\mathcal{O}$, *J. Symbolic Logic* **69** (2004) 585–611.

[71] S. S. Goncharov and B. Khoussainov, Complexity of theories of computable categorical models, *Algebra Logic* **43** (2004) 365–373 (English translation).

[72] S. S. Goncharov and J. F. Knight, Computable structure and nonstructure theorems, *Algebra Logic* **41** (2002) 351–373.

[73] S. Goncharov, S. Lempp and R. Solomon, The computable dimension of ordered abelian groups, *Adv. Math.* **175** (2003) 102–143.

[74] S. S. Goncharov and M. I. Marchuk, Index sets of constructive models that are autostable under strong constructivizations, *J. Math. Sci. (New York)* **205** (2015) 368–388.

[75] S. S. Goncharov and A. T. Nurtazin, Constructive models of complete decidable theories, *Algebra Logic* **12** (1974) 67–77 (English translation).

[76] N. Greenberg, A. Montalbán and T. A. Slaman, Relative to any non-hyperarithmetic set, *J. Math. Logic* **13** (2013) 1–26.

[77] V. Harizanov, Turing degrees of hypersimple relations on computable structures, *Ann. Pure Appl. Logic* **121** (2003) 209–226.

[78] V. Harizanov, Effectively nowhere simple relations on computable structures, in M. M. Arslanov and S. Lempp (eds.), *Recursion Theory and Complexity* (Walter de Gruyter, Berlin, 1999), pp. 59–70.

[79] V. S. Harizanov, Turing degrees of certain isomorphic images of computable relations, *Ann. Pure Appl. Logic* **93** (1998) 103–113.

[80] V. S. Harizanov, Some effects of Ash–Nerode and other decidability conditions on degree spectra, *Ann. Pure Appl. Logic* **55** (1991) 51–65.

[81] V. S. Harizanov, Degree spectrum of a recursive relation on a recursive structure, Ph.D. dissertation, University of Wisconsin, Madison (1987).

[82] V. S. Harizanov, J. F. Knight and A. S. Morozov, Sequences of n-diagrams, *J. Symbolic Logic* **67** (2002) 1227–1247.

[83] L. Harrington, Recursively presentable prime models, *J. Symbolic Logic* **39** (1974) 305–309.

[84] K. Harris, Δ_2^0-categorical Boolean algebras, preprint.

[85] J. Harrison, Recursive pseudo-well-orderings, *Trans. Amer. Math. Soc.* **131** (1968) 526–543.

[86] M. Harrison-Trainor, *Degree Spectra of Relations on a Cone*, Memoirs of the American Mathematical Society, Vol. 253 (American Mathematical Society, 2018).

[87] M. Harrison-Trainor, There is no classification of the decidably presentable structures, *J. Math. Logic* **18** (2018), 1850010, 41.

[88] M. Harrison-Trainor and M-C. Ho, On optimal sentences of finitely generated algebraic structures, *Proc. Amer. Math. Soc.* **146** (2018) 4473–4485.

[89] M. Harrison-Trainor, G. Igusa and J. F. Knight, Some new computable structures of high rank, *Proc. Amer. Math. Soc.* **146** (2018) 3097–3109.

[90] G. Hird, Recursive properties of intervals of recursive linear orders, in: J. N. Crossley, J. B. Remmel, R. A. Shore and M. E. Sweedler (eds.), *Logical Methods* (Birkhäuser, Boston, 1993), pp. 422–437.

[91] G. R. Hird, Recursive properties of relations on models, *Ann. Pure Appl. Logic* **63** (1993) 241–269.

[92] D. R. Hirschfeldt, Degree spectra of relations on computable structures in the presence of Δ_2^0 isomorphisms, *J. Symbolic Logic* **67** (2002) 697–720.

[93] D. R. Hirschfeldt, Prime models of theories of computable linear orderings, *Proc. Amer. Math. Soc.* **129** (2001) 3079–3083.

[94] D. R. Hirschfeldt, B. Khoussainov, R. A. Shore and A. M. Slinko, Degree spectra and computable dimensions in algebraic structures, *Ann. Pure Appl. Logic* **115** (2002) 71–113.

[95] D. Hirschfeldt, K. Kramer, R. Miller and A. Shlapentokh, Categoricity properties for computable algebraic fields, *Trans. Amer. Math. Soc.* **367** (2015) 3955–3980.

[96] M-C. Ho, Describing groups, *Proc. Amer. Math. Soc.* **145** (2017) 2233–2239.

[97] C. G. Jockusch and R. I. Soare, Degrees of orderings not isomorphic to recursive linear orderings, *Ann. Pure and Appl. Logic* **52** (1991) 39–64.

[98]   A. M. Kach and D. Turetsky, Δ_2^0-categoricity of equivalence structures, *New Zealand J. Math.* **39** (2009) 143–149.

[99]   O. Kharlampovich and A. Myasnikov, Elementary theory of free non-abelian groups, *J. Algebra* **302** (2006) 215–242.

[100] N. G. Khisamiev, Theory of abelian groups with constructive models, *Siberian Math. J.* 27 (1986) 572–585 (English translation).

[101] B. Khoussainov and A. Montalbán, A computable $\aleph_0$-categorical structure whose theory computes true arithmetic, *J. Symbolic Logic* **75** (2010) 728–740.

[102] B. Khoussainov and R. A. Shore, Computable isomorphisms, degree spectra of relations and Scott families, *Ann. Pure Appl. Logic* **93** (1998) 153–193.

[103] J. F. Knight, Minimality and completions of PA, *J. Symbolic Logic* **66** (2001) 1447–1457.

[104] J. Knight, Nonarithmetical $\aleph_0$-categorical theories with recursive models, *J. Symbolic Logic* **59** (1994) 106–112.

[105] J. F. Knight and C. McCoy, Index sets and Scott sentences, *Arch. Math. Logic* **53** (2014) 519–524.

[106] J. F. Knight and J. Millar, Computable structures of Scott rank ω_1^{CK}, *J. Math. Logic* **10** (2010) 31–43.

[107] J. F. Knight and V. Saraph, Scott sentences for certain groups, *Arch. Math. Logic* **57** (2017) 453–473.

[108] J. F. Knight and M. Stob, Computable Boolean algebras, *J. Symbolic Logic* **65** (2000) 1605–1623.

[109] K. Zh. Kudaibergenov, Effectively homogenous models, *Siberian Math. J.* **27** (1986) 180–182 (in Russian).

[110] O. Kudinov, An autostable 1-decidable model without a computable Scott family of $\exists$-formulas, *Algebra Logic* **35** (1996) 458–467.

[111] P. LaRoche, Recursively presented Boolean algebras, *Notices AMS* **24** (1977) A552–A553.

[112] S. Lempp, C. McCoy, R. Miller and R. Solomon, Computable categoricity of trees of finite height, *J. Symbolic Logic* **70** (2005) 151–215.

[113] M. Lerman and J. Schmerl, Theories with recursive models, *J. Symbolic Logic* **44** (1979) 59–76.

[114] A. I. Mal'cev, On recursive Abelian groups, *Soviet Math. Doklady* **3** (1962) 1431–1434 (English translation).

[115] M. Makkai, An example concerning Scott heights, *J. Symbolic Logic* **46** (1981) 301–318.

[116] M. Manasse, Techniques and counterexamples in almost categorical recursive model theory, Ph.D. dissertation, University of Wisconsin, Madison (1982).

[117] D. Marker and R. Miller, Turing degree spectra of differentially closed fields, *J. Symbolic Logic* **82** (2017) 1–25.

[118] L. B. Marshall, Computability-theoretic properties of partial injections, trees, and nested equivalences, Ph.D. dissertation, George Washington University (2015).

[119] C. F. D. McCoy, Δ_2^0-categoricity in Boolean algebras and linear orderings, *Ann. Pure Appl. Logic* **119** (2003) 85–120.

[120] C. F. D. McCoy, On Δ_3^0-categoricity for linear orders and Boolean algebras, *Algebra Logic* **41** (2002) 295–305 (English translation).

[121] C. McCoy and J. Wallbaum, Describing free groups, part II: Π_4^0-hardness and no Σ_2^0 basis, *Trans. Amer. Math. Soc.* **364** (2012) 5729–5734.

[122] J. Millar and G. E. Sacks, Atomic models higher up, *Ann. Pure Appl. Logic* **155** (2008) 225–241.

[123] T. Millar, Recursive categoricity and persistence, *J. Symbolic Logic* **51** (1986) 430–434.

[124] T. S. Millar, Homogeneous models and decidability, *Pacific J. Math.* **91** (1980) 407–418.

[125] T. S. Millar, Foundations of recursive model theory, *Ann. Math. Logic* **13** (1978) 45–72.

[126] A. Miller, The Borel classification of the isomorphism class of a countable model, *Notre Dame J. Formal Logic* **24** (1983) 22–34.

[127] D. E. Miller, The invariant Π_α^0 separation principle, *Trans. Amer. Math. Soc.* **242** (1978) 185–204.

[128] R. Miller, The computable dimension of trees of infinite height, *J. Symbolic Logic* **70** (2005) 111–141.

[129] R. Miller and A. Shlapentokh, Computable categoricity for algebraic fields with splitting algorithms, *Trans. Amer. Math. Soc.* **367** (2015) 3955–3980.

[130] R. Miller and H. Schoutens, Computably categorical fields via Fermat's Last Theorem, *Computability* **2** (2013) 51–65.

[131] M. Morley, Decidable models, *Israel J. Math.* **25** (1976) 233–240.

[132] M. Moses, Relations intrinsically recursive in linear orders, *Zeit. Math. Logik Grundlagen Math.* **32** (1986) 467–472.

[133] M. Nadel, Scott sentences and admissible sets, *Ann. Math. Logic* **7** (1974) 267–294.

[134] A. T. Nurtazin, Strong and weak constructivizations and computable families, *Algebra Logic* **13** (1974) 177–184 (English translation).

[135] E. N. Pavlovskii, An estimate for the algorithmic complexity of classes of computable models, *Siberian Math. J.* **49** (2008) 512–523 (English translation).

[136] M. G. Peretyat'kin, Criterion for strong constructivizability of a homogeneous model, *Algebra Logic* **17** (1978) 290–301 (English translation).

[137] J. B. Remmel, Recursively categorical linear orderings, *Proc. Amer. Math. Soc.* **83** (1981) 387–391.

[138] J. B. Remmel, Recursive isomorphism types of recursive Boolean algebras, *J. Symbolic Logic* **46** (1981) 572–594.

[139] J.-P. Ressayre, Models with compactness properties relative to an admissible language, *Ann. Math. Logic* **11** (1977) 31–55.

[140] H. Rogers, *Theory of Recursive Functions and Effective Computability* (McGraw-Hill, 1967).

[141] G. E. Sacks, *Higher Recursion Theory* (Springer, Berlin, 1990).

[142] G. E. Sacks, On the number of countable models, in C. T. Chong and M. J. Wicks (eds.), *Southeast Asian Conference on Logic* (North-Holland, 1983), pp. 185–195.

[143] V. L. Selivanov, Enumerations of families of general recursive functions, *Algebra Logic* **15** (1976) 128–141 (English translation).

[144] R. L. Smith, Two theorems on autostability in p-groups, in *Logic Year 1979–80*, University of Connecticut, Storrs, Lecture Notes in Mathematics, Vol. 859 (Springer, Berlin, 1981), pp. 302–311.

[145] R. I. Soare, *Turing Computability (Theory and Applications)* (Springer, Berlin, 2016).

[146] I. N. Soskov, Intrinsically Π_1^1 relations, *Math. Logic Quart.* **42** (1996) 109–126.

[147] I. N. Soskov, Intrinsically hyperarithmetical sets, *Math. Logic Quart.* **42** (1996) 469–480.

[148] A. I. Stukachev, A jump inversion theorem for the semilattices of Σ-degrees, *Sibirskie Élektronnye Mat. Izv.* **6** (2009) 182–190 (in Russian). (English translation in: *Siberian Adv. Math.* **20** (2010) 68–74.)

[149] R. L. Vaught, Sentences true in all constructive models, *J. Symbolic Logic* **25** (1960) 39–58.

[150] H. Walker, Computable isomorphisms of directed graphs and trees, Ph.D. dissertation, GWU (2017).

[151] H. Walker, Computable isomorphisms for certain classes of infinite graphs, *J. Knot Theory Ramifications* **27** (2018) 1841012, 17.

[152] W. White, Characterization for computable structures, Ph.D. dissertation, Cornell University (2000).

Author Index

Subject Index

A

μ-ary reduction, 158

almost always bounded, 203, 219–220, 224

almost always finite support, 211

almost always recursively bounded, 219–220, 226

almost included, 62

almost inclusion, 4, 20, 63

almost reduction, 3–5, 23, 26, 38, 44, 51–52

Answer Set Programming (ASP), 200

arithmetical, 289, 291–292, 304

arrow ultrafilters, 5–6, 48, 50

Ash–Nerode theorem, 330

B

barrier, 14, 16–17, 19–20

basically generated, 69, 78–79, 82–83

Bernoulli measure, 135

binary code, 113, 115–116, 121

binary net measure, 125–126

Boolean algebras, 294–295, 306, 319, 322–323, 344, 346

Borel reducibility, 155–156, 161, 171, 178, 194

C

canonical equivalence relations, 5, 12, 15–16, 20, 28–29, 33, 40, 42, 44, 48–49, 52

code-tree, 104–105, 107

coding theorem, 122, 133

cofinal map, 11–12, 20, 26–27, 32, 39, 46

complete, 166–169, 171, 182–184, 186, 190

complete combinatorics, 3–5, 9–10, 23, 25, 31–32, 34–35, 48, 50, 52–55, 59, 61

compressed, 121

compression ratio, 138–139

Printed in the United States
by Baker & Taylor Publisher Services